Antioxidants and Second Messengers of Free Radicals

Antioxidants and Second Messengers of Free Radicals

Special Issue Editor

Neven Zarkovic

MDPI • Basel • Beijing • Wuhan • Barcelona • Belgrade

MDPI

Special Issue Editor
Neven Zarkovic
Rudjer Boskovic Institute
Croatia

Editorial Office
MDPI
St. Alban-Anlage 66
4052 Basel, Switzerland

This is a reprint of articles from the Special Issue published online in the open access journal *Antioxidants* (ISSN 2076-3921) from 2018 (available at: https://www.mdpi.com/journal/antioxidants/special_issues/second_messengers_free_radicals).

For citation purposes, cite each article independently as indicated on the article page online and as indicated below:

LastName, A.A.; LastName, B.B.; LastName, C.C. Article Title. *Journal Name* **Year**, *Article Number*, Page Range.

ISBN 978-3-03897-533-5 (Pbk)
ISBN 978-3-03897-534-2 (PDF)

Contents

About the Special Issue Editor

Neven Zarkovic is a Senior Scientist (tenure) and the Head of the Laboratory for Oxidative Stress (LabOS) at the Rudjer Boskovic Institute in Zagreb, Croatia, where he acted as Associate Director for Science and Counsellor for International Affairs. He obtained his MD in 1984 from the Medical Faculty, Zagreb University. Following his MSc in biology in 1986 and PhD in 1989 he undertook a postdoctoral fellowship as a Lise Meitner awardee at the Institute of Biochemistry of the Karl Franz University of Graz, working mostly with Jörg Schaur and Hermann Esterbauer. His research focuses on oxidative stress and lipid peroxidation and the role of 4-hydroxynonenal (HNE) in the pathophysiology of stress and age-associated disorders. Currently he is coordinating the international offset project on metabolomics in cancer and PTSD patients supported by the Government of Croatia and the Finnish company Patria. He was a founding member of the International HNE-Club (SFRR-I) and now chairs its Steering Committee. Prof. Zarkovic was the proposer and co-ordinator of the European COST Action B35 on lipid peroxidation-associated disorders and was a member of the CMST Domain of COST and of the Board of Governors of EARTO, while he acts now as Study Director of the Interdisciplinary PhD Study Program in Molecular Biosciences and as visiting professor of the Medical University in Bialystok. He is a college professor of economy/management and holds university professorships in biology and in medicine. Professor Zarkovic has (co)authored more than 250 publications, which have been cited more than six thousand times (GS). In 1985, he received the national award for research in pathology, while in 2007 he received the national award for scientific achievements. He is a fellow of the Royal Society of Medicine (RSM, London) and a member of the European Order of St. Georg Knights.

Preface to "Antioxidants and Second Messengers of Free Radicals"

For decades, chemical species with unpaired electrons known as free radicals, in particular, oxygen free radicals, have been considered to act as enemies of living cells, damaging major bioactive molecules and thus causing degenerative and malignant diseases. Despite this, free radicals are live for very short periods and are highly reactive molecules that also play and important roles in the cellular metabolism. This paradox is further stressed by the fact that some products of the oxidative metabolism of lipids, especially reactive aldehydes, can mimic the bioactivity of free radicals, while living much longer and even acting on long distance. Therefore, there is a lack of general understanding of pathophysiological roles played by reactive aldehydes like malondialdehyde, 4-hydroxynoenal, 4-hydroxyhexenal, acrolein, etc., which are also considered to be "second messengers of free radicals". Being generated mostly by non-enzymatic lipid peroxidation, they often form bioactive adducts with macromolecules important for the pathophysiology of living cells, even in the absence of severe oxidative stress.

This book is based on the Special Issue of *Antioxidants* comprising original research papers and reviews on complex aspects of reactive aldehydes and their macromolecular adducts (especially with proteins) generated during lipid peroxidation and their interference with natural and synthetic antioxidants in the physiology of cells and in the pathophysiology of various diseases studied by modern bioanalytical methods applied in translational and clinical medicine.

Taken together, these scientific papers suggest that understanding the pathophysiology of reactive aldehydes might indeed be crucial to a better understanding of major human diseases, while monitoring their production and controlling them by using efficient antioxidants might help the development of the modern, interdisciplinary life sciences and integrative biomedicine.

Neven Zarkovic
Special Issue Editor

antioxidants

MDPI

Editorial

Antioxidants and Second Messengers of Free Radicals

Neven Zarkovic

Laboratory for Oxidative Stress (LabOS), Institute "Rudjer Boskovic", HR-10000 Zagreb, Croatia; zarkovic@irb.hr

Received: 1 November 2018; Accepted: 1 November 2018; Published: 6 November 2018

In the recent years, numerous research on the pathology of oxidative stress has been completed by intense studies on redox signaling implementing various experimental models and clinical trials. Nonetheless, there is still a lack of general understanding of pathophysiological roles played by reactive aldehydes like malondialdehyde, 4-hydroxynonenal, 4-hydroxyhexenal, acrolein, etc., which are considered as "second messengers of free radicals" [1,2]. Mostly being generated by lipid peroxidation, reactive aldehydes often form bioactive adducts with macromolecules that are important for the pathophysiology of living cells, thus, mimicking the effects of reactive oxygen species (ROS) even in the absence of severe oxidative stress [3–5]. Accordingly, we lack understanding on the complex effects of antioxidants that might be active in the regulation of toxic and/or hormetic effects of reactive aldehydes. This knowledge is necessary for better understanding of human physiology from the earliest days of life as well and for prevention and treatment of various stress- and age-associated diseases, which require integrative medicine treatment protocols [6,7].

This Special Issue collected original research papers and reviews on complex aspects of reactive aldehydes and their protein adducts generated during lipid peroxidation and their interference with natural and synthetic antioxidants in the physiology of cell and in the pathophysiology of various diseases, studied by modern bioanalytical methods applied in translational and integrative medicine.

Hence, focusing on the damaging effects of 4-hydroxynonenal (HNE) to the earliest events in our lives, i.e., the process of sperm-egg recognition, the Australian scientists reviewed the negative effects of HNE affecting the function and the stability of several germline proteins. Additionally, the authors pointed to the arachidonate 15-lipoxygenase (ALOX15) as a potential therapeutic target that could be exploited to protect human spermatozoa against oxidative stress [8]. They also prepared a very informative list of antioxidants tested for the improvement of male fertility summarizing their efficiencies.

On the other hand, focusing on the other end of the time-scale of human life, the Italian researchers wrote a very informative review on the relevance of reactive aldehydes in age-related disorders [9]. Reviewing the current knowledge on these complex topics of major relevance for modern biomedicine, the authors suggest that a major fraction of the toxic effects of oxidative stress observed in age-related disorders could depend on the formation of aldehyde-protein adducts (in particular, protein adducts of HNE and malondialdehyde (MDA). They also stressed the relevance of novel redox-proteomic approaches, which might reveal aldehydic modifications of distinct cellular proteins targeted in and after the course of oxidative stress, aiming to pave the way to targeted therapeutic strategies for age-associated disorders.

The importance of novel analytical approaches of redox-proteomics was also shown by researchers from Austria who described in their original research paper the method for detection of lipid-modified proteins that does not require an a priori knowledge on the chemical structure of lipid oxidation products or identification of their target proteins [10]. The method is based on the change of electrophoretic mobility of lipid-modified proteins, which is induced by conformational changes and cross-linking with other proteins. The authors have applied this method to successfully study the effects of oxidized palmitoyl-arachidonoyl-phosphatidylcholine (OxPAPC) on endothelial cells, identifying several known but also many new OxPAPC-binding proteins, thus presenting an important analytical breakthrough. This supports previous research by the Austrian pioneers in the field as Hermann Esterbauer and collaborators who discovered HNE, thus, constructing the fundaments for the modern scientific arena of lipid peroxidation [11].

The pathophysiological aspects of lipid peroxidation were further reviewed from two complementary aspects; by summarizing findings on HNE in redox homeostasis of gastrointestinal mucosa with possible implications for the stomach in health and in gastrointestinal diseases [12] and by reviewing options for modulation of oxidative stress and lipid peroxidation by endocannabinoids and their lipid analogues [13]. In the former article, the authors point to pathophysiological relevance of the HNE-protein adducts in digestive system of humans, especially stressing increased accumulation of HNE-modified proteins in gastric mucosa during infection and even after eradication of *H. pillory* infection. However, the authors of the later review paper suggest that a link between the endocannabinoid system (ECS) and redox homeostasis impairment could be crucial for cellular and tissue damages occurring in redox-dependent processes involving reactive oxygen and nitrogen species as well as lipid peroxidation-derived reactive aldehydes including acrolein, MDA and HNE.

Consistent with that are the findings on the bioactivities or natural and synthetic antioxidants targeting reactive aldehydes as second messengers of free radicals in vitro or in vivo presented in the remaining papers of this Special Issue [14–17]. The authors of one review and two original papers were studying the structure-activity relations of particular plant extracts on their chemical composition. This might help us to better understand their activity principles [14–16], while in the last article of this Special Issue, the authors studied the relationship between antioxidant and growth regulating effects of synthetic chemical substances, notably of 1,4-dihydropyridine derivatives (DHS) [17]. Namely, various DHPs are known for their pleiotropic activity, some also act as antioxidants that are already used for UV-protection or as antihypertensive agents. In their original in vitro study using several well-known or newly synthesized DHPs to treat human osteoblast-like cells, the authors revealed some DHPs as possible therapeutic agents for osteoporosis. However, further research is needed to elucidate their bioactivity mechanisms in respect to signaling pathways involving HNE and related second messengers of free radicals [17].

Similarly, although working on a very different in vitro model of human skin cells treated with sea buckthorn seed oil, another group analyzed the effects of the particular oil on the redox balance and lipid metabolism in UV irradiated skin cells. This research aimed to examine whether the plant oil can have the UV-protective effect [16]. By doing so, the authors found beneficial effects of the buckthorn seed oil, which decreased the production of lipid peroxidation products (including HNE) simultaneously decreasing the cannabinoid receptor expression in UV-irradiated keratinocytes and fibroblasts.

Another in vitro study used several cell lines to test if HNE might be a relevant factor of beneficial effects of the widely used *Aloe vera* extracts (AV) [15]. This study found that the cell-type specific effects of AV, by itself was not toxic for any type of cells, while it modulated the cellular response to oxidative stress induced by hydrogen peroxide. Of particular relevance, it was found that high antioxidant levels of the AV did not interfere with enhanced cellular accumulation of the HNE-protein adducts in human endothelial cells, as revealed by the genuine cell-based ELISA specific for HNE-His, which was used for the first time. The authors concluded that these findings might help in understanding the activity principles of AV, particularly if used for the promotion of wound healing and/or for adjuvant cancer treatments.

Some options for the modulation of lipid peroxidation pathophysiology by plant extracts reach in antioxidants were eventually summarized in the review on the relationship between biological activities of such extracts and their chemical composition in the article focusing on the evening primrose extracts [14]. The authors of this review point to the biomedical use of the evening primrose oil (EPO) rich in linoleic acid (70–74%) and linolenic acid (8–10%), which are precursors of anti-inflammatory eicosanoids. Thus, EPO supplementation may result in an increase in plasma levels of linolenic acid and its metabolite dihomo-linolenic acid, which is oxidized by lipoxygenase (15-LOX) to 15-hydroxyeicosatrienoic acid (15-HETrE) or can be, under the influence of cyclooxygenase (COX), metabolized to series 1 prostaglandins, which exert anti-inflammatory and anti-proliferative properties. In addition, linolenic acid itself may suppress the production of inflammatory cytokines. Since linoleic acid is also a major source of HNE, one may assume that lipid peroxidation generating HNE could be also important for the multiple biological effects of EPO, as suggested for the *Aloe vera* extract in the paper described above [15].

In conclusion, more research is needed to evaluate, using advanced analytical methods and translation models, how the natural and/or synthetic antioxidants interfere with the pathophysiology of lipid peroxidation. Yet, by doing so, we could increase not only our understanding of this important field but also support the development of the modern integrative biomedicine for which both antioxidants and second messengers of free radicals, represented by HNE, are of highest importance.

Conflicts of Interest: The author declares no conflict of interest.

References

1. Wonisch, W.; Kohlwein, S.; Schaur, J.; Tatzber, F.; Guttenberger, H.; Žarković, N.; Winkler, R.; Esterbauer, H. Treatment of the budding yeast (*Saccharomyces cerevisiae*) with the lipid peroxidation product 4-HNE provokes a temperorary cell cycle arrest in G1 phase. *Free Radic. Biol. Med.* **1998**, *25*, 682–687. [CrossRef]
2. Fedorova, M.; Zarkovic, N. Preface to the Special Issue on 4-Hydroxynonenal and Related Lipid Oxidation Products. *Free Radic. Biol. Med.* **2017**, *111*, 1. [CrossRef] [PubMed]
3. Žarković, K.; Uchida, K.; Kolenc, D.; Hlupic, L.J.; Žarković, N. Tissue distribution of lipid peroxidation product acrolein in human colon carcinogenesis. *Free Radic. Res.* **2006**, *40*, 543–552. [CrossRef] [PubMed]
4. Zarkovic, K.; Jakovcevic, A.; Zarkovic, N. Contribution of the HNE-immunohistochemistry to modern pathological concepts of major human diseases. *Free Radic. Biol. Med.* **2017**, *111*, 110–125. [CrossRef] [PubMed]
5. Poli, G.; Zarkovic, N. Editorial Introduction to the Special Issue on 4-Hydroxynonenal and Related Lipid Oxidation Products. *Free Radic. Biol. Med.* **2017**, *111*, 2–5. [CrossRef] [PubMed]
6. Gverić-Ahmetašević, S.; Borović Šunjić, S.; Skala, H.; Andrišić, L.; Štroser, M.; Žarković, K.; Škreblin, S.; Tatzber, F.; Čipak, A.; Jaganjac, M.; et al. Oxidative stress in small-for-gestational age (SGA) term newborns and their mothers. *Free Radic. Res.* **2009**, *43*, 376–384. [CrossRef] [PubMed]
7. Milkovic, L.; Siems, W.; Siems, R.; Zarkovic, N. Oxidative stress and antioxidants in carcinogenesis and integrative therapy of cancer. *Curr. Pharm. Des.* **2014**, *20*, 6529–6542. [CrossRef] [PubMed]
8. Walters, J.L.H.; De Iuliis, G.D.; Nixon, B.; Bromfield, E.G. Oxidative Stress in the Male Germline: A Review of Novel Strategies to Reduce 4-Hydroxynonenal Production. *Antioxidants* **2018**, *7*, 132. [CrossRef] [PubMed]
9. Barrera, G.; Pizzimenti, S.; Daga, M.; Dianzani, C.; Arcaro, A.; Cetrangolo, G.P.; Giordano, G.; Cucci, M.A.; Graf, M.; Gentile, F. Lipid Peroxidation-Derived Aldehydes, 4-Hydroxynonenal and Malondialdehyde in Aging-Related Disorders. *Antioxidants* **2018**, *7*, 102. [CrossRef] [PubMed]
10. Gesslbauer, B.; Kuerzl, D.; Valpatic, N.; Bochkov, V. Unbiased Identification of Proteins Covalently Modified by Complex Mixtures of Peroxidized Lipids Using a Combination of Electrophoretic Mobility Band Shift with Mass Spectrometry. *Antioxidants* **2018**, *7*, 116. [CrossRef] [PubMed]
11. Esterbauer, H.; Schaur, R.J.; Zollner, H. Chemistry and Biochemistry of 4-hydroxynonenal, malonaldehyde and related aldehydes. *Free Radic. Biol. Med.* **1991**, *11*, 81–128. [CrossRef]
12. Cherkas, A.; Zarkovic, N. 4-Hydroxynonenal in Redox Homeostasis of Gastrointestinal Mucosa: Implications for the Stomach in Health and Diseases. *Antioxidants* **2018**, *7*, 118. [CrossRef]
13. Gallelli, C.A.; Calcagnini, S.; Romano, A.; Koczwara, J.B.; De Ceglia, M.; Dante, D.; Villani, R.; Giudetti, A.M.; Cassano, T.; Gaetani, S. Modulation of the Oxidative Stress and Lipid Peroxidation by Endocannabinoids and Their Lipid Analogues. *Antioxidants* **2018**, *7*, 93. [CrossRef] [PubMed]
14. Timoszuk, M.; Bielawska, K.; Skrzydlewska, E. Evening Primrose (*Oenothera biennis*) Biological Activity Dependent on Chemical Composition. *Antioxidants* **2018**, *7*, 108. [CrossRef] [PubMed]

15. Cesar, V.; Jozić, I.; Begović, L.; Vuković, T.; Mlinarić, S.; Lepeduš, H.; Borović Šunjić, S.; Žarković, N. Cell-Type-Specific Modulation of Hydrogen Peroxide Cytotoxicity and 4-Hydroxynonenal Binding to Human Cellular Proteins In Vitro by Antioxidant *Aloe vera* Extract. *Antioxidants* **2018**, *7*, 125. [CrossRef] [PubMed]

16. Gęgotek, A.; Jastrząb, A.; Jarocka-Karpowicz, I.; Muszyńska, M.; Skrzydlewska, E. The Effect of Sea Buckthorn (*Hippophae rhamnoides* L.) Seed Oil on UV-Induced Changes in Lipid Metabolism of Human Skin Cells. *Antioxidants* **2018**, *7*, 110. [CrossRef] [PubMed]

17. Milkovic, L.; Vukovic, T.; Zarkovic, N.; Tatzber, F.; Bisenieks, E.; Kalme, Z.; Bruvere, I.; Ogle, Z.; Poikans, J.; Velena, A.; et al. Antioxidative 1,4-Dihydropyridine Derivatives Modulate Oxidative Stress and Growth of Human Osteoblast-Like Cells In Vitro. *Antioxidants* **2018**, *7*, 123. [CrossRef] [PubMed]

antioxidants

MDPI

Review

Oxidative Stress in the Male Germline: A Review of Novel Strategies to Reduce 4-Hydroxynonenal Production

Jessica L. H. Walters, Geoffry N. De Iuliis, Brett Nixon [†] and Elizabeth G. Bromfield *,[†]

Priority Research Centre for Reproductive Science, School of Environmental and Life Sciences, Discipline of Biological Sciences, University of Newcastle, Callaghan, NSW 2380, Australia; jwalters1@uon.edu.au (J.L.H.W.); geoffry.deiuliis@newcastle.edu.au (G.N.D.I.); Brett.nixon@newcastle.edu.au (B.N.)
* Correspondence: Elizabeth.bromfield@newcastle.edu.au; Tel.: +61-2-4921-6267
† These authors contributed equally to this work.

Received: 16 August 2018; Accepted: 26 September 2018; Published: 3 October 2018

Abstract: Germline oxidative stress is intimately linked to several reproductive pathologies including a failure of sperm-egg recognition. The lipid aldehyde 4-hydroxynonenal (4HNE) is particularly damaging to the process of sperm-egg recognition as it compromises the function and the stability of several germline proteins. Considering mature spermatozoa do not have the capacity for de novo protein translation, 4HNE modification of proteins in the mature gametes has uniquely severe consequences for protein homeostasis, cell function and cell survival. In somatic cells, 4HNE overproduction has been attributed to the action of lipoxygenase enzymes that facilitate the oxygenation and degradation of ω-6 polyunsaturated fatty acids (PUFAs). Accordingly, the arachidonate 15-lipoxygenase (ALOX15) enzyme has been intrinsically linked with 4HNE production, and resultant pathophysiology in various complex conditions such as coronary artery disease and multiple sclerosis. While ALOX15 has not been well characterized in germ cells, we postulate that ALOX15 inhibition may pose a new strategy to prevent 4HNE-induced protein modifications in the male germline. In this light, this review focuses on (i) 4HNE-induced protein damage in the male germline and its implications for fertility; and (ii) new methods for the prevention of lipid peroxidation in germ cells.

Keywords: male fertility; oxidative stress; 4-hydroxynonenal (4HNE); arachidonate 15-lipoxygenase (ALOX15); lipid peroxidation; reactive oxygen species (ROS)

---

1. Introduction: Fertility and Oxidative Stress

A decline in fertility rates is becoming an increasingly prevalent issue worldwide, with current estimates indicating that 1 in every 6 couples experience issues with conception [1]. Furthermore, the contribution of male factor infertility accounts for up to half of these cases [2]. The leading cause of male infertility stems from a loss of sperm function, ultimately resulting in a loss of fertilization potential [3]. This loss in function is causatively linked to oxidative stress within the cell [4,5] driven by the presence and/or overproduction of intracellular reactive oxygen species (ROS). Reactive oxygen species are oxygen-containing molecules that can contain unpaired electrons (radicals) or be non-radical oxidizing agents [6]. The consequences of ROS are realized through redox reactions with a great number of biological substrates, producing either further reactive products or oxidized biomolecules. Within spermatozoa, low levels of ROS are essential for promoting key stages of development. For instance, ROS actively participate in metabolic pathways during sperm activation, which leads to cholesterol efflux, cyclic adenosine monophosphate (cAMP) production and tyrosine phosphorylation, important events that contribute to fertilization competence [5,7–9]. However, if intracellular ROS production escalates beyond the buffering antioxidant capacity of the cell in a state of oxidative stress, the redox biochemistry leads to damaging effects such as lipid peroxidation, organelle degradation, DNA damage and eventually cell death [10,11]. Typically, antioxidants, which counteract and protect against oxidative stress, are housed within the cytoplasm and mitochondria of somatic cells [12,13]. However, spermiogenesis, a process that gives rise to the unique architecture of mature spermatozoa, results in significant cytoplasmic depletion [14,15], thereby diminishing antioxidant capacity in the spermatozoon [16]. Furthermore, during testicular maturation, there is an enrichment of long chain poly-unsaturated fatty acids (PUFAs) in the sperm plasma membrane, which can serve as important substrates for lipid peroxidation [10]. Indeed, PUFAs such as arachidonic acid, linoleic acid and docosahexaenoic acid are enriched within the sperm plasma membrane [17,18], and can be broken down into cytotoxic lipid aldehydes that promote cellular damage and the dysregulation of cell function [19]. Common metabolites of lipid peroxidation within spermatozoa include reactive aldehyde compounds such as 4-hydroxynonenal (4HNE) and malondialdehyde (MDA) [19–21]. Herein, we review literature pertaining to the reactivity, production and prevention of these cytotoxic lipid peroxidation products in the male germline.

2. Aldehydes in the Male Germline

In developing male germ cells and mature spermatozoa, two of the primary aldehyde products of lipid peroxidation that have been reported to cause cellular damage are MDA and 4HNE [19,22]. Increased levels of MDA are linked to a reduction in sperm concentration, normal morphology and motility [23,24]. Similarly, MDA is present at higher levels within the sperm of infertile men and is thought to initiate a loss of motility, reduction in sperm concentration and atypical morphology [24]. The levels of 4HNE within spermatozoa are positively correlated with mitochondrial superoxide formation [10], suggesting that elevated levels of 4HNE place sperm cells under increased levels of oxidative stress. Accordingly, the presence of 4HNE has been linked to numerous adverse effects on sperm function including a decline in motility, morphology, the capacity to acrosome react, and to engage in interactions with the zona pellucida of oocytes [19,25,26]. Specifically, the exposure of biomolecules to 4HNE stimulates an upregulation of mitochondrial ROS, generating a cascade of oxidative stress within human spermatozoa [19], as depicted in Figure 1.

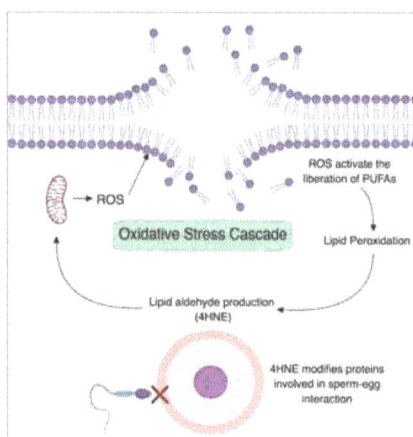

Figure 1. The cascade of oxidative stress in human spermatozoa. Mitochondrial reactive oxygen species (ROS) are produced and initiate the breakdown of the lipid plasma membrane. This promotes lipid peroxidation and the production of cytotoxic lipid aldehydes such as 4-hydroxynonenal (4HNE). In turn, 4HNE upregulates ROS production while causing an overall decline in cell function, ultimately impairing sperm-egg interaction. Figure created with BioRender.

Overproduction of 4HNE within sperm cells is linked to a reduction in sperm motility [26] and sperm-zona pellucida (ZP) interaction mediated by the molecular chaperone heat shock protein A2 (HSPA2) [25], and an increase in cell death [19]. There are several non-enzymatic pathways for aldehyde production, the best characterized being Fenton reactions, whereby ferrous iron (Fe^{2+}) within the cell is able to interact with lipids (LOOH) allowing the formation of lipid hydroperoxides ($LO^{\bullet}$) as shown in Equation (1) [27] and the production of aldehydes (as reviewed by Spiteller and Ayala et al.) [20,27].

$$LOOH + Fe^{2+} \rightarrow LO\bullet + Fe^{3+} + OH\bullet \tag{1}$$

Importantly, 4HNE is also produced via enzymatic pathways involving lipoxygenases such as arachidonate 15-lipoxygenase (ALOX15), with several studies highlighting that key metabolites such as 13-HpODE lead to the production of 4HNE [20,28], while MDA appears to be synthesized independent of lipoxygenase activity [29]. 4-hydroxynonenal is considered to be the most toxic lipid aldehyde produced within the cell [30]. This is due, at least in part, to its reactivity and subsequent capacity to alkylate proteins, generate DNA damage and ultimately cause cell death [19,25,26,31]. The reactivity of 4HNE lies in its ability to form Schiff bases and/or participate in Michael reactions. The preferential biological targets for these reactions are proteins, specifically primary amines such as lysine, but reactions with cysteine and histidine amino acid residues are also common [32,33]. A particular target for 4HNE adduction is succinate dehydrogenase (SDH) [19], a key protein in the electron transport chain within the mitochondria. Excess 4HNE has been shown to form adducts with SDH, which result in a loss of function. This ultimately facilitates electron leakage to electron acceptors in an unregulated fashion, increasing the production of ROS and eventually precipitating a state of oxidative stress within the cell [19]. Another such example in human spermatozoa is the molecular chaperone HSPA2 [34], which is also targeted for adduction by 4HNE [25]. Such modifications of HSPA2 results in a loss of its chaperoning ability and thus significantly attenuates the ability of the protein to coordinate the expression of receptors on the sperm surface; a maturational event that is critical for sperm-egg recognition [25]. Ultimately, this sequence of events culminates in a severely reduced capacity for fertilization [25,26].

Overall, the production of 4HNE has been shown to have a direct effect on the function of its protein targets, leading to cellular damage in the male germline as well as other cell types. Therefore, targeting the lipoxygenases responsible for the production of these reactive aldehydes may be an important strategy to both counter the onset of oxidative stress and reduce the cellular damage generated by 4HNE. Here, we investigate in more detail the involvement of lipoxygenase proteins in the enzymatic production of 4HNE.

3. Mechanisms for the Generation of 4HNE: A Focus on Lipoxygenase Proteins

Lipoxygenase proteins are a highly conserved family of enzymes that are ubiquitously found in plants [35,36], fungi [37] and mammals [38], but are rarely found in lower eukaryotes and prokaryotes and are absent in archaea and viruses [38–40]. Mammalian lipoxygenases typically consist of singular polypeptide chains, two functional domains and a molecular mass of ~75–80 kDa [41–43]. The C terminus contains the catalytic domain, while the N terminus is involved in processes governing membrane binding and interaction with substrates [42]. The catalytic pocket of the enzyme coordinates a single, non-heme containing iron atom per molecule [41,44], which is actively involved in the redox reactions necessary to facilitate the selective peroxidation of PUFAs [41,45]. However, this two domain structure is not conserved across all prokaryotes [46], and the presence of manganese replaces iron in the catalytic site of some fungal lipoxygenases [47–49]. The classification system of lipoxygenases (ALOX-n) defines the carbon position where oxygenation takes place along the PUFA chain. Table 1 indicates the known paralogs of human lipoxygenases, their substrates and metabolic products.

PUFA substrates for ALOX15 include ω-6 fatty acids such as arachidonic and linoleic acid and the ω-3 fatty acid, docosahexaenoic acid [50]. The mechanisms underpinning lipoxygenase function are still not entirely understood. However, it is clear that the iron center can alternate between ferric (Fe^{3+}/active) and ferrous (Fe^{2+}/inactive) forms [43] and this redox activity assists in hydrogen abstraction (L-H→L) of PUFAs when the iron atom undergoes a reduction ($Fe^{3+}→Fe^{2+}$) [41,51]. This reaction mechanism anticipates that the enzyme is converted back to its active form through oxidation of the iron center ($Fe^{2+}→Fe^{3+}$) and oxygenation (L→LOO) of the PUFA [41,43]. Importantly, recent studies assessing the enzymatic action of ALOX15 have identified binding sites for allosteric inhibition, which will allow for further insight into its specific activity [52,53].

Table 1. Paralogs and metabolites of the family of human lipoxygenase enzymes.

Lipoxygenase Enzyme [1]	Substrates [2]	Metabolic Products	References
ALOX5	AA LA EPA	5-HpETE, 5-HETE and DGLA, Leukotrienes	[43,54]
ALOX12	AA LA EPA DGLA	12-HpETE, 12-HETE, 12-HPETre, 12-HEPE, 12-HPOTrE	[50,54,55]
ALOX15	AA LA DHA	15-HpETE, 15-HETE, 13-HpODE, 13-HODE, 17-HpDHA	[50,54,56,57]
ALOX12B	AA LA LωHC	12R-HpETE, 12R-HETE, 9R-HpODE, 9HωHC	[50,54,58]
ALOX15B	AA	15-HpETE, 15-HETE	[50,54,59]
ALOXE3	12(R)HpETE 9HωHC	Epoxyalchohols (metabolism of 12(R)-HpETE) 9TEHωHC	[54,60] [3]

[1,2] Paralogs of the lipoxygenase family are shown along with their corresponding substrates of arachidonic acid (AA, red), linoleic acid (LA, green), eicosapentanoic acid (EPA) and docosahexaenoic acid (DHA). Abbreviations: arachidonate lipoxygenase (ALOX), epidermal type lipoxygenase (ALOXE), hydroperoxyeicosatetraenoic acid (HpETE), hydroxyeicosatetraenoic acid (HETE), 12-hydroxyeicosapentaenoic acid (HEPE), Hydroperoxyeicosatrienoic acid (HPEtrE) hydroperoxyoctadecadienoic (HpODE), hydroxyoctadecadienoic (HODE), 12-hydroxy-9Z,13E,15-octadecatrienoic (12-HPOTrE) hydroperoxydocosahexaenoic acid (HpDHA) and Dihomo-γ-linoleic acid (DGLA), Linoleyl-ω-hydroxy ceramide (LωHC), 9(R)-hydroperoxyllinoleoyl-ω-hydroxy ceramide (9HωHC), 9(R)-10(R)-trans-epoxy-11E-13(R)-hydroxylinoleoyl-ω-hydroxy ceramide (9TEHωHC). [3] It is noted that under normoxic conditions ALOXE3 does not exhibit lipoxygenase activity [60].

Numerous studies have focused on the possible pathogenic implications of the lipoxygenase family, with a key focus on ALOX5 due to its role in the biosynthesis of leukotrienes, which are inflammatory mediators [61]. Leukotrienes can cause pathological inflammatory responses in diseases such as cystic fibrosis [62], inflammatory bowel disease [63] and asthma [64], thereby presenting a relationship between lipoxygenase activity and immune responses. Chronic inflammation has the potential to place cells under stress, which in turn can promote cell death or abnormal cell differentiation [65]; the latter of these, in turn, has the potential to promote tumorigenesis [66]. In the case of ALOX15, several studies have implicated this protein in the inflammation pathway of diseases such as colorectal cancer [67], prostate cancer [68] and chronic myeloid leukemia [69]. However, while the formation of 14,15-leukotrienes from ALOX15 has been proposed [70], the biological relevance of these specific compounds has not yet been explored.

Interestingly, ALOX15 activity has also been linked to obesity as the enzyme is highly expressed in omental tissue compared to the subcutaneous fat layer of obese patients [71]. Accordingly, analysis of ALOX15 transgenic mice supports a link between inflammation, obesity and insulin resistance [72]. Indeed, this study proposes that an overexpression of ALOX15 stimulates the production of pro-inflammatory mediators, which promote insulin resistance induced through a high fat diet [72]. In turn, insulin resistance results in an overall increased risk in developing type 2 diabetes and obesity [72]. It is now well established that obesity can have detrimental impacts on both maternal and paternal fertility, as well as embryo health and development [73,74]. Obesity in males, is linked with an increased time to conception and a decrease in sperm function [73]. With these lines of evidence, the activity of ALOX15 may have a systemic and indirect effect on male infertility through obesity, alongside the direct effects it may have within the male germline through 4HNE production. The imperative for understanding mechanisms of male infertility is further supported by the growing evidence that male fertility status may in fact be an effective indicator of general health of the individual [75–77]. Specifically, studies assessing the fertility of more than 40,000 males have revealed that important semen parameters such as volume, cell count, and morphology are directly correlated with life expectancy [76]. A similar link has also been observed in the context of the prevalence of infertility in diseased men experiencing inflammatory bowel disease [78], obesity [79–81], diabetes [82], hypertension and also sexually transmitted diseases such as chlamydia [83], human immunodeficiency virus (HIV), and hepatitis C [84]. Such data suggest that drivers of poor fertility may originate in systemic issues rather than being restricted to the male reproductive tract, again emphasizing the importance of gaining a better understanding of the fundamental aspects of infertility and its origins.

At this time, literature on ALOX15 in the male germline is very scarce. Nevertheless, analysis of ALOX15 within mature spermatozoa has indicated a putative role for the enzyme within the cytoplasmic droplet of mammalian species such as boar [85] and mouse [86]. These studies suggest that ALOX15 works in concert with the ubiquitin pathway to cause organelle degradation, assisting in the removal of the cytoplasmic droplet [85]. Additionally, the production of an ALOX15 knockout mouse model has shown that the loss of this enzyme does not compromise sperm production per se. However, the spermatozoa produced from null males exhibited atypical cytoplasmic droplet degradation during epididymal transit [85]. Earlier work provided an indication that the bull sperm acrosome reaction may be suppressed following lipoxygenase inhibition [87]. However, these data must be interpreted with caution owing to the use of non-specific lipoxygenase inhibitors, and the absence of substantiating evidence to illuminate the direct role of ALOX15 in the induction of acrosomal exocytosis. Recent studies have suggested a possible link between this lipoxygenase enzyme and oxidative stress propagation in human spermatozoa [88] and in mouse germ cells [89,90]. Using an immortalized spermatocyte cell line [GC-2spd(ts)], we have demonstrated that the treatment of these cells with an ALOX15 inhibitor resulted in significant reductions in 4HNE protein modifications and subsequent oxidative stress cascades [89]. However, direct evidence of the ability of PD146176 to inhibit ALOX15 function is yet to be established and further work is required to verify the function of ALOX15 in rodent models [91]. Despite these shortcomings, using a double knockout study, Brütsch and colleagues have established a clear link between ALOX15 activity and a key antioxidant, glutathione peroxidase 4 (GPX4), in mouse germ cells [90]. In this study, the inactivation of *Gpx4* (genotype $Gpx4^{+/-}$) led to significant sperm defects, including marked reductions in sperm motility (total, rapid and progressive). These $Gpx4^{+/-}$ mice correspondingly exhibited significantly reduced litter sizes compared to wild type mice. However, both the motility attributes and the litter sizes of the animals were significantly improved following a simultaneous knockout of the *Alox15* gene (i.e., genotype $Gpx4^{+/-}/Alox15^{-/-}$), thus implicating ALOX15 in the mediation of oxidative damage in the mouse [90].

In addition to these animal studies, we have recently reported on a possible role for ALOX15 in human spermatozoa using the selective ALOX15 inhibitor 6,11-dihydro[1]benzothiopyrano[4,3-b]indole (PD146176, Tocris). This (PD146176) inhibits ALOX15 through non-competitive and non-antioxidant means [92,93] and has previously been shown to reduce the production of specific ALOX15 metabolites such as 15-HPETE [94] and 13-HODE [95]. Though minimal studies have used PD146176 in spermatozoa, the use of this inhibitor in conjunction with an oxidative challenge has been documented to give rise to significant ROS reductions in neuronal cells [96]. This is consistent with our findings in human spermatozoa, that under oxidative stress conditions ALOX15 inhibition significantly decreased ROS production and lipid peroxidation levels while also improving the functional competence of sperm populations including their motility, acrosome reaction rates and ability to undergo sperm-egg interaction processes [88]. Importantly, such studies are also consistent with those completed in the context of neurological disorders such as Alzheimer's disease, where disease progression often relies on oxidative stress and the production of 4HNE. This lipid peroxide end product has been shown to promote the production of amyloid beta plaques and neuronal death [97–99]. Strikingly, these studies have demonstrated reduced amyloid plaque production with significant improvements in memory deficits through the inclusion of the same ALOX15 inhibitor, PD146176 [100,101]. These data provide further evidence for the use of

PD146176 as a potential therapeutic means to prevent pathologies induced through oxidative stress and lipid peroxidation.

4. Protecting the Germline from 4HNE-Induced Damage

There are increasing numbers of couples using assisted reproductive technologies (ART) to achieve conception. This has led to more than 5 million births since the invention of this technology [102]. While ART has undoubtedly changed the lives of many, such technologies are highly expensive and have a live birth success rate of no more than ~30% [103,104]. There may also be a level of risk associated with assisted conception where numerous studies have confirmed that higher levels of DNA damage are present in men with subfertility [105]. This presents the possibility that ARTs could be inadvertently using damaged sperm cells which may elevate the risk of adverse health outcomes for the offspring conceived through assisted reproduction [106,107]. Additionally, the lack of selection pressure on the gametes may eventually propagate further fertility issues for future generations. A major origin of sperm cell damage arises through the onset of oxidative stress. Congruent with DNA fragmentation, markers of oxidative stress are also elevated in the infertile population [108]. It is therefore without surprise that antioxidant supplementation is an extensively studied area for the mitigation of male infertility. Table 2 summarizes numerous studies that have examined male fertility following antioxidant supplementation and their corresponding reproductive outcome. This table was collated through examination of external literature as well as analysis of a variety of detailed reviews [109–111]. Interestingly, only 5 out of the 28 investigated studies presented improvements to pregnancy and live birth rates following antioxidant supplementation, with positive effects associated with astaxanthin [112], L-carnitine + L-acetyl carnitine [113], Menevit® [114], vitamin E [115] and zinc sulphate [116]. While some studies observed very high levels of variability when measuring semen parameters, the studies focusing on L-glutathione, lycopene, N-acetylcysteine + selenium, ubiquinone, selenium and zinc sulphate, consistently presented improvements in at least one or more semen parameter [116–122]. Other studies using antioxidants such as Co-enzyme Q10, folic acid + zinc sulphate, lycopene and L-carnitine + L-acetyl carnitine showed variation in effects between trials, with some studies reporting improvements to semen parameters [113,118,123–128], while others showing no positive effects [129–131]. Some of this variation may be attributed to intrinsic variations within each trial, such as dose regimes, methodology and the duration of treatments. Nonetheless, this variability, combined with a lack of clinical success in terms of increased pregnancy rates and live birth rates, highlights a clear need for further investigation into effective alternative strategies to prevent, or at least limit, ROS production in the male germline to improve a large subset of male fertility issues.

In seeking to account for the lack of consistent clinical success using regimens based on single antioxidant supplementation as a means for combating male infertility, it is possible that the scavenging nature of antioxidants [132] fails to provide direct protection against the cascades of lipid peroxidation and 4HNE production that ensue under conditions of oxidative stress. Interestingly, nucleophiles such as penicillamine have been shown to successfully reduce cellular ROS in both human spermatozoa and in oocytes [19,25,133,134]; effects that manifest in the recovery of sperm-oocyte interaction in vitro [25]. However, this antioxidant has serious off-target toxicity concerns [135], and thus investigation into the clinical utility of penicillamine is not possible. Another novel antioxidant formulation therapy in the male germline is Fertilix® (Cell Oxess, Ewing, USA), which has been shown to protect against DNA damage in antioxidant deficient mice [136]. However, clinical trials have yet to be performed to establish whether this therapy is an appropriate method for treating infertile men. Among alternative methods that have shown promise in protecting somatic cells from diseases linked to lipid peroxidation-dependent mechanisms [137–139], is the stabilization of the lipid membrane through deuteration [140]. Such success provides an important precedent to investigate the efficacy of this strategy to protect sperm membranes.

Table 2. Benefits of antioxidant supplementation for male fertility. A summary of tested antioxidants and their relative success for the improvement of male fertility as reviewed by Ahmadi et al., Ross et al., and Majzoub and Agarwal [109–111].

Antioxidant	Outcomes	References
Astaxanthin	Increased pregnancy rates	[112]
	Reduced oxidative stress	[112]
Co-enzyme Q10	Improved sperm motility	[123]
	Improved sperm concentration and morphology	[124]
	Altered antioxidant enzyme activity	[124,129]
	No improvements to sperm motility, concentration or morphology	[129]
Folic Acid + Zinc Sulphate	Improved sperm concentration	[125,126]
	No improvements to sperm motility, concentration or morphology	[130]
L-Glutathione	Improved motility	[117]
L-Carnitine + L-acetyl carnitine	Increased motility (progressive and total)	[127,128]
	No changes to motility or concentration	[131]
	Increased pregnancy rates and improved sperm concentration, motility and morphology	[113]
Lycopene	Improved sperm motility and concentration	[118]
Menevit	Improved pregnancy rates	[114]
N-acetylcysteine	Increased sperm concentration	[141]
	No significant increase in spontaneous pregnancies	[141]
	Improved sperm volume, motility and viscosity	[142]
	Reduced oxidative stress	[142]
N-acetylcysteine + Selenium	Improved sperm motility, concentration and morphology	[119]
Ubiquinone	Improved sperm motility, concentration and morphology	[143]
Vitamin E	Improved sperm motility	[115]
	Improved pregnancy rates	[115]
	Decreased lipid peroxidation products	[115]
Vitamin E + Vitamin C	No changes to motility or concentration	[144,145]
	Reduced DNA damage	[144]
	Improved ICSI outcomes	[146]
Vitamin E + Selenium	Improved morphology	[147]
	Improved sperm motility	[148]
	Decreased lipid peroxidation products	[148]
Selenium	Improved sperm motility	[120]
	No changes to sperm concentration	[120]
Zinc Sulphate	Improved semen volume, sperm motility and concentration	[116,121]
	Improved live birth rate	[116]
	Altered antioxidant enzyme activity	[122]

5. Conclusions

In this review we discuss strategies to alleviate oxidative stress in males suffering from fertility issues (summarized in Figure 2). Here we provide new perspectives on the lipoxygenase–lipid peroxidation pathway and discuss the merit of ALOX15 as a potential therapeutic target that could be exploited to protect human spermatozoa against oxidative stress, a key origin of poor cell function. Overall, this review highlights the importance of correct lipid metabolism in the maintenance of sperm function and fertility and provides the impetus to explore targeted, lipid-based antioxidant approaches to prevent lipid-peroxidation induced changes in the male germline.

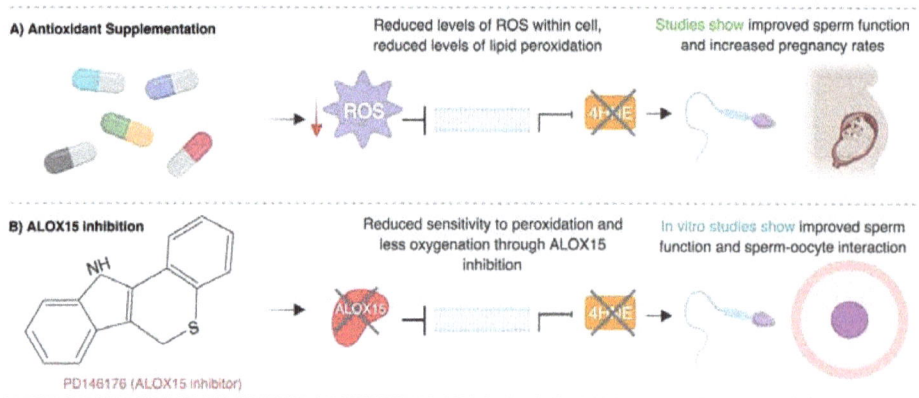

Figure 2. Potential strategies to protect human spermatozoa against oxidative stress. This model explores two strategies to protect against oxidative stress: antioxidant supplementation and arachidonate 15-lipoxygenase (ALOX15) inhibition. (**A**) Antioxidant supplementation has been shown to reduce levels of ROS, hence lipid peroxidation may be prevented, or its products scavenged, allowing sperm function to be improved. (**B**) ALOX15 inhibition in human sperm has been demonstrated to reduce lipid peroxidation and improve sperm function and sperm-oocyte interaction in vitro. Figure created with BioRender.

Author Contributions: J.L.H.W., G.N.D.I., B.N. and E.G.B. contributed to this review as follows; conceptualization, J.L.H.W. and E.G.B.; Investigation and data curation, J.L.H.W.; Writing—original draft preparation, J.L.H.W.; Writing—Review and editing, E.G.B., B.N. and G.N.D.; Funding acquisition, B.N. Project Administration, B.N., G.N.D.I. and E.G.B.

Funding: This study was funded through an NHMRC CJ Martin Fellowship (APP1138701) to E.G.B. and an NHMRC Project Grant (APP1101953) to B.N. J.L.H.W. is the recipient of an Australian Government Research Training Program (RTP) PhD Scholarship.

Conflicts of Interest: The authors declare that there is no conflict of interest.

References

1. Trussell, J. Optimal diagnosis and medical treatment of male infertility. *Semin. Reprod. Med.* **2013**, *31*, 235–236. [CrossRef] [PubMed]
2. Kumar, N.; Singh, A.K. Trends of male factor infertility, an important cause of infertility: A review of literature. *J. Hum. Reprod. Sci.* **2015**, *8*, 191–196. [CrossRef] [PubMed]
3. Liu, D.; Baker, H. Defective sperm–zona pellucida interaction: A major cause of failure of fertilization in clinical in-vitro fertilization. *Hum. Reprod.* **2000**, *15*, 702–708. [CrossRef] [PubMed]
4. Guthrie, H.; Welch, G. Effects of reactive oxygen species on sperm function. *Theriogenology* **2012**, *78*, 1700–1708. [CrossRef] [PubMed]
5. Aitken, R.J.; Baker, M.A.; Nixon, B. Are sperm capacitation and apoptosis the opposite ends of a continuum driven by oxidative stress? *Asian J. Androl.* **2015**, *17*, 633–639. [CrossRef] [PubMed]
6. Bayr, H. Reactive oxygen species. *Crit. Care Med.* **2005**, *33*, S498–S501. [CrossRef]
7. Aitken, J.R.; Nixon, B. Sperm capacitation: A distant landscape glimpsed but unexplored. *Mol. Hum. Reprod.* **2013**, *19*, 785–793. [CrossRef] [PubMed]
8. O'Flaherty, C.; Beorlegui, N.; Beconi, M. Reactive oxygen species requirements for bovine sperm capacitation and acrosome reaction. *Theriogenology* **1999**, *52*, 289–301. [CrossRef]
9. Aitken, R.J.; Gordon, E.; Harkiss, D.; Twigg, J.P.; Milne, P.; Jennings, Z.; Irvine, D.S. Relative impact of oxidative stress on the functional competence and genomic integrity of human spermatozoa. *Biol. Reprod.* **1998**, *59*, 1037–1046. [CrossRef] [PubMed]
10. Aitken, R.J.; Clarkson, J.S.; Fishel, S. Generation of reactive oxygen species, lipid peroxidation, and human sperm function. *Biol. Reprod.* **1989**, *41*, 183–197. [CrossRef] [PubMed]

11. Marnett, L.J. Oxy radicals, lipid peroxidation and DNA damage. *Toxicology* **2002**, *181*, 219–222. [CrossRef]

12. Chaves, F.J.; Mansego, M.L.; Blesa, S.; Gonzalez-Albert, V.; Jiménez, J.; Tormos, M.C.; Espinosa, O.; Giner, V.; Iradi, A.; Saez, G.; et al. Inadequate cytoplasmic antioxidant enzymes response contributes to the oxidative stress in human hypertension. *Am. J. Hypertens.* **2007**, *20*, 62–69. [CrossRef] [PubMed]

13. Birben, E.; Sahiner, U.M.; Sackesen, C.; Erzurum, S.; Kalayci, O. Oxidative stress and antioxidant defense. *World Allergy Organ. J.* **2012**, *5*, 9. [CrossRef] [PubMed]

14. Fischer, M.A.; Willis, J.; Zini, A. Human sperm DNA integrity: Correlation with sperm cytoplasmic droplets. *Urology* **2003**, *61*, 207–211. [CrossRef]

15. Cooper, T.G.; Yeung, C.H.; Fetic, S.; Sobhani, A.; Nieschlag, E. Cytoplasmic droplets are normal structures of human sperm but are not well preserved by routine procedures for assessing sperm morphology. *Hum. Reprod.* **2004**, *19*, 2283–2288. [CrossRef] [PubMed]

16. Sabeti, P.; Pourmasumi, S.; Rahiminia, T.; Akyash, F.; Talebi, A.R. Etiologies of sperm oxidative stress. *Int. J. Reprod. Biomed.* **2016**, *14*, 231–240. [CrossRef]

17. Lenzi, A.; Picardo, M.; Gandini, L.; Dondero, F. Lipids of the sperm plasma membrane: From polyunsaturated fatty acids considered as markers of sperm function to possible scavenger therapy. *Hum. Reprod. Update* **1996**, *2*, 246–256. [CrossRef] [PubMed]

18. Sanocka, D.; Kurpisz, M. Reactive oxygen species and sperm cells. *Reprod. Biol. Endocrinol.* **2004**, *2*, 12. [CrossRef] [PubMed]

19. Aitken, R.J.; Whiting, S.; De Iuliis, G.N.; McClymont, S.; Mitchell, L.A.; Baker, M.A. Electrophilic aldehydes generated by sperm metabolism activate mitochondrial reactive oxygen species generation and apoptosis by targeting succinate dehydrogenase. *J. Biol. Chem.* **2012**, *287*, 33048–33060. [CrossRef] [PubMed]

20. Ayala, A.; Muñoz, M.F.; Argüelles, S. Lipid peroxidation: Production, metabolism, and signaling mechanisms of malondialdehyde and 4-hydroxy-2-nonenal. *Oxid. Med. Cell. Longev.* **2014**, *2014*, 360438. [CrossRef] [PubMed]

21. Pizzimenti, S.; Ciamporcero, E.S.; Daga, M.; Pettazzoni, P.; Arcaro, A.; Cetrangolo, G.; Minelli, R.; Dianzani, C.; Lepore, A.; Gentile, F.; et al. Interaction of aldehydes derived from lipid peroxidation and membrane proteins. *Front. Physiol.* **2013**, *4*, 242. [CrossRef] [PubMed]

22. Moazamian, R.; Polhemus, A.; Connaughton, H.; Fraser, B.; Whiting, S.; Gharagozloo, P.; Aitken, R.J. Oxidative stress and human spermatozoa: Diagnostic and functional significance of aldehydes generated as a result of lipid peroxidation. *Mol. Hum. Reprod.* **2015**, *21*, 502–515. [CrossRef] [PubMed]

23. Hsieh, Y.-Y.; Chang, C.-C.; Lin, C.-S. Seminal malondialdehyde concentration but not glutathione peroxidase activity is negatively correlated with seminal concentration and motility. *Int. J. Biol. Sci.* **2006**, *2*, 23–29. [CrossRef] [PubMed]

24. Mehrotra, A.; Katiyar, D.K.; Agarwal, A.; Das, V.; Pant, K.K. Role of total antioxidant capacity and lipid peroxidation in fertile and infertile men. *Biomed. Res.* **2013**, *24*, 347–352.

25. Bromfield, E.G.; Aitken, R.J.; Anderson, A.L.; McLaughlin, E.A.; Nixon, B. The impact of oxidative stress on chaperone-mediated human sperm–egg interaction. *Hum. Reprod.* **2015**, *30*, 2597–2613. [CrossRef] [PubMed]

26. Aitken, R.J.; Gibb, Z.; Mitchell, L.A.; Lambourne, S.R.; Connaughton, H.S.; De Iuliis, G.N. Sperm motility is lost in vitro as a consequence of mitochondrial free radical production and the generation of electrophilic aldehydes but can be significantly rescued by the presence of nucleophilic thiols. *Biol. Reprod.* **2012**, *87*, 110. [CrossRef] [PubMed]

27. Spiteller, G. Is lipid peroxidation of polyunsaturated acids the only source of free radicals that induce aging and age-related diseases? *Rejuv. Res.* **2010**, *13*, 91–103. [CrossRef] [PubMed]

28. Schneider, C.; Tallman, K.A.; Porter, N.A.; Brash, A.R. Two distinct pathways of formation of 4-hydroxynonenal mechanisms of nonenzymatic transformation of the 9-and 13-hydroperoxides of linoleic acid to 4-hydroxyalkenals. *J. Biol. Chem.* **2001**, *276*, 20831–20838. [CrossRef] [PubMed]

29. Morisaki, N.; Lindsey, J.A.; Stitts, J.M.; Zhang, H.; Cornwell, D.G. Fatty acid metabolism and cell proliferation. V. Evaluation of pathways for the generation of lipid peroxides. *Lipids* **1984**, *19*, 381–394. [CrossRef] [PubMed]

30. Shoeb, M.; Naseem, H.A.; Satish, K.S.; Kota, V.R. 4-Hydroxynonenal in the pathogenesis and progression of human diseases. *Curr. Med. Chem.* **2014**, *21*, 230–237. [CrossRef] [PubMed]

31. Aitken, R.J.; Koppers, A.J. Apoptosis and DNA damage in human spermatozoa. *Asian J. Androl.* **2011**, *13*, 36–42. [CrossRef] [PubMed]

32. Rauniyar, N.; Prokai, L. Detection and identification of 4-hydroxy-2-nonenal Schiff-base adducts along with products of Michael addition using data-dependent neutral loss-driven MS3 acquisition: Method evaluation through an in vitro study on cytochrome c oxidase modifications. *Proteomics* **2009**, *9*, 5188–5193. [CrossRef] [PubMed]

33. Doorn, J.A.; Petersen, D.R. Covalent modification of amino acid nucleophiles by the lipid peroxidation products 4-hydroxy-2-nonenal and 4-oxo-2-nonenal. *Chem. Res. Toxicol.* **2002**, *15*, 1445–1450. [CrossRef] [PubMed]

34. Redgrove, K.A.; Nixon, B.; Baker, M.A.; Hetherington, L.; Baker, G.; Liu, D.Y.; Aitken, R.J. The molecular chaperone HSPA2 plays a key role in regulating the expression of sperm surface receptors that mediate sperm-egg recognition. *PLoS ONE* **2012**, *7*, e50851. [CrossRef] [PubMed]

35. Grechkin, A. Recent developments in biochemistry of the plant lipoxygenase pathway. *Prog. Lipid Res.* **1998**, *37*, 317–352. [CrossRef]

36. Shibata, D.; Axelrod, B. Plant lipoxygenases. *J. Lipid Med. Cell Signal.* **1995**, *12*, 213–228. [CrossRef]

37. Heshof, R.; Jylhä, S.; Haarmann, T.; Jørgensen, A.L.; Dalsgaard, T.K.; de Graaff, L.H. A novel class of fungal lipoxygenases. *Appl. Microbiol. Biotechnol.* **2014**, *98*, 1261–1270. [CrossRef] [PubMed]

38. Horn, T.; Adel, S.; Schumann, R.; Sur, S.; Kakularam, K.R.; Polamarasetty, A.; Redanna, P.; Kuhn, H.; Heydeck, D. Evolutionary aspects of lipoxygenases and genetic diversity of human leukotriene signaling. *Prog. Lipid Res.* **2015**, *57*, 13–39. [CrossRef] [PubMed]

39. Vance, R.E.; Hong, S.; Gronert, K.; Serhan, C.N.; Mekalanos, J.J. The opportunistic pathogen Pseudomonas aeruginosa carries a secretable arachidonate 15-lipoxygenase. *Proc. Natl. Acad. Sci. USA* **2004**, *101*, 2135–2139. [CrossRef] [PubMed]

40. Hansen, J.; Garreta, A.; Benincasa, M.; Fusté, M.C.; Busquets, M.; Manresa, A. Bacterial lipoxygenases, a new subfamily of enzymes? A phylogenetic approach. *Appl. Microbiol. Biotechnol.* **2013**, *97*, 4737–4747. [CrossRef] [PubMed]

41. Ivanov, I.; Kuhn, H.; Heydeck, D. Structural and functional biology of arachidonic acid 15-lipoxygenase-1 (ALOX15). *Gene* **2015**, *573*, 1–32. [CrossRef] [PubMed]

42. Kuhn, H.; Banthiya, S.; van Leyen, K. Mammalian lipoxygenases and their biological relevance. *Biochim. Biophys. Acta (BBA)-Mol. Cell Biol. Lipids* **2015**, *1851*, 308–330. [CrossRef] [PubMed]

43. Brash, A.R. Lipoxygenases: Occurrence, functions, catalysis, and acquisition of substrate. *J. Biol. Chem.* **1999**, *274*, 23679–23682. [CrossRef] [PubMed]

44. Chasteen, N.D.; Grady, J.K.; Skorey, K.I.; Neden, K.J.; Riendeau, D.; Percival, M.D. Characterization of the non-heme iron center of human 5-lipoxygenase by electron paramagnetic resonance, fluorescence, and ultraviolet-visible spectroscopy: Redox cycling between ferrous and ferric states. *Biochemistry* **1993**, *32*, 9763–9771. [CrossRef] [PubMed]

45. Feussner, I.; Wasternack, C. The lipoxygenase pathway. *Ann. Rev. Plant Biol.* **2002**, *53*, 275–297. [CrossRef] [PubMed]

46. Garreta, A.; Val-Moraes, S.P.; García-Fernández, Q.; Busquets, M.; Juan, C.; Oliver, A.; Ortiz, A.; Gaffney, B.J.; Fita, I.; Manresa, À.; et al. Structure and interaction with phospholipids of a prokaryotic lipoxygenase from Pseudomonas aeruginosa. *FASEB J.* **2013**, *27*, 4811–4821. [CrossRef] [PubMed]

47. Wennman, A.; Oliw, E.H.; Karkehabadi, S.; Chen, Y. Crystal structure of manganese lipoxygenase of the rice blast fungus Magnaporthe oryzae. *J. Biol. Chem.* **2016**. [CrossRef] [PubMed]

48. Wennman, A.; Jernerén, F.; Magnuson, A.; Oliw, E.H. Expression and characterization of manganese lipoxygenase of the rice blast fungus reveals prominent sequential lipoxygenation of α-linolenic acid. *Arch. Biochem. Biophys.* **2015**, *583*, 87–95. [CrossRef] [PubMed]

49. Su, C.; Oliw, E.H. Manganese lipoxygenase purification and characterization. *J. Biol. Chem.* **1998**, *273*, 13072–13079. [CrossRef] [PubMed]

50. Dobrian, A.D.; Lieb, D.C.; Cole, B.K.; Taylor-Fishwick, D.A.; Chakrabarti, S.K.; Nadler, J.L. Functional and pathological roles of the 12-and 15-lipoxygenases. *Prog. Lipid Res.* **2011**, *50*, 115–131. [CrossRef] [PubMed]

51. Ivanov, I.; Heydeck, D.; Hofheinz, K.; Roffeis, J.; O'Donnell, V.B.; Kuhn, H.; Walther, M. Molecular enzymology of lipoxygenases. *Arch. Biochem. Biophys.* **2010**, *503*, 161–174. [CrossRef] [PubMed]

52. Meng, H.; Dai, Z.; Zhang, W.; Liu, Y.; Lai, L. Molecular mechanism of 15-lipoxygenase allosteric activation and inhibition. *Phys. Chem. Chem. Phys.* **2018**, *20*, 14785–14795. [CrossRef] [PubMed]

53. Meng, H.; McClendon, C.L.; Dai, Z.; Li, K.; Zhang, X.; He, S.; Shang, E.; Liu, Y.; Lai, L. Discovery of novel 15-lipoxygenase activators to shift the human arachidonic acid metabolic network toward inflammation resolution. *J. Med. Chem.* **2015**, *59*, 4202–4209. [CrossRef] [PubMed]

54. Wisastra, R.; Dekker, F.J. Inflammation, cancer and oxidative lipoxygenase activity are intimately linked. *Cancers* **2014**, *6*, 1500–1521. [CrossRef] [PubMed]

55. Tersey, S.A.; Bolanis, E.; Holman, T.R.; Maloney, D.J.; Nadler, J.L.; Mirmira, R.G. Minireview: 12-lipoxygenase and islet β-cell dysfunction in diabetes. *Mol. Endocrinol.* **2015**, *29*, 791–800. [CrossRef] [PubMed]

56. Vangaveti, V.; Baune, B.T.; Kennedy, R.L. Hydroxyoctadecadienoic acids: Novel regulators of macrophage differentiation and atherogenesis. *Ther. Adv. Endocrinol. MeTable* **2010**, *1*, 51–60. [CrossRef] [PubMed]

57. Chang, J.; Jiang, L.; Wang, Y.; Yao, B.; Yang, S.; Zhang, B.; Zhang, M.Z. 12/15 lipoxygenase regulation of colorectal tumorigenesis is determined by the relative tumor levels of its metabolite 12-HETE and 13-HODE in animal models. *Oncotarget* **2015**, *6*, 2879. [CrossRef] [PubMed]

58. Muñoz-Garcia, A.; Thomas, C.P.; Keeney, D.S.; Zheng, Y.; Brash, A.R. The importance of the lipoxygenase-hepoxilin pathway in the mammalian epidermal barrier. *Biochim. Biophys. Acta (BBA)-Mol. Cell Biol. Lipids* **2014**, *1841*, 401–408. [CrossRef] [PubMed]

59. Brash, A.R.; Boeglin, W.E.; Chang, M.S. Discovery of a second 15S-lipoxygenase in humans. *Proc. Natl. Acad. Sci. USA* **1997**, *94*, 6148–6152. [CrossRef] [PubMed]

60. Yu, Z.; Schneider, C.; Boeglin, W.E.; Marnett, L.J.; Brash, A.R. The lipoxygenase gene ALOXE3 implicated in skin differentiation encodes a hydroperoxide isomerase. *Proc. Natl. Acad. Sci. USA* **2003**, *100*, 9162–9167. [CrossRef] [PubMed]

61. Haeggstrom, J.Z.; Funk, C.D. Lipoxygenase and leukotriene pathways: Biochemistry, biology, and roles in disease. *Chem. Rev.* **2011**, *111*, 5866–5898. [CrossRef] [PubMed]

62. Lawrence, R.; Sorrell, T. Eicosapentaenoic acid in cystic fibrosis: Evidence of a pathogenetic role for leukotriene B. 4. *Lancet* **1993**, *342*, 465–469. [CrossRef]

63. Sharon, P.; Stenson, W.F. Enhanced Synthesis of Leukotriene B4 by Colonic Mucosa in Inflammatory Bowel. *Gastroenterology* **1984**, *86*, 453–460. [PubMed]

64. O'Byrne, P.M. Leukotrienes in the pathogenesis of asthma. *Chest J.* **1997**, *111*, 27S–34S. [CrossRef]

65. Rock, K.L.; Kono, H. The inflammatory response to cell death. *Ann. Rev. Pathol.* **2008**, *3*, 99–126. [CrossRef] [PubMed]

66. Okada, F. Inflammation-related carcinogenesis: Current findings in epidemiological trends, causes and mechanisms. *Yonago Acta Med.* **2014**, *57*, 65–72. [PubMed]

67. Mao, F.; Wang, M.; Wang, J.; Xu, W.R. The role of 15-LOX-1 in colitis and colitis-associated colorectal cancer. *Inflamm. Res.* **2015**, *64*, 661–669. [CrossRef] [PubMed]

68. Kelavkar, U.P.; Harya, N.S.; Hutzley, J.; Bacich, D.J.; Monzon, F.A.; Chandran, U.; Dhir, R.; O'Keefe, D.S. DNA methylation paradigm shift: 15-lipoxygenase-1 upregulation in prostatic intraepithelial neoplasia and prostate cancer by atypical promoter hypermethylation. *Prostaglandins Other Lipid Med.* **2007**, *82*, 185–197. [CrossRef] [PubMed]

69. Chen, Y.; Peng, C.; Abraham, S.A.; Shan, Y.; Guo, Z.; Desouza, N.; Cheloni, G.; Li, D.; Holyoake, T.L.; Li, S. Arachidonate 15-lipoxygenase is required for chronic myeloid leukemia stem cell survival. *J. Clin. Investig.* **2014**, *124*, 3847–3862. [CrossRef] [PubMed]

70. Bryant, R.W.; Schewe, T.; Rapoport, S.M.; Bailey, J.M. Leukotriene formation by a purified reticulocyte lipoxygenase enzyme. Conversion of arachidonic acid and 15-hydroperoxyeicosatetraenoic acid to 14, 15-leukotriene A4. *J. Biol. Chem.* **1985**, *260*, 3548–3555. [PubMed]

71. Dobrian, A.D.; Lieb, D.C.; Ma, Q.; Lindsay, J.W.; Cole, B.K.; Ma, K.; Chakrabarti, S.K.; Kuhn, N.S.; Wohlgemuth, S.D.; Fontana, M.; et al. Differential expression and localization of 12/15 lipoxygenases in adipose tissue in human obese subjects. *Biochem. Biophys. Res. Commun.* **2010**, *403*, 485–490. [CrossRef] [PubMed]

72. Sears, D.D.; Miles, P.D.; Chapman, J.; Ofrecio, J.M.; Almazan, F.; Thapar, D.; Miller, Y.I. 12/15-lipoxygenase is required for the early onset of high fat diet-induced adipose tissue inflammation and insulin resistance in mice. *PLoS ONE* **2009**, *4*, e7250. [CrossRef] [PubMed]

73. Palmer, N.O.; Bakos, H.W.; Fullston, T.; Lane, M. Impact of obesity on male fertility, sperm function and molecular composition. *Spermatogenesis* **2012**, *2*, 253–263. [CrossRef] [PubMed]

74. Leddy, M.A.; Power, M.L.; Schulkin, J. The impact of maternal obesity on maternal and fetal health. *Rev. Obstet. Gynecol.* **2008**, *1*, 170–178. [PubMed]

75. Eisenberg, M.L.; Li, S.; Behr, B.; Cullen, M.R.; Galusha, D.; Lamb, D.J.; Lipshultz, L.I. Semen quality, infertility and mortality in the USA. *Hum. Reprod.* **2014**, *29*, 1567–1574. [CrossRef] [PubMed]

76. Jensen, T.K.; Jacobsen, R.; Christensen, K.; Nielsen, N.C.; Bostofte, E. Good Semen Quality and Life Expectancy: A Cohort Study of 43,277 Men. *Am. J. Epidemiol.* **2009**, *170*, 559–565. [CrossRef] [PubMed]

77. Eisenberg, M.L.; Li, S.; Behr, B.; Pera, R.R.; Cullen, M.R. Relationship between semen production and medical comorbidity. *Fertil. Steril.* **2015**, *103*, 66–71. [CrossRef] [PubMed]

78. Rossato, M.; Foresta, C. Antisperm antibodies in inflammatory bowel disease. *Arch. Intern. Med.* **2004**, *164*, 2281–2283. [CrossRef] [PubMed]

79. Sallmén, M.; Sandler, D.P.; Hoppin, J.A.; Blair, A.; Baird, D.D. Reduced fertility among overweight and obese men. *Epidemiology* **2006**, *17*, 520–523. [CrossRef] [PubMed]

80. Nguyen, R.H.; Wilcox, A.J.; Skjærven, R.; Baird, D.D. Men's body mass index and infertility. *Hum. Reprod.* **2007**, *22*, 2488–2493. [CrossRef] [PubMed]

81. Ramlau-Hansen, C.H.; Thulstrup, A.M.; Nohr, E.A.; Bonde, J.P.; Sørensen, T.I.; Olsen, J. Subfecundity in overweight and obese couples. *Hum. Reprod.* **2007**, *22*, 1634–1637. [CrossRef] [PubMed]

82. Agbaje, I.M.; Rogers, D.A.; McVicar, C.M.; McClure, N.; Atkinson, A.B.; Mallidis, C.; Lewis, S.E. Insulin dependant diabetes mellitus: Implications for male reproductive function. *Hum. Reprod.* **2007**, *22*, 1871–1877. [CrossRef] [PubMed]

83. Mazzoli, S.; Cai, T.; Addonisio, P.; Bechi, A.; Mondaini, N.; Bartoletti, R. Chlamydia trachomatis infection is related to poor semen quality in young prostatitis patients. *Eur. Urol.* **2010**, *57*, 708–714. [CrossRef] [PubMed]

84. Lorusso, F.; Palmisano, M.; Chironna, M.; Vacca, M.; Masciandaro, P.; Bassi, E.; Luigi, S.L.; Depalo, R. Impact of chronic viral diseases on semen parameters. *Andrologia* **2010**, *42*, 121–126. [CrossRef] [PubMed]

85. Fischer, K.A.; Van Leyen, K.; Lovercamp, K.W.; Manandhar, G.; Sutovsky, M.; Feng, D.; Safranski, T.; Sutovsky, P. 15-Lipoxygenase is a component of the mammalian sperm cytoplasmic droplet. *Reproduction* **2005**, *130*, 213–222. [CrossRef] [PubMed]

86. Moore, K.; Lovercamp, K.; Feng, D.; Antelman, J.; Sutovsky, M.; Manandhar, G.; van Leyen, K.; Safranski, T.; Sutovsky, P. Altered epididymal sperm maturation and cytoplasmic droplet migration in subfertile male Alox15 mice. *Cell Tissue Res.* **2010**, *340*, 569–581. [CrossRef] [PubMed]

87. Lax, Y.; Grossman, S.; Rubinstein, S.; Magid, N.; Breitbart, H. Role of lipoxygenase in the mechanism of acrosome reaction in mammalian spermatozoa. *Biochim. Biophys. Acta (BBA)-Lipids Lipid MeTable* **1990**, *1043*, 12–18. [CrossRef]

88. Walters, J.L.; De Iuliis, G.N.; Dun, M.D.; Aitken, R.J.; McLaughlin, E.A.; Nixon, B.; Bromfield, E.G. Pharmacological inhibition of arachidonate 15-lipoxygenase (ALOX15) protects human spermatozoa against oxidative stress. *Biol. Reprod.* **2018**, *98*, 784–794. [CrossRef] [PubMed]

89. Bromfield, E.G.; Mihalas, B.P.; Dun, M.D.; Aitken, R.J.; McLaughlin, E.A.; Walters, J.L.; Nixon, B. Inhibition of arachidonate 15-lipoxygenase prevents 4-hydroxynonenal-induced protein damage in male germ cells. *Biol. Reprod.* **2017**, *96*, 598–609. [CrossRef] [PubMed]

90. Brütsch, S.H.; Wang, C.C.; Li, L.; Stender, H.; Neziroglu, N.; Richter, C.; Kuhn, H.; Borchert, A. Expression of inactive glutathione peroxidase 4 leads to embryonic lethality, and inactivation of the alox15 gene does not rescue such knock-in mice. *Antioxid. Redox Signal.* **2015**, *22*, 281–293. [CrossRef] [PubMed]

91. Gregus, A.M.; Dumlao, D.S.; Wei, S.C.; Norris, P.C.; Catella, L.C.; Meyerstein, F.G.; Buczynski, M.W.; Steinauer, J.J.; Fitzsimmons, B.L.; Yaksh, T.L.; et al. Systematic analysis of rat 12/15-lipoxygenase enzymes reveals critical role for spinal eLOX3 hepoxilin synthase activity in inflammatory hyperalgesia. *FASEB J.* **2013**, *27*, 1939–1949. [CrossRef] [PubMed]

92. Sadeghian, H.; Jabbari, A. 15-Lipoxygenase inhibitors: A patent review. *Expert Opin. Ther. Pat.* **2016**, *26*, 65–88. [CrossRef] [PubMed]

93. Sendobry, S.M.; Cornicelli, J.A.; Welch, K.; Bocan, T.; Tait, B.; Trivedi, B.K.; Colbry, N.; Dyer, R.D.; Feinmark, S.J.; Daugherty, A. Attenuation of diet-induced atherosclerosis in rabbits with a highly selective 15-lipoxygenase inhibitor lacking significant antioxidant properties. *Bri. J. Pharmacol.* **1997**, *120*, 1199–1206. [CrossRef] [PubMed]

94. Sordillo, L.M.; Weaver, J.A.; Cao, Y.Z.; Corl, C.; Sylte, M.J.; Mullarky, I.K. Enhanced 15-HPETE production during oxidant stress induces apoptosis of endothelial cells. *Prostaglandins Other Lipid Med.* **2005**, *76*, 19–34. [CrossRef] [PubMed]

95. Bocan, T.M.; Rosebury, W.S.; Mueller, S.B.; Kuchera, S.; Welch, K.; Daugherty, A.; Cornicelli, J.A. A specific 15-lipoxygenase inhibitor limits the progression and monocyte–macrophage enrichment of hypercholesterolemia-induced atherosclerosis in the rabbit. *Atherosclerosis* **1998**, *136*, 203–216. [CrossRef]

96. Tobaben, S.; Grohm, J.; Seiler, A.; Conrad, M.; Plesnila, N.; Culmsee, C. Bid-mediated mitochondrial damage is a key mechanism in glutamate-induced oxidative stress and AIF-dependent cell death in immortalized HT-22 hippocampal neurons. *Cell Death Differ.* **2011**, *18*, 282–292. [CrossRef] [PubMed]

97. Zhou, L.; Qian, J.; Liu, J.; Zhao, R.; Li, B.; Wang, R. Identification of the sites of 4-hydroxy-2-nonenal and neprilysin adduction using a linear trap quadrupole Velos Pro-Orbitrap Elite mass spectrometer. *Eur. J. Mass Spectrom.* **2016**, *22*, 133–139. [CrossRef] [PubMed]

98. Tsirulnikov, K.; Abuladze, N.; Bragin, A.; Faull, K.; Cascio, D.; Damoiseaux, R.; Schibler, M.J.; Pushkin, A. Inhibition of aminoacylase 3 protects rat brain cortex neuronal cells from the toxicity of 4-hydroxy-2-nonenal mercapturate and 4-hydroxy-2-nonenal. *Toxicol. Appl. Pharmacol.* **2012**, *263*, 303–314. [CrossRef] [PubMed]

99. Siegel, S.J.; Bieschke, J.; Powers, E.T.; Kelly, J.W. The oxidative stress metabolite 4-hydroxynonenal promotes Alzheimer protofibril formation. *Biochemistry* **2007**, *46*, 1503–1510. [CrossRef] [PubMed]

100. Chu, J.; Li, J.G.; Giannopoulos, P.F.; Blass, B.E.; Childers, W.; Abou-Gharbia, M.; Pratico, D. Pharmacologic blockade of 12/15-lipoxygenase ameliorates memory deficits, Aβ and tau neuropathology in the triple-transgenic mice. *Mol. Psychiatry* **2015**, *20*, 1329. [CrossRef] [PubMed]

101. Di Meco, A.; Li, J.G.; Blass, B.E.; Abou-Gharbia, M.; Lauretti, E.; Praticò, D. 12/15-Lipoxygenase inhibition reverses cognitive impairment, brain amyloidosis, and tau pathology by stimulating autophagy in aged triple transgenic mice. *Biol. Psychiatry.* **2017**, *81*, 92–100. [CrossRef] [PubMed]

102. Okhovati, M.; Zare, M.; Zare, F.; Bazrafshan, M.S.; Bazrafshan, A. Trends in Global Assisted Reproductive Technologies Research: A Scientometrics study. *Electr. Phys.* **2015**, *7*, 1597–1601. [CrossRef] [PubMed]

103. Dyer, S.; Chambers, G.M.; de Mouzon, J.; Nygren, K.G.; Zegers-Hochschild, F.; Mansour, R.; Ishihara, O.; Banker, M.; Adamson, G.D. International Committee for Monitoring Assisted Reproductive Technologies world report: Assisted reproductive technology 2008, 2009 and 2010. *Hum. Reprod.* **2016**, *31*, 1588–1609. [CrossRef] [PubMed]

104. Mansour, R.; Ishihara, O.; Adamson, G.D.; Dyer, S.; de Mouzon, J.; Nygren, K.G.; Sullivan, E.; Zegers-Hochschild, F. International Committee for Monitoring Assisted Reproductive Technologies world report: Assisted reproductive technology 2006. *Hum. Reprod.* **2014**, *29*, 1536–1551. [CrossRef] [PubMed]

105. Rex, A.; Aagaard, J.; Fedder, J. DNA fragmentation in spermatozoa: A historical review. *Andrology* **2017**, *5*, 622–630. [CrossRef] [PubMed]

106. Gao, J.; He, X.; Cai, Y.; Wang, L.; Fan, X. Association between assisted reproductive technology and the risk of autism spectrum disorders in the offspring: A meta-analysis. *Sci. Rep.* **2017**, *7*, 46207.

107. Hart, R.; Norman, R.J. The longer-term health outcomes for children born as a result of IVF treatment. Part II–Mental health and development outcomes. *Hum. Reprod. Update* **2013**, *19*, 244–250. [CrossRef] [PubMed]

108. Aitken, R.J.; De Iuliis, G.N.; Finnie, J.M.; Hedges, A.; McLachlan, R.I. Analysis of the relationships between oxidative stress, DNA damage and sperm vitality in a patient population: Development of diagnostic criteria. *Hum. Reprod.* **2010**, *25*, 2415–2426. [CrossRef] [PubMed]

109. Ahmadi, S.; Bashiri, R.; Ghadiri-Anari, A.; Nadjarzadeh, A. Antioxidant supplements and semen parameters: An evidence based review. *Int. J. Reprod. Biomed.* **2016**, *14*, 729–736. [CrossRef]

110. Ross, C.; Morriss, A.; Khairy, M.; Khalaf, Y.; Braude, P.; Coomarasamy, A.; El-Toukhy, T. A systematic review of the effect of oral antioxidants on male infertility. *Reprod. Biomed. Online* **2010**, *20*, 711–723. [CrossRef] [PubMed]

111. Majzoub, A.; Agarwal, A. Systematic review of antioxidant types and doses in male infertility: Benefits on semen parameters, advanced sperm function, assisted reproduction and live-birth rate. *Arab J. Urol.* **2018**, *16*, 113–124. [CrossRef] [PubMed]

112. Comhaire, F.H.; Garem, Y.E.; Mahmoud, A.H.; Eertmans, F.; Schoonjans, F.R. Combined conventional/antioxidant "Astaxanthin" treatment for male infertility: A double blind, randomized trial. *Asian J. Androl.* **2005**, *7*, 257–262. [CrossRef] [PubMed]

113. Cavallini, G.; Ferraretti, A.P.; Gianaroli, L.; Biagiotti, G.; Vitali, G. Cinnoxicam and L-carnitine/acetyl-L-carnitine treatment for idiopathic and varicocele-associated oligoasthenospermia. *J. Androl.* **2004**, *25*, 761–770. [CrossRef] [PubMed]

114. Tremellen, K.; Miari, G.; Froiland, D.; Thompson, J. A randomised control trial examining the effect of an antioxidant (Menevit) on pregnancy outcome during IVF-ICSI treatment. *Aust. N. Z. J. Obstet. Gynaecol.* **2007**, *47*, 216–221. [CrossRef] [PubMed]

115. Suleiman, S.A.; Ali, M.E.; Zaki, Z.M.; El-Malik, E.M.; Nasr, M.A. Lipid peroxidation and human sperm motility: Protective role of vitamin E. *J. Androl.* **1996**, *17*, 530–537. [PubMed]

116. Omu, A.E.; Dashti, H.; Al-Othman, S. Treatment of asthenozoospermia with zinc sulphate: Andrological, immunological and obstetric outcome. *Eur. J. Obstet. Gynecol. Reprod. Biol.* **1998**, *79*, 179–184. [CrossRef]

117. Lenzi, A.; Culasso, F.; Gandini, L.; Lombardo, F.; Dondero, F. Andrology: Placebo-controlled, double-blind, cross-over trial of glutathione therapy in male infertility. *Hum. Reprod.* **1993**, *8*, 1657–1662. [CrossRef] [PubMed]

118. Gupta, N.P.; Kumar, R. Lycopene therapy in idiopathic male infertility-a preliminary report. *Int. Urol. Nephrol.* **2002**, *34*, 369–372. [CrossRef] [PubMed]

119. Safarinejad, M.R.; Safarinejad, S. Efficacy of selenium and/or N-acetyl-cysteine for improving semen parameters in infertile men: A double-blind, placebo controlled, randomized study. *J. Urology* **2009**, *181*, 741–751. [CrossRef] [PubMed]

120. Scott, R.; MacPherson, A.; Yates, R.W.; Hussain, B.; Dixon, J. The effect of oral selenium supplementation on human sperm motility. *Br. J. Urol.* **1998**, *82*, 76–80. [CrossRef] [PubMed]

121. Hadwan, M.H.; Almashhedy, L.A.; Alsalman, A.R.S. Oral zinc supplementation restore high molecular weight seminal zinc binding protein to normal value in Iraqi infertile men. *BMC Urol.* **2012**, *12*, 32. [CrossRef] [PubMed]

122. Hadwan, M.H.; Almashhedy, L.A.; Alsalman, A.R.S. Study of the effects of oral zinc supplementation on peroxynitrite levels, arginase activity and NO synthase activity in seminal plasma of Iraqi asthenospermic patients. *Reprod. Biol. Endocrinol.* **2014**, *12*, 1. [CrossRef] [PubMed]

123. Balercia, G.; Buldreghini, E.; Vignini, A.; Tiano, L.; Paggi, F.; Amoroso, S.; Ricciardo-Lamonica, G.; Boscaro, M.; Lenzi, A.; Littarru, G. Coenzyme Q10 treatment in infertile men with idiopathic asthenozoospermia: A placebo-controlled, double-blind randomized trial. *Fertil. Steril.* **2009**, *91*, 1785–1792. [CrossRef] [PubMed]

124. Nadjarzadeh, A.; Shidfar, F.; Amirjannati, N.; Vafa, M.R.; Motevalian, S.A.; Gohari, M.R.; Nazeri Kakhki, S.A.; Akhondi, M.M.; Sadeghi, M.R. Effect of Coenzyme Q10 supplementation on antioxidant enzymes activity and oxidative stress of seminal plasma: A double-blind randomised clinical trial. *Andrologia* **2014**, *46*, 177–183. [CrossRef] [PubMed]

125. Ebisch, I.M.; Pierik, F.H.; De Jong, F.H.; Thomas, C.M.; Steegers-Theunissen, R.P. Does folic acid and zinc sulphate intervention affect endocrine parameters and sperm characteristics in men? *Int. J. Androl.* **2006**, *29*, 339–345. [CrossRef] [PubMed]

126. Wong, W.Y.; Merkus, H.M.; Thomas, C.M.; Menkveld, R.; Zielhuis, G.A.; Steegers-Theunissen, R.P. Effects of folic acid and zinc sulfate on male factor subfertility: A double-blind, randomized, placebo-controlled trial. *Fertil. Steril.* **2002**, *77*, 491–498. [CrossRef]

127. Lenzi, A.; Sgro, P.; Salacone, P.; Paoli, D.; Gilio, B.; Lombardo, F.; Santulli, M.; Agarwal, A.; Gandini, L. A placebo-controlled double-blind randomized trial of the use of combined l-carnitine and l-acetyl-carnitine treatment in men with asthenozoospermia. *Fertil. Steril.* **2004**, *81*, 1578–1584. [CrossRef] [PubMed]

128. Balercia, G.; Regoli, F.; Armeni, T.; Koverech, A.; Mantero, F.; Boscaro, M. Placebo-controlled double-blind randomized trial on the use of L-carnitine, L-acetylcarnitine, or combined L-carnitine and L-acetylcarnitine in men with idiopathic asthenozoospermia. *Fertil. Steril.* **2005**, *84*, 662–671. [CrossRef] [PubMed]

129. Nadjarzadeh, A.; Sadeghi, M.R.; Amirjannati, N.; Vafa, M.R.; Motevalian, S.A.; Gohari, M.R.; Akhondi, M.A.; Yavari, P.; Shidfar, F. Coenzyme Q10 improves seminal oxidative defense but does not affect on semen parameters in idiopathic oligoasthenoteratozoospermia: A randomized double-blind, placebo controlled trial. *J. Endocrinol. Investig.* **2011**, *34*, e224–e228.

130. Raigani, M.; Yaghmaei, B.; Amirjannti, N.; Lakpour, N.; Akhondi, M.M.; Zeraati, H.; Hajihosseinal, M.; Sadeghi, M.R. The micronutrient supplements, zinc sulphate and folic acid, did not ameliorate sperm functional parameters in oligoasthenoteratozoospermic men. *Andrologia* **2014**, *46*, 956–962. [CrossRef] [PubMed]

131. Sigman, M.; Glass, S.; Campagnone, J.; Pryor, J.L. Carnitine for the treatment of idiopathic asthenospermia: A randomized, double-blind, placebo-controlled trial. *Fertil. Steril.* **2006**, *85*, 1409–1414. [CrossRef] [PubMed]

132. Adewoyin, M.; Ibrahim, M.; Roszaman, R.; Isa, M.; Alewi, N.; Rafa, A.; Anuar, M. Male infertility: The effect of natural antioxidants and phytocompounds on seminal oxidative stress. *Diseases* **2017**, *5*, 9. [CrossRef] [PubMed]

133. Mihalas, B.P.; Iuliis, G.N.; Redgrove, K.A.; McLaughlin, E.A.; Nixon, B. The lipid peroxidation product 4-hydroxynonenal contributes to oxidative stress-mediated deterioration of the ageing oocyte. *Sci. Rep.* **2017**, *7*, 6247. [CrossRef] [PubMed]

134. Lord, T.; Martin, J.H.; Aitken, R.J. Accumulation of electrophilic aldehydes during postovulatory aging of mouse oocytes causes reduced fertility, oxidative stress, and apoptosis. *Biol. Reprod.* **2015**, *92*, 1–13. [CrossRef] [PubMed]

135. Walshe, J. Penicillamine neurotoxicity: An hypothesis. *ISRN neurology* **2011**, *2011*, 464572. [CrossRef] [PubMed]

136. Gharagozloo, P.; Gutiérrez-Adán, A.; Champroux, A.; Noblanc, A.; Kocer, A.; Calle, A.; Pérez-Cerezales, S.; Pericuesta, E.; Polhemus, A.; Moazamian, A.; et al. A novel antioxidant formulation designed to treat male infertility associated with oxidative stress: Promising preclinical evidence from animal models. *Hum. Reprod.* **2016**, *31*, 252–262. [CrossRef] [PubMed]

137. Elharram, A.; Czegledy, N.M.; Golod, M.; Milne, G.L.; Pollock, E.; Bennett, B.M.; Shchepinov, M.S. Deuterium-reinforced polyunsaturated fatty acids improve cognition in a mouse model of sporadic Alzheimer's disease. *FEBS J.* **2017**, *284*, 4083–4095. [CrossRef] [PubMed]

138. Berbée, J.F.; Mol, I.M.; Milne, G.L.; Pollock, E.; Hoeke, G.; Lütjohann, D.; Monaco, C.; Rensen, P.C.; van der Ploeg, L.H.; Shchepinov, M.S. Deuterium-reinforced polyunsaturated fatty acids protect against atherosclerosis by lowering lipid peroxidation and hypercholesterolemia. *Atherosclerosis* **2017**, *264*, 100–107. [CrossRef] [PubMed]

139. Cotticelli, M.G.; Crabbe, A.M.; Wilson, R.B.; Shchepinov, M.S. Insights into the role of oxidative stress in the pathology of Friedreich ataxia using peroxidation resistant polyunsaturated fatty acids. *Redox Biol.* **2013**, *1*, 398–404. [CrossRef] [PubMed]

140. Hill, S.; Lamberson, C.R.; Xu, L.; To, R.; Tsui, H.S.; Shmanai, V.V.; Bekish, A.V.; Awad, A.M.; Marbois, B.N.; Cantor, C.R.; et al. Small amounts of isotope-reinforced polyunsaturated fatty acids suppress lipid autoxidation. *Free Radic. Biol. Med.* **2012**, *53*, 893–906. [CrossRef] [PubMed]

141. Galatioto, G.P.; Gravina, G.L.; Angelozzi, G.; Sacchetti, A.; Innominato, P.F.; Pace, G.; Ranieri, G.; Vicentini, C. May antioxidant therapy improve sperm parameters of men with persistent oligospermia after retrograde embolization for varicocele? *World J. Urol.* **2008**, *26*, 97–102. [CrossRef] [PubMed]

142. Ciftci, H.; Verit, A.; Savas, M.; Yeni, E.; Erel, O. Effects of N-acetylcysteine on semen parameters and oxidative/antioxidant status. *Urology* **2009**, *74*, 73–76. [CrossRef] [PubMed]

143. Safarinejad, M.R.; Safarinejad, S.; Shafiei, N.; Safarinejad, S. Effects of the reduced form of coenzyme Q10 (ubiquinol) on semen parameters in men with idiopathic infertility: A double-blind, placebo controlled, randomized study. *J. Urol.* **2012**, *188*, 526–531. [CrossRef] [PubMed]

144. Greco, E.; Iacobelli, M.; Rienzi, L.; Ubaldi, F.; Ferrero, S.; Tesarik, J. Reduction of the incidence of sperm DNA fragmentation by oral antioxidant treatment. *J. Androl.* **2005**, *26*, 349–353. [CrossRef] [PubMed]

145. Rolf, C.; Cooper, T.G.; Yeung, C.H.; Nieschlag, E. Antioxidant treatment of patients with asthenozoospermia or moderate oligoasthenozoospermia with high-dose vitamin C and vitamin E: A randomized, placebo-controlled, double-blind study. *Hum. Reprod.* **1999**, *14*, 1028–1033. [CrossRef] [PubMed]

146. Greco, E.; Romano, S.; Iacobelli, M.; Ferrero, S.; Baroni, E.; Minasi, M.G.; Ubaldi, F.; Rienzi, L.; Tesarik, J. ICSI in cases of sperm DNA damage: Beneficial effect of oral antioxidant treatment. *Hum. Reprod.* **2005**, *20*, 2590–2594. [CrossRef] [PubMed]

147. Moslemi, M.K.; Tavanbakhsh, S. Selenium–vitamin E supplementation in infertile men: Effects on semen parameters and pregnancy rate. *Int. J. Gen. Med.* **2011**, *4*, 99–104. [CrossRef] [PubMed]

148. Keskes-Ammar, L.; Feki-Chakroun, N.; Rebai, T.; Sahnoun, Z.; Ghozzi, H.; Hammami, S.; Zghal, K.; Fki, H.; Damak, J.; Bahloul, A. Sperm oxidative stress and the effect of an oral vitamin E and selenium supplement on semen quality in infertile men. *Arch. Androl.* **2003**, *49*, 83–94. [CrossRef] [PubMed]

antioxidants

MDPI

Review

Lipid Peroxidation-Derived Aldehydes, 4-Hydroxynonenal and Malondialdehyde in Aging-Related Disorders

Giuseppina Barrera [1,*]**, Stefania Pizzimenti** [1]**, Martina Daga** [1]**, Chiara Dianzani** [2]**,**
Alessia Arcaro [3]**, Giovanni Paolo Cetrangolo** [3]**, Giulio Giordano** [4]**, Marie Angele Cucci** [1]**,**
Maria Graf [4] **and Fabrizio Gentile** [3] iD

[1]   Dipartimento di Scienze Cliniche e Biologiche, Università di Torino, 10124 Turin, Italy;
       stefania.pizzimenti@unito.it (S.P.); martina.daga@unito.it (M.D.); marieangele.cucci@unito.it (M.A.C.)
[2]   Dipartimento di Scienze e Tecnologia del Farmaco, Università di Torino, 10124 Turin, Italy;
       chiara.dianzani@unito.it
[3]   Dipartimento di Medicina e Scienze della Salute "V. Tiberio", Università del Molise,
       86100 Campobasso, Italy; alessia.arcaro@unimol.it (A.A.); gianpaolo.cet@tiscali.it (G.P.C.);
       gentilefabrizio@unimol.it (F.G.)
[4]   Presidio Ospedaliero "A. Cardarelli", Azienda Sanitaria Regione Molise, 86100 Campobasso, Italy;
       giuliogiordano@hotmail.com (G.G.); mariagraf@tiscali.it (M.G.)
*   Correspondence: giuseppina.barrera@unito.it

Received: 29 June 2018; Accepted: 27 July 2018; Published: 30 July 2018

Abstract: Among the various mechanisms involved in aging, it was proposed long ago that a prominent role is played by oxidative stress. A major way by which the latter can provoke structural damage to biological macromolecules, such as DNA, lipids, and proteins, is by fueling the peroxidation of membrane lipids, leading to the production of several reactive aldehydes. Lipid peroxidation-derived aldehydes can not only modify biological macromolecules, by forming covalent electrophilic addition products with them, but also act as second messengers of oxidative stress, having relatively extended lifespans. Their effects might be further enhanced with aging, as their concentrations in cells and biological fluids increase with age. Since the involvement and the role of lipid peroxidation-derived aldehydes, particularly of 4-hydroxynonenal (HNE), in neurodegenerations, inflammation, and cancer, has been discussed in several excellent recent reviews, in the present one we focus on the involvement of reactive aldehydes in other age-related disorders: osteopenia, sarcopenia, immunosenescence and myelodysplastic syndromes. In these aging-related disorders, characterized by increases of oxidative stress, both HNE and malondialdehyde (MDA) play important pathogenic roles. These aldehydes, and HNE in particular, can form adducts with circulating or cellular proteins of critical functional importance, such as the proteins involved in apoptosis in muscle cells, thus leading to their functional decay and acceleration of their molecular turnover and functionality. We suggest that a major fraction of the toxic effects observed in age-related disorders could depend on the formation of aldehyde-protein adducts. New redox proteomic approaches, pinpointing the modifications of distinct cell proteins by the aldehydes generated in the course of oxidative stress, should be extended to these age-associated disorders, to pave the way to targeted therapeutic strategies, aiming to alleviate the burden of morbidity and mortality associated with these disturbances.

Keywords: aldehydes; osteopenia; sarcopenia; myelodysplastic syndromes; immunosenescence

1. Introduction

In recent years, increased life expectancy due to an improved quality of life and decline in mortality rates, is leading to a society in which the aging population is growing more rapidly than the entire population. Life expectancy is projected to increase in industrialized countries. In 2016, 27.3 million very old adults were living in the European Union, and in the UK, 2.4% of the population (1.6 million) were aged 85 and over [1]. It has been calculated that there is more than a 50% probability that, by 2030, female life expectancy will exceed the 90-year barrier, a level that was deemed unattainable at the turn of the 21st century [2]. The functional disturbances appearing in old age are referred to as the "aging process", which entails changes in body composition, imbalances in energy production and use, homeostatic dysregulation, neurodegeneration and loss of neuroplasticity [3]. The scientific community has formulated over 300 theories to explain the driving forces behind aging [4], but none has proven, so far, to be universally applicable. The free radical theory of aging has gained widespread acceptance, thus becoming one of the leading explanations of the aging process at the molecular level. It is commonly believed that the aging process is related to an imbalance favoring pro-oxidant over antioxidant molecules and a consequent increase of oxidative stress [5]. Oxidative stress entails elevated intracellular levels of reactive oxygen species (ROS), which can cause damage to proteins, lipids, and DNA. The mitochondria are the primary sources of intracellular ROS (~1–5%), due to the electron leakage primarily resulting from the electron transport chain [6]. Indeed, mitochondrial dysfunction has long been considered a major contributor to aging and age-related diseases. In elderly subjects, mitochondria are characterized by functional impairment, such as lowered oxidative capacity, reduced oxidative phosphorylation, decreased adenosine triphosphate production, significantly increased ROS generation, and diminished antioxidant defense [7]. Depending on their concentration in the cells, ROS are either physiological signals essential for cell life or toxic species which damage cell structure and functions. In particular, ROS cause the oxidation of polyunsaturated fatty acids in membrane lipid bilayers, leading eventually to the formation of aldehydes, which have been considered as toxic messengers of oxidative stress, able to propagate and amplify oxidative injury [8].

Among the aldehydes produced by lipid peroxidation (LPO), malonaldehyde (MDA) and 4-hydroxynonenal (HNE) have gained most attention, since MDA is produced at high levels during LPO, so that it is commonly used as a measure of oxidative stress, and HNE has been shown to be endowed with the highest biological activity. The production, metabolism, and signaling mechanisms of two main omega-6 fatty acids lipid peroxidation products, MDA and HNE, have been extensively studied and reported in an excellent review [9]. Briefly, MDA and HNE originate from the peroxidation of polyunsaturated fatty acids. LPO can be described generally as a process under which oxidant, free radical or non-radical chemical species attack lipids containing carbon–carbon double bond(s). This process involves hydrogen abstraction from a carbon atom and oxygen insertion, resulting in the formation of peroxyl radicals and lipid hydroperoxides, as illustrated in Figure 1.

Since aldehydes have high chemical reactivities, mammals have evolved a full set of enzymes converting them into less reactive chemical species and contributing to the control of their intracellular concentrations, which reflect the steady-state between the rates of formation by LPO and catabolism into less reactive compounds. Once formed, MDA and HNE can be reduced to alcohols by aldo-keto reductases or alcohol dehydrogenases, or can be oxidized to acids by aldehyde dehydrogenases [10]. Moreover, HNE, which is the most reactive among aldehydes, easily reacts with low-molecular-weight compounds, such as glutathione. This reaction can occur spontaneously or can be catalyzed by glutathione-S-transferases [10]. HNE and MDA are able to affect several signaling processes. Most of these effects depend on their ability to bind covalently to proteins and DNA.

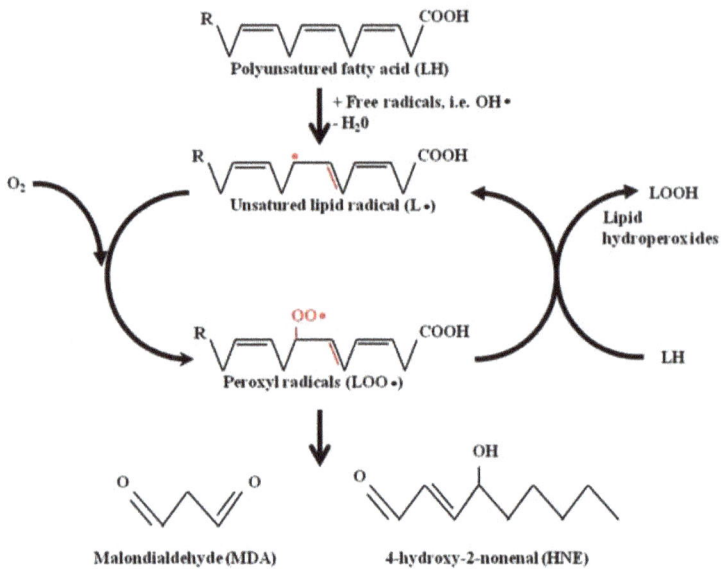

Figure 1. Malondialdehyde (MDA) and 4-hydroxynonenal (HNE) formation from polyunsaturated fatty acids.

Aldehyde–Protein Adduct in Human Disease

LPO-derived aldehydes easily react with proteins, generating a wide variety of intra- and inter-molecular covalent adducts. Depending on their structural features, they can form Schiff bases and/or Michael adducts with the free amine group of lysine, the imidazoline group of histidine, the guanidine group of arginine and the thiol group of cysteine [11]. HNE can affect cell functions, through its ability to form adducts with proteins involved in signal transduction and gene expression, including receptors, kinases, phosphatases and transcription factors [11]. MDA can form protein adducts by specifically modifying the lysyl residues of proteins [12], and modified autologous biomolecules can generate neoepitopes from self epitopes, capable of inducing undesired innate and adaptive immune responses, including atherosclerosis [13]. HNE adducts can accumulate progressively in the vascular system, leading to cellular dysfunctions and tissue damaging effects, which are involved in the progression of atherosclerosis. Moreover, HNE, by forming HNE-apoB adducts, contributes to the atherogenicity of oxidized low-density lipoproteins (LDL), leading to the formation of foam cells [14]. The presence of HNE-protein adducts has been detected in inflammation-related diseases, such as alcoholic liver disorders and chronic alcoholic pancreatitis, in which the increased formation of HNE-protein adducts was evidenced in acinar cells adjacent to the interlobular connective tissue [15]. Moreover, HNE-protein adducts have been detected in brain tissues and body fluids in several neurodegenerative diseases, such as Alzheimer's disease, Huntington disease, Parkinson's disease, amyotrophic lateral sclerosis, and Down syndrome [10].

The involvement of MDA and HNE in age-related diseases has been supported by the observations that the concentrations of MDA and HNE, in human erythrocytes and blood plasma, increased during the aging process [16]. The role of reactive aldehydes in age-related pathologies, such as atherosclerosis [13], neurodegenerations [17], inflammation, and cancer [10] has been discussed by many excellent reviews. Other age-related disorders, such as osteopenia/osteoporosis, sarcopenia, immunosenescence and myelodysplastic syndromes have received less attention. In this review, we focus our attention on to recent advances in the role of LPO-derived aldehydes and aldehyde-protein adducts in these age-related disorders, all of which are characterized by increases of oxidative stress and consequent increases in the generation of LPO-derived aldehydes.

2. Osteopenia/Osteoporosis

Bone remodeling is a highly dynamic physiological process: osteoblasts (bone-forming cells) and osteoclasts (bone-resorbing cells) work simultaneously to maintain bone density and strength [18]. During aging, bone density decreases and the measure of bone mineral density is commonly employed to evaluate whether or not a patient is affected by osteopenia or osteoporosis. Osteopenia is the thinning of bone mass. Such a decrease in bone mass is not usually "severe", but is considered a very serious risk factor for the development of osteoporosis. Osteoporosis is characterized by profound losses of skeletal mass, coupled with architectural deterioration, which increase bone fragility and susceptibility to fractures. Several recent papers have provided insights into the current prevalence of osteoporosis in specific older populations. The current National Osteoporosis Foundation guidelines for the USA, coupled with the most recently available population data from the National Health and Nutrition Survey, show that the eligibility for osteoporosis treatment increases exponentially with age; roughly 10% of both men and women meet criteria for treatment at age 50 years, whereas 48% of men and 79% of women over 80 years meet treatment guidelines [19]. Current treatments for osteoporosis, which increase bone density and reduce fracture risk, include anti-resorptive medications (bisphosphonates and denosumab), which primarily increase endocortical bone and cortical thickness, and anabolic medications (teriparatide and abaloparatide), which increase the periosteal and endosteal perimeters, without causing large changes in cortical thickness [20].

Increases of ROS levels have been observed consistently in osteopenia and osteoporosis. In aging people, the number and activity of individual osteoblasts decrease and osteoblast apoptosis increases, in association with oxidative stress-induced osteoporosis [21]. Some studies explored the relationships between the use of antioxidants and bone metabolism. Indeed, a marked decrease in plasma antioxidants was found in aged or osteoporotic rats and in aged or osteoporotic women [22–24]. The loss of antioxidant capacity leads to accelerated bone loss through the activation of a tumor necrosis factor alpha (TNF-α)-dependent signalling pathway [25]. In turn, the administration of antioxidants such as vitamin C, E, N-acetyl-cysteine and linoleic acid, had beneficial effects in individuals with osteoporosis [26].

Osteoblast apoptosis was induced by oxidative stress through the activation of the c-Jun N-terminal kinase (JNK) pathway [27] and the NF-kB pathway [28]. In unstimulated cells, NF-κB proteins are sequestered in the cytoplasm because of their association with IκB (inhibitor of κ light gene enhancer in B cells) proteins. Phosphorylation and degradation of IκB disrupt this association and allow the translocation of NF-κB proteins into the nucleus. ROS induce the oxidation of critical cysteines and enhance the activity of several cytoplasmic kinases, which promote IκB phosphorylation and degradation, including IκB kinase and the PKC family of serine/threonine kinases [29]. Additionally, ROS-induced modifications control key steps in the nuclear phase of the NF-κB program, including recruitment of coactivators, chromatin remodeling, and DNA binding. The activation of NF-kB in osteoblastic cells increased the phosphorylation of p66 (shc) protein, which amplified mitochondrial ROS generation and stimulated apoptosis [29]. The increase of ROS in osteoblastic MC3T3-E1 cells induced phosphorylation of MAPKs (mitogen-activated protein kinases), which subsequently triggered the intrinsic apoptosis pathway. Moreover, the advanced oxidation protein products induced osteoblast apoptosis in aged Sprague–Dawley rats [30].

The NF-κB pathway is involved in osteoclast activity, as well. Indeed, ROS stimulated osteoclast differentiation and activity through activation of the NF-κB pathway, and these effects were reversed upon NF-κB suppression [31].

Other than by the direct actions mentioned above, ROS can affect bone turnover by enhancing the production of LPO-derived reactive intermediates, thereby causing protein damage and inflammatory responses [32]. In vitro experiments have proved that HNE, the most biologically active aldehyde produced by LPO, can induce intense oxidative stress, inflammatory reactions, and apoptosis in osteoblasts, via the induction of protein phosphatase 2A activity, which has earned a reputation as a mediator of oxidative stress-induced apoptosis in these cells [21]. The role of LPO-derived aldehydes in osteoblast apoptosis is also supported by studies on aldehyde dehydrogenases (ALDH), a family of enzymes involved in aldehyde degradation [33]. ALDH inhibition by disulfiram resulted in bone loss in rats [34]. Additional confirmation of the role of ALDH activity in osteoporosis comes from the results obtained with transgenic mice expressing the Aldh2*2 (Aldh2*2 Tg) dominant-negative form of ALDH2. These transgenic mice exhibited severe osteoporosis, indicating that ALDH2 regulates physiological bone homeostasis. Moreover, the Aldh2*2 transgene or treatment with acetaldehyde induced the accumulation of HNE and the expression of peroxysome proliferator-activated receptor γ (PPAR-γ) a transcription factor that promotes adipogenesis and inhibits osteoblastogenesis [35]. On the contrary, the activation of ALDH2 by *N*-(1,3-benzodioxol-5-ylmethyl)-2,6-dichlorobenzamide (alda-1) had an osteogenic effect, involving increased production of bone morphogenetic protein-2 by osteoblasts [36]. Taken together, these experimental data seem to delineate a role for the products of LPO in the induction of osteoporosis. However, the measurements of MDA as an indicator of LPO in post-menopausal osteoporotic women gave conflicting results. One study disclosed significant increases in plasma MDA concentrations in post-menopausal osteoporotic women, compared with control subjects [37], whereas another one failed to detect changes of MDA concentrations between osteoporotic and non-osteoporotic post-menopausal women [38]. Thus, further studies are needed to precisely define the role played by LPO-derived aldehydes and to validate the possible inclusion of targeted therapies aimed to reducing the effects of reactive aldehydes in the development and progression of age-related osteopenia/osteoporosis.

3. Sarcopenia

Sarcopenia is a geriatric syndrome characterized by a progressive and generalized loss of skeletal muscle mass, together with low muscle strength and/or poor physical performance in the elderly [39]. A recent study in Brazilians aged 60 years or older demonstrated that the overall prevalence of sarcopenia in older Brazilians was 17.0%. Sensitivity analysis showed rates of 20.0% in women and 12.0% in men [40]. In order to counteract the age-related muscle decline several interventions have been explored, including protein supplementation, testosterone replacement in men, estrogen replacement in women, growth hormone replacement, and treatment with vitamin D. To date, adequate protein intake and physical exercise are the most promising interventions aiming to prevent and/or delay the decline of muscle mass [41].

Oxidative stress and inflammation are implicated in the pathogenesis of sarcopenia [42]. It has been suggested that an increase of oxidative stress might activate apoptosis, leading to the loss of skeletal muscle fibers, thus contributing to the progression of sarcopenia [43]. Skeletal muscle tissue is unique, with respect to apoptosis, because muscle cells are one of the three cell types, along with osteoclasts and cytotrophoblasts, that are multinucleated [44]. The process by which nuclei are eliminated from multinucleated muscle fibers appears to be similar to apoptosis, since it involves chromatin condensation and DNA fragmentation. Moreover, it has been suggested that oxidative stress and the consequent mitochondrial genotoxic damage might play a causal role in the numerical loss of muscle fibers with aging [45].

Deregulation of redox homeostasis has emerged in recent years as a common pathogenetic mechanism and potential therapeutic target in collagen VI-related congenital muscular dystrophies, and in Duchenne muscular dystrophy, as well as in other more prevalent processes, such as age-related muscle loss [46]. Moreover, sarcopenia can be associated with obesity. This association is defined as sarcopenic obesity and represents a chronic condition, whose increase in prevalence has been related to parallel increases in the mean age of the population, the prevalence of obesity, and the changes in lifestyle during the last several decades [47]. In obese individuals, adipose tissue secretes both bioactive molecules, called "adipokines", and ROS, which might represent the mechanistic link between obesity and its associated metabolic complications, including sarcopenia [48].

The aging of human skeletal muscle cells is marked by a progressive functional decline of mitochondria, resulting in the accumulation of ROS [43]. $Sod1^{-}/^{-}$ mice lack the superoxide dismutase [Cu-Zn] 1 (CuZnSOD1) enzyme and undergo accelerated sarcopenia, i.e., display the characteristics of aging muscle in an accelerated manner. In $Sod1^{-}/^{-}$ mice, muscle loss was accompanied by a progressive decline of mitochondrial function, with an increased mitochondrial generation of ROS and faster induction of mitochondrial-mediated apoptosis and loss of myonuclei. $Sod1^{-}/^{-}$ mice also exhibited a strikingly increased number of dysfunctional mitochondria near neuromuscular junctions [49]. Increased ROS production is a necessary response to exercise of a sufficient intensity [50]. ROS and RNS (reactive nitrogen species, such as NO and the peroxynitrite anion) play important roles in the function of skeletal muscle, as they may mediate muscle adaptive responses, e.g., to physical exercise [51], by facilitating glucose uptake or inducing mitochondrial biogenesis [52,53]. However, during extended disuse periods, redox imbalance contributes to deleterious muscle remodeling via myonuclear apoptosis and atrophy, with the mediation of a redox-sensitive transcriptional regulator, NF-κB [54,55]. ROS can stimulate the production of TNF-α and IL-1β, through the activation of NF-κB-mediated pathways [56]. Systemic inflammation, impaired responses to stressors, and weakened regenerative capacity are all associated with sarcopenia. In Wistar rats, aging (40 weeks) was accompanied by reduced mitochondrial respiratory chain complex activities in saponin-skinned soleus muscle fibers, increased production of ROS and decreased transcription of the genes encoding mitochondrial superoxide dismutase 2 (*SOD2*), PPAR-γ coactivator-1β (*PGC-1β*) and sirtuin 1, in comparison with young (16 weeks) rats. Chronic intake of polyphenols normalized V_{max}, decreased ROS production and enhanced *SOD2* and *PGC-1β* expression, in comparison with age-matched untreated rats [57].

Notably, sarcopenia is also accompanied by fat infiltration in the skeletal muscle, which can not only affect muscle quality and functional performance [58], but might render muscle tissue prone to oxidative stress-induced LPO. In disease processes marked by increased ROS production, novel 2,4-dinitrophenylhydrazine (DNPH)-reactive carbonyl groups in proteins, susceptible to spectrometric or antibody-mediated detection, might be either produced by the direct metal-catalyzed oxidation of aminoacyl side chains or introduced by stable adduct formation via the reaction of the latter with reducing sugars or reactive carbonyl species, including MDA and HNE. Age-related increases in protein carbonyl levels were detected in skeletal muscle by derivatization with DNPH, followed by fluorescent anti-DNPH antibodies, in descending ranking order, in the extracellular space, subsarcolemmal mitochondria, intermyofibrillar mitochondria and cytoplasm [59]. By the use of quantitative proteomics, a number of mitochondrial proteins susceptible to carbonylation in a muscle type-dependent (slow- vs. fast-twitch) and age-dependent manner were identified in Fischer 344 rat skeletal muscle. Fast-twitch muscle revealed twice as many carbonylated mitochondrial proteins than slow-twitch muscle, with 22 proteins showing significant changes (mostly increases) in carbonylation state with age. Ingenuity pathway analysis revealed that these proteins belonged to functional classes and pathways known to be impaired in muscle aging, including cellular function and maintenance, fatty acid metabolism and citric acid cycle. Although proof was not provided that carbonylation was responsible for any functional changes, these data delineated a catalog of protein targets deserving investigation, because of their potential implication in muscle aging [60]. In the gastrocnemius muscle of male C57B1/6 mice, a distinct increase of HNE adducts was noticed, upon the progression of age from 5 to 25 months. This was accompanied by increased expression of inducible nitric oxide synthase, decreased expression of G6PDH, activation of JNK, caspase 2, caspase 9, and inactivation of BCL-2, through phosphorylation at Ser70 [61]. Other observations have been collected in transgenic mice expressing a dominant-negative form of ALDH2 (ALDH2*2 Tg mice), which indicate a role of HNE as an inducer of apoptosis. Mice, in whose muscles ALDH activity was selectively attenuated, exhibited small body size, muscle atrophy, decreased fat content, osteopenia, and kyphosis, accompanied by increased muscular HNE levels [62].

Oxidative stress, in sarcopenic patients, can lead to increased LPO production. The mechanistic importance of apoptosis in the loss of muscle cells in age-related sarcopenia and the involvement of LPO-derived aldehydes in this process are supported by several studies. The abundance of protein carbonyl adducts was determined within skeletal muscle sarcoplasmic, myofibrillar, and mitochondrial protein subfractions from musculus *vastus lateralis* biopsies, using immunoblotting techniques, in two groups of 16 old males ("old" and "old sarcopenic") [63]. Concentrations of cytoprotective proteins (e.g., heat shock proteins, αβ-crystallin) were also assayed. Aging was associated with increased mitochondrial (but not myofibrillar or sarcoplasmic) protein carbonyl adducts, independent of (stage-I) sarcopenia. Mitochondrial protein carbonyl abundance negatively correlated with muscle strength, but not muscle mass. According to this study, mitochondrial protein carbonylation increased moderately with age, and this increase might impact upon skeletal muscle function, but was not a hallmark of sarcopenia, per se. It should be considered, though, that the subjects under study were affected by low-grade (stage-I) sarcopenia [63].

A recent study, conducted with an urban Spanish cohort of elderly people (≥70 years), demonstrated a significant increase of MDA and HNE in blood samples of sarcopenic subjects, with respect to sarcopenia-exempt control subjects [64]. Notably, among several parameters examined in this study, only MDA and HNE levels were significantly associated with sarcopenia. These observations have been confirmed by studies demonstrating the presence of plasma MDA/HNE protein adducts in sarcopenic patients [65]. Interestingly, proteomic analysis of muscle extracts from adult and aged post-menopausal women demonstrated that ALDH2 and AKR1A1 were up-regulated in aged women, suggesting that the scavenging of reactive aldehyde products in skeletal muscle cells of the elderly was enhanced [66]. These results may be interpreted as a demonstration of the existence of a mechanism of adaptation of muscle cells to increased oxidative stress and to the consequently increased production of LPO-derived aldehydes.

Taken together, these results support a pathogenic link between oxidative stress, LPO-derived aldehydes, aldehyde-protein adducts and age-related sarcopenia in humans.

4. Immunosenescence

Immunosenescence takes a relevant part in the aging phenotype, and there is growing consensus on the idea that it might result from a progressive imbalance, favoring inflammatory over anti-inflammatory mechanisms, which some authors define as "inflammaging". According to this view, continuing antigenic stimulation in the course of life, with accompanying oxidative stress, steadily leads, as enzymatic and non-enzymatic antioxidant defences fade off with age, to an up-regulation of inflammatory responses and a remodeling of immune responses, as revealed by the increased serum levels of pro-inflammatory cytokines, such as IL-6 and TNF-α, and a loss of efficacy of adaptive responses. Pro-inflammatory cytokines stimulate further the generation of ROS and cytotoxic LPO products.

As already well documented in other tissues and organ systems, accumulation of ROS and LPO products might cause significant damage to cell lipids, nucleic acids and crucial cell proteins also in cells of the immune system. This pro-inflammatory condition negatively affects the overall chances for good health, self-sufficiency and extended survival of the elderly and is a hallmark of the so-called "fragility" of the elderly [67,68]. Moreover, the remodeling of immunity strongly contributes to a number of age-associated diseases (infectious diseases, autoimmunity, cancer, metabolic, vascular and neurodegenerative diseases). Immunosenescence is best revealed by the changes in the modulation of survival of T cells, which become more resistant to damage-induced apoptosis. As a consequence, CD28-senescent T cells increase in number (with effector/memory cells, most of the CD8+ CD45RO+ CD25+ phenotype, prevailing over naïve cells), dysfunctional cells accumulate, and the immunological space in lymphoid tissues is reduced. Increased activation-induced cell death (AICD), in response to TNF-α, causes a progressive depletion of lymphoid niches, particularly of the naïve T cell compartment. At the same time, the generation of immunocompetent T cells in thymus declines with age [68–72].

The age-dependent shortening of DNA telomeres, which is associated with increased mortality in individuals over 65 years of age, contributes to the reduction of the potential for clonal expansion and differentiation of naïve T cells, which is revealed by the dramatic decay of T-cell-receptor excision circles (TREC). The overall outcome is a reduction in the repertoire of clonal antigenic specificities, which is reflected in a decreased ability to enact recall immune responses to formerly encountered antigens (e.g., cytomegalovirus, CMV) and to mount adaptive responses to novel antigenic stimuli, combined with an increased tendency to autoimmunity [68–72]. The inversion of the CD4+/CD8+ cell number ratio, the increased fraction of effector-memory cells and the seropositivity for CMV identify an immune risk phenotype in elderly patients [73]. Reduced B cell function with age reflects the decreases both of CD19+ cell number and of T helper-mediated cooperation with B cell responses. Age-associated changes in innate immunity include reductions of antigen uptake by DCs, phagocytosis and production of lymphocyte-derived chemotactic factor by macrophages, FCγRIII (CD16)-stimulated production of superoxide anion by neutrophils, and IFN-γ secretion and expression of activating NKp receptors by NK cells.

Thymus involution is associated with a drastic reduction in organ volume and replacement of functional cortical and medullary areas with adipose tissue. This process, which starts early in life, is almost complete by the age of 40–50 years. Increasing formation of lipid-laden cells with aging has also been observed in lympho-hematopoietic organs, including the thymus [74,75]. Age-related thymic involution appears to reflect the compound effects of increased rates of thymocyte death in the thymus and decreased thymic differentiation and output of T cells. The former might result from intrathymic inflammation and lipotoxicity, the latter from the failure of stromal cell-dependent thymocyte maturation and survival (Figure 2) [76].

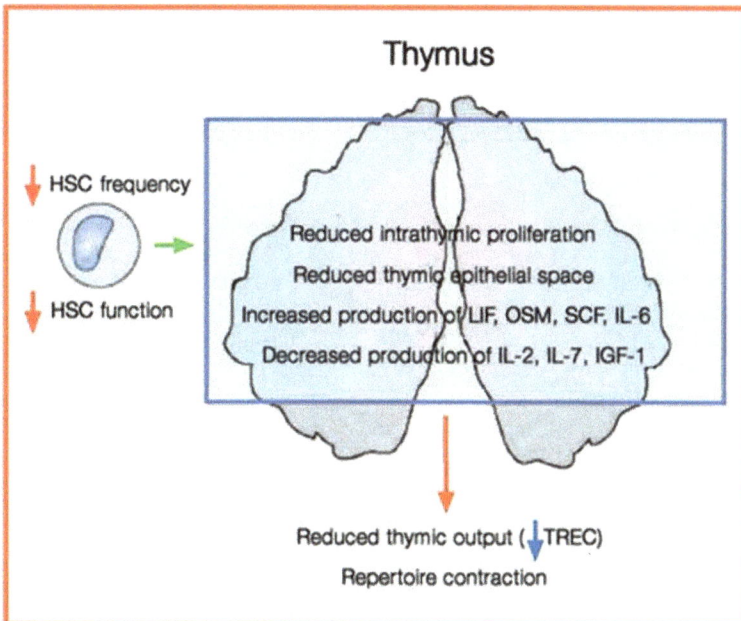

Figure 2. Main aspects of thymic involution. The decrease of hematopoietic cells causes a decrease of production of T cells from the thymus. HSC: hematopoietic stem cells; LIF: leukemia- inhibitory factor; OSM: Oncostatin M; SCF: stem cell factor; IL-2: interleukin 2; IL-6 interleukin 6; IL-7: interleukin 7; IGF-1: insulin-like growth factor 1; TREC: T cell receptor excision circles (modified from [68]).

The levels of pro-inflammatory chemokines (MIP-1α, MIP-1β, RANTES) secreted by activated DC, macrophages and endothelial cells and recognized by CCR5 increase in the aging thymus, just like intrathymic lipid-laden multilocular and adipose cells (LLC), whose number is inversely related with thymic function [76]. It was suggested that LLC derive from CCR5-expressing, perithymic and perivascular preadipocytes, migrating into the aging thymus [77]. They produce pro-inflammatory cytokines (LIF, Oncostatin M, IL-6, TNF-α), thus creating a pro-oxidative, cytotoxic intrathymic *milieu*, which might favor thymocyte death [77]. The age-related alterations of lipid metabolism and redox balance were studied in mouse thymus and isolated thymocytes. An evaluation of the lipidomics profile of the whole thymus, between the ages of 4 weeks (young age) and 18 months (old age), revealed increased amounts of triacylglycerides, cholesterol, HNE, sulfatide ceramide and ganglioside GD1a in the aged thymus. Increased levels of cholesterol esters and HNE adducts were also found in isolated thymocytes from elder mice, compared with younger individuals. Increased levels of TNF-α and increased expression of CD204, a scavenger receptor of oxidized LDL, were detected in the thymic parenchyma of older individuals, as compared to younger ones. GH reduced thymic levels of TNF-α and HNE and increased the number of thymocytes, in accordance with several observations that indicated growth hormone as a powerful immunomodulating agent, stimulating thymopoiesis and limiting the number of adipocytes and fat locules in rodent and human thymus [74]. The levels of MDA and protein carbonyls (PC), as well as of oxidized and reduced glutathione and the activities of several antioxidant enzymes were measured in peripheral blood lymphocytes from 100 individuals, equally subdivided into groups of ages varying from 11–20 to 51–60 years. This study evidenced distinct, steady increases in MDA and PC levels with age, and decreasing glutathione levels and antioxidant enzyme activities, documenting a progressive redox imbalance with aging [78]. When rats subjected to ovariectomy, which undergo premature aging of the immune system, were subjected to dietary supplementation with polyphenolic antioxidants (soybean and green tea polyphenols), beneficial effects were observed on several parameters of the immune function (macrophage phagocytosis, chemotaxis and ROS production; lymphocyte mitogenic responses; NK activity) and redox balance (catalase and glutathione peroxidase activity, oxidized versus reduced glutathione ratio, MDA levels) of peritoneal leukocytes [60].

5. Myelodysplastic Syndromes

Myelodysplastic syndromes (MDS) include a heterogeneous range of stem cell disorders characterized by peripheral cytopenias and increased risk of progression to acute myeloid leukemia. The incidence of MDS increases markedly with age. It is conservatively estimated that >10,000 new cases of MDS occur in the United States annually [79]. In MDS, next-generation sequencing allowed the identification of molecular mutations in nearly 90% of the patients. Consequently, molecular mutation markers were integrated into current classification systems. The growing insights into molecular aspects of the pathogenesis of MDS may help to predict the possible evolution towards leukemia [80].

Most MDS patients have anemia and many develop transfusion dependence and iron overload (IOL). IOL is considered a negative independent prognostic factor, associated with a higher risk of leukemic transformation and shorter survival [81]. Serologic and molecular markers of oxidative stress have been evidenced in many patients with MDS [82], with the concentration of the LPO product MDA being significantly increased in patients with MDS and IOL, compared with IOL-exempt patients and control subjects. Moreover, both plasma nitrite and MDA concentrations were positively correlated with the ferritin levels, suggesting a relationship between IOL and the increased production of ROS in MDS [83]. In regard, it has to be noticed that the impact of iron on the redox balance of cells and body fluids is strictly dependent on transferrin saturation (TSAT). In the presence of normal TSAT, the M1 polarization of macrophages is decreased and the production of pro-inflammatory cytokines, ROS and LPO products are inhibited [84]. Conversely, patients with IOL and high TSAT show higher O_2% saturation levels than patients at the time of diagnosis and normal controls. Antioxidant systems, with the exception of SOD, exhibit significant activity changes in IOL patients, compared with controls.

Moreover, iron chelation treatment with deferoxamine has also been shown to reduce cytopenia in patients with MDS. In a study of 11 patients with MDS who were receiving deferoxamine for up to 60 months, increases in platelet and neutrophil counts were observed in 64% and 78% of patients, respectively [85]. Furthermore, mitochondrial dysfunction has been detected only in IOL cases, but not in control subjects [86]. High-level ROS production in MDS can be responsible for genotoxic damage to nuclear and mitochondrial DNA, resulting in genomic instability, which contributes to disease progression and leukemia onset [87,88].

Even though the role of ROS in MDS is well established, the role played by LPO products in the disease and its progression has not been completely assessed. It was suggested that the loss of ALDHs, which are mainly responsible for the metabolism of reactive aldehydes, might lead to the alteration of various cell processes, which may foster MDS progression to leukemic transformation [89]. Recent data indicate that ALDH1A1-defective cell lines, as well as primary leukemic cells, are sensitive to the treatment with drugs that directly or indirectly generate toxic ALDH substrates, including HNE. On the contrary, normal HSCs are relatively resistant to these compounds, suggesting that LPO-derived aldehydes could selectively kill ALDH1A1-defective leukemic cells [90]. According to these observations, in human lymphoid leukemic CDM-NKR cells, high concentrations of HNE caused significant cytotoxic effects on DNA synthesis and mitochondrial activity, whereas no significant toxicity of HNE was detected in normal hematopoietic precursor cells [91]. In addition, murine erythroleukemia (MEL) cells from HNE-treated mice exhibit a higher degree of differentiation, in comparison with dimethyl sulfoxide-treated MEL cells. These findings indicate that HNE, at concentrations physiologically detected in many normal tissues and in plasma, induces MEL cell differentiation, by modulating the expression of specific genes [92].

However, the role of HNE and other reactive aldehydes in the development of leukemia is controversial. The ALDH1A1 and ALDH3A1 isoforms are important for the metabolism of reactive aldehydes and ROS, and are expressed at high levels in HSCs. Indeed, the loss of these two isoforms resulted in a variety of effects on HSC biology, such as increased DNA damage and increased rates of leukemic transformation [93]. On the other hand, ALDH activity in human leukemic cells also mediates resistance to a number of drugs [94], and high levels of ALDH activity are predictors of poor therapeutic outcomes [95]. These conclusions are supported by the identification, by the in silico screening of the Gene Expression Omnibus database, using the *PTL* gene signature as a template, of 2 new agents, celastrol and HNE, which were able to eradicate acute myeloid leukemia (AML) at the bulk, progenitor, and stem cell level [96].

Serologic and molecular markers of oxidative stress in patients with MDS include increased concentrations of the LPO product MDA and the presence of oxidized bases in CD34+ cells. Potential mechanisms of oxidative stress include mitochondrial dysfunction via IOL and mitochondrial DNA mutation, systemic inflammation, and bone marrow stromal defects [82]. MDA levels in plasma correlated moderately with serum ferritin and free iron levels and were significantly higher in MDS patients with iron overload, when compared to healthy blood donors, once more emphasizing the role of oxidative stress in the development of MDS [97]. Moreover, a cytoprotective effect was reported of erythropoietin on the plasma membrane of erythroid cells in MDS, which was closely reminiscent of effects detected in certain conditions of impaired glucose metabolism, which were associated with increased LPO-dependent cell stress in the elderly [98]. The selective cytotoxic activities exhibited by LPO products towards transformed hematopoietic cells might be partly responsible for the effectiveness of hypomethylating drugs, widely employed in the chemotherapy of AML and MDS. In fact, the use of these chemotherapeutic agents is associated with the generation of enormous amounts of ROS. Antioxidant supplementation in these patients must be approached with caution, because of the high probability that it might result in significant reductions of the therapeutic efficacy of hypomethylating drugs, whose cytotoxic effect is probably mediated by plasma MDA concentrations, which increase significantly during the 14-day post-chemotherapy period [99].

In conclusion, it appears that the overall effects of LPO products within the context of MDS are strictly dependent on their concentrations, the degree of inflammation and the disease phase. Physiological HNE and MDA concentrations in an early phase of MDS, with low-grade inflammation and normal iron concentrations, seem to favor blast apoptosis and normal hematopoietic cell differentiation in bone marrow niches, whereas high HNE and MDA concentrations in later phases of MDS, with high-grade inflammation and higher than normal iron levels, might favor normal cell death and awry maturation of bone marrow cells [99].

6. Conclusions

From the overall body of evidence presented, the pathogenic involvement of oxidative stress and lipid peroxidation appears to be firmly grounded in all of the senile pathological and dysfunctional conditions examined in this review. In turn, increases in the production of ROS and reactive aldehydes from lipid peroxidation concur to induce apoptosis, which has been observed during aging in osteoblasts, muscle cells, thymocytes and hematopoietic cells. The use of antioxidants as an adjuvant therapy to counteract ROS increases in these disorders gave interesting results [26,87,100,101]. However, the efficacy of antioxidants in reducing the concentrations of aldehydes and/or protein-aldehyde adducts in blood or in tissues has not yet been measured. It is our conception that osteoporosis/osteopenia, sarcopenia, immunosenescence and myelodysplastic syndromes may all represent concurring expressions of the age-related decay of molecular turnover and repair capabilities in post-mitotic cells, altogether expressing themselves as the progressive multiorgan/multisystem failure of senescence. A few systematic redox proteomic approaches, pinpointing the modifications of distinct cell proteins with LPO products generated in the course of oxidative stress, have been conducted to date, mostly in the skeletal muscle of rodents. It is warranted that studies of this kind be extended to the other organ systems undergoing age-associated decay, and be complemented by functional studies as well. We anticipate that continuing investigation in this field may pave the way, in the end, to targeted therapeutic strategies aiming to alleviate the burden of morbidity and mortality associated with these disturbances.

Author Contributions: G.B. and F.G. wrote the paper, S.P., M.D., C.D., A.A, G.P.C., G.G., M.A.C. and M.G. were involved in the previous work on these topics in our group and/or substantively revised it.

Funding: This work was supported by the University of Turin and the Department of Clinical and Biological Sciences (Local Funds ex-60%).

Conflicts of Interest: The authors declare no conflict of interest.

Abbreviations

ALDH	aldehyde dehydrogenase
AML	acute myeloid leukemia
apoB	apolipoprotein B
CMV	cytomegalovirus
DNPH	dinitrophenylhydrazine
HNE	4-hydroxynonenal
HSC	hematopoietic stem cell
IkB	inihibitor of κ light gene enhancer in B cells
IGF-1	insulin-like growth factor 1
IOL	iron overload
JNK	c-jun N-terminal kinase
LDL	low-density lipoprotein
LIF	leukemia-inhibitory factor
LLC	lipid-laden multilocular and adipose cell
LPO	lipid peroxidation
MAPK	mitogen-activated protein kinase
MDA	malondialdehyde

MDS	myelodysplastic syndromes
MEL	murine erythroleukemia
NK-κB	nuclear factor of κ light gene enhancer in B cells
OSM	oncostatin M
PC	protein carbonyl
PPAR-γ	peroxysome proliferator-activated receptor γ
SCF	stem cell factor
TNF-α	tumor necrosis factor alpha
TREC	T-cell receptor excision circle
TSAT	transferrin saturation

References

1. Granic, A.; Mendonça, N.; Hill, T.R.; Jagger, C.; Stevenson, E.J.; Mathers, J.C.; Sayer, A.A. Nutrition in the Very Old. *Nutrients* **2018**, *10*, 269. [CrossRef] [PubMed]
2. Kontis, V.; Bennett, J.E.; Mathers, C.D.; Li, G.; Foreman, K.; Ezzati, M. Future life expectancy in 35 industrialised countries: Projections with a Bayesian model ensemble. *Lancet* **2017**, *389*, 1323–1335. [CrossRef]
3. Bektas, A.; Schurman, S.H.; Sen, R.; Ferrucci, L. Aging, inflammation and the environment. *Exp. Gerontol.* **2018**, *105*, 10–18. [CrossRef] [PubMed]
4. Vina, J.; Borras, C.; Miquel, J. Theories of ageing. *IUBMB Life* **2007**, *59*, 249–254. [CrossRef] [PubMed]
5. Sohal, R.S.; Orr, W.C. The redox stress hypothesis of aging. *Free Radic. Biol. Med.* **2012**, *52*, 539–555. [CrossRef] [PubMed]
6. Boveris, A.; Chance, B. The mitochondrial generation of hydrogen peroxide. General properties and effect of hyperbaric oxygen. *Biochem. J.* **1973**, *134*, 707–716. [CrossRef] [PubMed]
7. Chistiakov, D.A.; Sobenin, I.A.; Revin, V.V.; Orekhov, A.N.; Bobryshev, Y.V. Mitochondrial aging and age-related dysfunction of mitochondria. *BioMed. Res. Int.* **2014**, *2014*, 238463. [CrossRef] [PubMed]
8. Esterbauer, H.; Schaur, R.J.; Zollner, H. Chemistry and Biochemistry of 4-hydroxynonenal, malonaldehyde and related aldehydes. *Free Radic. Biol. Med.* **1991**, *11*, 81–128. [CrossRef]
9. Ayala, A.; Muñoz, M.F.; Argüelles, S. Lipid peroxidation: Production, metabolism, and signaling mechanisms of malondialdehyde and 4-hydroxy-2-nonenal. *Oxid. Med. Cell. Longev.* **2014**, *2014*, 360438. [CrossRef] [PubMed]
10. Barrera, G.; Pizzimenti, S.; Ciamporcero, E.S.; Daga, M.; Ullio, C.; Arcaro, A.; Cetrangolo, G.P.; Ferretti, C.; Dianzani, C.; Lepore, A.; et al. Role of 4-hydroxynonenal-protein adducts in human diseases. *Antioxid. Redox Signal.* **2015**, *22*, 1681–1702. [CrossRef] [PubMed]
11. Domingues, R.M.; Domingues, P.; Melo, T.; Pérez-Sala, D.; Reis, A.; Spickett, C.M. Lipoxidation adducts with peptides and proteins: Deleterious modifications or signaling mechanisms? *J. Proteom.* **2013**, *92*, 110–131. [CrossRef] [PubMed]
12. Esterbauer, H.; Zollern, H. Methods for determination of aldehydic lipid peroxidation products. *Free Radic. Biol. Med.* **1989**, *7*, 197–203. [CrossRef]
13. Papac-Milicevic, N.; Busch, C.J.; Binder, C.J. Malondialdehyde Epitopes as Targets of Immunity and the Implications for Atherosclerosis. *Adv. Immunol.* **2016**, *131*, 1–59. [CrossRef] [PubMed]
14. Nègre-Salvayre, A.; Garoby-Salom, S.; Swiader, A.; Rouahi, M.; Pucelle, M.; Salvayre, R. Proatherogenic effects of 4-hydroxynonenal. *Free Radic. Biol. Med.* **2017**, *111*, 127–139. [CrossRef] [PubMed]
15. Casini, A.; Galli, A.; Pignalosa, P.; Frulloni, L.; Grappone, C.; Milani, S.; Pederzoli, P.; Cavallini, G.; Surrenti, C. Collagen type I synthesized by pancreatic periacinar stellate cells (PSC) co-localizes with lipid peroxidation-derived aldehydes in chronic alcoholic pancreatitis. *J. Pathol.* **2000**, *192*, 81–89. [CrossRef]
16. Gil, L.; Siems, W.; Mazurek, B.; Gross, J.; Schroeder, P.; Voss, P.; Grune, T. Age-associated analysis of oxidative stress parameters in human plasma and erythrocytes. *Free Radic. Res.* **2006**, *40*, 495–505. [CrossRef] [PubMed]
17. Sultana, R.; Perluigi, M.; Allan Butterfield, D. Lipid peroxidation triggers neurodegeneration: A redox proteomics view into the Alzheimer disease brain. *Free Radic. Biol. Med.* **2013**, *62*, 157–169. [CrossRef] [PubMed]
18. Manolagas, S.C. Birth and death of bone cells: Basic regulatory mechanisms and implications for the pathogenesis and treatment of osteoporosis. *Endocr. Rev.* **2000**, *2*, 115–137. [CrossRef]

19. Dawson-Hughes, B.; Looker, A.C.; Tosteson, A.N.; Johansson, H.; Kanis, J.A.; Melton, L.J., III. The potential impact of the National Osteoporosis Foundation guidance on treatment eligibility in the USA: An update in NHANES 2005–2008. *Osteoporos. Int.* **2012**, *23*, 811–820. [CrossRef] [PubMed]

20. Choksi, P.; Jepsen, K.J.; Clines, G.A. The challenges of diagnosing osteoporosis and the limitations of currently available tools. *Clin. Diabetes Endocrinol.* **2018**, *4*, 12. [CrossRef] [PubMed]

21. Huang, C.X.; Lv, B.; Wang, Y. Protein Phosphatase 2A Mediates Oxidative Stress Induced Apoptosis in Osteoblasts. *Mediat. Inflamm.* **2015**, *2015*, 804260. [CrossRef] [PubMed]

22. Sendur, O.F.; Turan, Y.; Tastaban, E.; Serter, M. Antioxidant status in patients with osteoporosis: A controlled study. *Joint Bone Spine* **2009**, *76*, 514–518. [CrossRef] [PubMed]

23. Almeida, M.; Han, L.; Martin-Millan, M.; Plotkin, L.I.; Stewart, S.A.; Roberson, P.K.; Kousteni, S.; O'Brien, C.A.; Bellido, T.; Parfitt, A.M.; et al. Skeletal involution by age-associated oxidative stress and its acceleration by loss of sex steroids. *J. Biol. Chem.* **2007**, *282*, 27285–27297. [CrossRef] [PubMed]

24. Maggio, D.; Barabani, M.; Pierandrei, M.; Polidori, M.C.; Catani, M.; Mecocci, P.; Senin, U.; Pacifici, R.; Cherubini, A. Marked decrease in plasma antioxidants in aged osteoporotic women: Results of a cross-sectional study. *J. Clin. Endocrinol. Metab.* **2003**, *88*, 1523–1527. [CrossRef] [PubMed]

25. Lean, J.M.; Jagger, C.J.; Kirstein, B.; Fuller, K.; Chambers, T.J. Hydrogen peroxide is essential for estrogen-deficiency bone loss and osteoclast formation. *Endocrinology* **2005**, *146*, 728–735. [CrossRef] [PubMed]

26. Domazetovic, V.; Marcucci, G.; Iantomasi, T.; Brandi, M.L.; Vincenzini, M.T. Oxidative stress in bone remodeling: Role of antioxidants. *Clin. Cases Miner. Bone Metab.* **2017**, *14*, 209–216. [CrossRef] [PubMed]

27. Li, X.; Han, Y.; Guan, Y.; Zhang, L.; Bai, C.; Li, Y. Aluminum induces osteoblast apoptosis through the oxidative stress-mediated JNK signaling pathway. *Biol. Trace Elem. Res.* **2012**, *150*, 502–508. [CrossRef] [PubMed]

28. Pantano, C.; Reynaert, N.L.; van der Vliet, A.; Janssen-Heininger, Y.M. Redox-sensitive kinases of the nuclear factor-κB signaling pathway. *Antioxid. Redox Signal.* **2006**, *8*, 1791–1806. [CrossRef] [PubMed]

29. Almeida, M.; Han, L.; Ambrogini, E.; Bartell, S.M.; Manolagas, S.C. Oxidative stress stimulates apoptosis and activates NF-κB in osteoblastic cells via a PKCβ/p66shc signaling cascade: Counter regulation by estrogens or androgens. *Mol. Endocrinol.* **2010**, *24*, 2030–2037. [CrossRef] [PubMed]

30. Zhu, S.Y.; Zhuang, J.S.; Wu, Q.; Liu, Z.Y.; Liao, C.R.; Luo, S.G.; Chen, J.T.; Zhong, Z.M. Advanced oxidation protein products induce pre-osteoblast apoptosis through a nicotinamide adenine dinucleotide phosphate oxidase-dependent, mitogen-activated protein kinases-mediated intrinsic apoptosis pathway. *Aging Cell* **2018**, e12764. [CrossRef] [PubMed]

31. Wang, X.; Chen, B.; Sun, J.; Jiang, Y.; Zhang, H.; Zhang, P.; Fei, B.; Xu, Y. Iron-induced oxidative stress stimulates osteoclast differentiation via NF-κB signaling pathway in mouse model. *Metabolism* **2018**, *83*, 167–176. [CrossRef] [PubMed]

32. Rachner, T.D.; Khosla, S.; Hofbauer, L.C. Osteoporosis: Now and the future. *Lancet* **2011**, *377*, 1276–1287. [CrossRef]

33. Muzio, G.; Maggiora, M.; Paiuzzi, E.; Oraldi, M.; Canuto, R.A. Aldehyde dehydrogenases and cell proliferation. *Free Radic. Biol. Med.* **2012**, *52*, 735–746. [CrossRef] [PubMed]

34. Mittal, M.; Khan, K.; Pal, S.; Porwal, K.; China, S.P.; Barbhuyan, T.K.; Baghel, K.S.; Rawat, T.; Sanyal, S.; Bhadauria, S.; et al. The thiocarbamate disulphide drug, disulfiram induces osteopenia in rats by inhibition of osteoblast function due to suppression of acetaldehyde dehydrogenase activity. *Toxicol. Sci.* **2014**, *139*, 257–270. [CrossRef] [PubMed]

35. Hoshi, H.; Hao, W.; Fujita, Y.; Funayama, A.; Miyauchi, Y.; Hashimoto, K.; Miyamoto, K.; Iwasaki, R.; Sato, Y.; Kobayashi, T.; et al. Aldehyde-stress resulting from Aldh2 mutation promotes osteoporosis due to impaired osteoblastogenesis. *J. Bone Miner. Res.* **2012**, *27*, 2015–2023. [CrossRef] [PubMed]

36. Mittal, M.; Pal, S.; China, S.P.; Porwal, K.; Dev, K.; Shrivastava, R.; Raju, K.S.; Rashid, M.; Trivedi, A.K.; Sanyal, S.; et al. Pharmacological activation of aldehyde dehydrogenase 2 promotes osteoblast differentiation via bone morphogenetic protein-2 and induces bone anabolic effect. *Toxicol. Appl. Pharmacol.* **2017**, *316*, 63–73. [CrossRef] [PubMed]

37. Akpolat, V.; Bilgin, H.M.; Celik, M.Y.; Erdemoglu, M.; Isik, B. An evaluation of nitric oxide, folate, homocysteine levels and lipid peroxidation in postmenopausal osteoporosis. *Adv. Clin. Exp. Med.* **2013**, *22*, 403–409. [PubMed]

38. Wu, Q.; Zhong, Z.M.; Pan, Y.; Zeng, J.H.; Zheng, S.; Zhu, S.Y.; Chen, J.T. Advanced Oxidation Protein Products as a Novel Marker of Oxidative Stress in Postmenopausal Osteoporosis. *Med. Sci. Monit.* **2015**, *21*, 2428–2432. [CrossRef] [PubMed]

39. Cruz-Jentoft, J.P.; Baeyens, J.M.; Bauer, Y.; Boirie, T.; Cederholm, F.; Landi, F.C.; Martin, J.P.; Michel, Y.; Rolland, S.M.; Schneider, E.; et al. European consensus on definition and diagnosis: Report of the European Working Group on Sarcopenia in Older People. *Age Ageing* **2010**, *39*, 412–423. [CrossRef] [PubMed]

40. Diz, J.B.; Leopoldino, A.A.; Moreira, B.S.; Henschke, N.; Dias, R.C.; Pereira, L.S.; Oliveira, V.C. Prevalence of sarcopenia in older Brazilians: A systematic review and meta-analysis. *Geriatr. Gerontol. Int.* **2017**, *17*, 5–16. [CrossRef] [PubMed]

41. Zanandrea, V.; Giua, R.; Costanzo, L.; Vellas, B.; Zamboni, M.; Cesari, M. Interventions against sarcopenia in older persons. *Curr. Pharm. Des.* **2014**, *20*, 5983–6006. [PubMed]

42. Marzetti, E.; Calvani, R.; Bernabei, R.; Leeuwenburgh, C. Apoptosis in skeletal myocytes: A potential target for interventions against sarcopenia and physical frailty—A mini-review. *Gerontology* **2012**, *58*, 99–106. [CrossRef] [PubMed]

43. Buonocore, D.; Rucci, S.; Vandoni, M.; Negro, M.; Marzatico, F. Oxidative system in aged skeletal muscle. *Muscles Ligaments Tendons J.* **2011**, *1*, 85–90. [PubMed]

44. Dupont-Versteegden, E.E. Apoptosis in muscle atrophy: Relevance to sarcopenia. *Exp. Gerontol.* **2005**, *40*, 473–481. [CrossRef] [PubMed]

45. Aiken, J.; Bua, E.; Cao, Z.; Lopez, M.; Wanagat, J.; McKenzie, D.; McKiernan, S. Mitochondrial DNA deletion mutations and sarcopenia. *Ann. N. Y. Acad. Sci.* **2002**, *959*, 412–423. [CrossRef] [PubMed]

46. Moulin, M.; Ferreiro, A. Muscle redox disturbances and oxidative stress as pathomechanisms and therapeutic targets in early-onset myopathies. *Semin. Cell Dev. Biol.* **2017**, *64*, 213–223. [CrossRef] [PubMed]

47. Polyzos, S.A.; Margioris, A.N. Sarcopenic obesity. *Hormones* **2018**. [CrossRef] [PubMed]

48. Lefranc, C.; Friederich-Persson, M.; Palacios-Ramirez, R.; Nguyen Dinh Cat, A. Mitochondrial oxidative stress in obesity: Role of the mineralocorticoid receptor. *J. Endocrinol.* **2018**, *238*, R143–R159. [CrossRef] [PubMed]

49. Jang, Y.C.; Lustgarten, M.S.; Liu, Y.; Muller, F.L.; Bhattacharya, A.; Liang, H.; Salmon, A.B.; Brooks, S.V.; Larkin, L.; Hayworth, C.R.; et al. Increased superoxide in vivo accelerates age-associated muscle atrophy through mitochondrial dysfunction and neuromuscular junciton degeneration. *FASEB J.* **2010**, *24*, 1376–1390. [CrossRef] [PubMed]

50. Vollaard, N.B.; Shearmann, J.P.; Cooper, C.E. Exercise induced oxidative stress: Myths, realities and physiological relevance. *Sports Med.* **2005**, *35*, 1045–1062. [CrossRef] [PubMed]

51. Zuo, L.; Pannell, B.K. Redox characterization of functioning skeletal muscle. *Front. Physiol.* **2015**, *6*, 338. [CrossRef] [PubMed]

52. Merry, T.L.; Steinberg, G.R.; Lynch, G.S.; McConell, G.K. Skeletal muscle uptake during contraction is regulated by nitric oxide and ROS independently of AMPK. *Am. J. Physiol. Endocrinol. Metab.* **2009**, *298*, E577–E585. [CrossRef] [PubMed]

53. Powers, S.K.; Talbert, E.E.; Adhihetty, P.J. Reactive oxygen and nitrogen species as intracellular signals in skeletal muscle. *J. Physiol.* **2011**, *589*, 2129–2138. [CrossRef] [PubMed]

54. Bassel-Duby, R.; Olson, E.N. Signaling pathways in skeletal muscle remodeling. *Annu. Rev. Biochem.* **2006**, *75*, 19–37. [CrossRef] [PubMed]

55. Powers, S.K.; Duarte, J.; Kavazis, A.N.; Talbert, E.E. Reactive oxygen species are signalling molecules for skeletal muscle adaptation. *Exp. Physiol.* **2010**, *95*, 1–9. [CrossRef] [PubMed]

56. Malik, V.; Rodino-Klapac, L.R.; Mendell, J.R. Emerging drugs for Duchenne muscular dystrophy. *Expert Opin. Emerg. Drugs* **2012**, *17*, 261–277. [CrossRef]

57. Charles, A.-L.; Meyer, A.; Dal-Ros, S.; Auger, C.; Keller, N.; Ramamoorthy, T.G.; Zoll, J.; Metzger, D.; Schini-Kerth, V.; Geny, B. Polyphenols prevent ageing-related impairment in skeletal muscle mitochondrial function through decreased reactive oxygen species production. *Exp. Physiol.* **2013**, *98*, 536–545. [CrossRef] [PubMed]

58. Visser, S.B.; Kritchevsky, B.H.; Goodpaster, A.B.; Newman, M.; Nevitt, E. Leg muscle mass and composition in relation to lower extremity performance in men and women aged 70–79: The health, aging and body composition study. *J. Am. Geriatr. Soc.* **2002**, *50*, 897–904. [CrossRef] [PubMed]

59. Feng, J.; Navratil, M.; Thompson, L.V.; Arriaga, E.A. Estimating relative carbonyl levels in muscle microstructures by fluorescence imaging. *Anal. Bioanal. Chem.* **2008**, *391*, 2591–2598. [CrossRef] [PubMed]

60. Feng, J.; Xie, H.; Meany, D.L.; Thompson, L.V.; Arriaga, E.A.; Griffin, T.J. Quantitative proteomic profiling of muscle type-dependent and age-dependent protein carbonylation in rat skeletal muscle mitochondria. *J. Gerontol. A Biol. Sci. Med.Sci.* **2008**, *63*, 1137–1152. [CrossRef] [PubMed]

61. Braga, M.; Sinha Hikim, A.P.; Datta, S.; Ferrini, M.G.; Brown, D.; Kovacheva, E.L.; Gonzalez-Cadavid, N.F.; Sinha-Hikim, I. Involvement of oxidative stress and caspase 2-mediated intrinsic pathway signaling in age-related increase in muscle cell apoptosis in mice. *Apoptosis* **2008**, *13*, 822–832. [CrossRef] [PubMed]

62. Nakashima, Y.; Ohsawa, I.; Nishimaki, K.; Kumamoto, S.; Maruyama, I.; Suzuki, Y.; Ohta, S. Preventive effects of Chlorella on skeletal muscle atrophy in muscle-specific mitochondrial aldehyde dehydrogenase 2 activity-deficient mice. *BMC Complement. Altern. Med.* **2014**, *14*, 390. [CrossRef] [PubMed]

63. Beltran Valls, M.R.; Wilkinson, D.J.; Narici, M.V.; Smith, K.; Phillips, B.E.; Caporossi, D.; Atherton, P.J. Protein carbonylation and heat shock proteins in human skeletal muscle: Relationships to age and sarcopenia. *J. Gerontol. A Biol. Sci. Med. Sci.* **2014**, *70*, 174–181. [CrossRef] [PubMed]

64. Coto Montes, A.; Boga, J.A.; Bermejo Millo, C.; Rubio González, A.; Potes Ochoa, Y.; Vega Naredo, I.; Martínez Reig, M.; Romero Rizos, L.; Sánchez Jurado, P.M.; Solano, J.J.; et al. Potential early biomarkers of sarcopenia among independent older adults. *Maturitas* **2017**, *104*, 117–122. [CrossRef] [PubMed]

65. Bellanti, F.; Romano, A.D.; Lo Buglio, A.; Castriotta, V.; Guglielmi, G.; Greco, A.; Serviddio, G.; Vendemiale, G. Oxidative stress is increased in sarcopenia and associated with cardiovascular disease risk in sarcopenic obesity. *Maturitas* **2018**, *109*, 6–12. [CrossRef] [PubMed]

66. Gueugneau, M.; Coudy-Gandilhon, C.; Gourbeyre, O.; Chambon, C.; Combaret, L.; Polge, C.; Taillandier, D.; Attaix, D.; Friguet, B.; Maier, A.B.; et al. Proteomics of muscle chronological ageing in post-menopausal women. *BMC Genom.* **2014**, *15*, 1165. [CrossRef] [PubMed]

67. Franceschi, C.; Capri, M.; Monti, D.; Giunta, S.; Olivieri, F.; Sevini, F.; Panourgia, M.P.; Invidia, L.; Celani, L.; Scurti, M.; et al. Inflammaging and anti-inflammaging: A systemic perspective on aging and longevity emerged from studies in humans. *Mech. Ageing Dev.* **2007**, *128*, 92–105. [CrossRef] [PubMed]

68. Ventura, M.T.; Casciaro, M.; Gangemi, S.; Buquicchio, R. Immunosenescence in aging: Between immune cells depletion and cytokines. *Clin. Mol. Allergy* **2017**, *15*, 21. [CrossRef] [PubMed]

69. Goronzy, J.J.; Weyand, C.M. T cell development and receptor diversity during aging. *Curr. Opin. Immunol.* **2005**, *17*, 468–475. [CrossRef] [PubMed]

70. Taub, D.D.; Longo, D.L. Insight into thymic aging and regeneration. *Immunol. Rev.* **2005**, *205*, 72–93. [CrossRef] [PubMed]

71. Herndler-Brandstetter, D.; Cioca, D.P.; Grubeck-Loebensteiner, B. Immunizations in the elderly: Do they live up to their promise? *Wien. Med. Wochenschr.* **2006**, *156*, 130–141. [CrossRef] [PubMed]

72. Gruver, A.L.; Hudson, L.L.; Sempowski, G.D. Immunosenescence of ageing. *J. Pathol.* **2007**, *211*, 144–156. [CrossRef] [PubMed]

73. Wikby, A.; Ferguson, F.; Forsey, R.; Thompson, J.; Strindhall, J.; Lofgren, S.; Nilsson, B.O.; Ernerudh, J.; Pawelec, G.; Johansson, B. An immune risk phenotype, cognitive impairment, and survival in very late life: Impact of allostatic load in Swedish octogenarian and nonagenarian humans. *J. Gerontol. A Biol. Sci. Med. Sci.* **2005**, *60*, 556–565. [CrossRef] [PubMed]

74. De Mello-Coelho, V.; Cutler, R.G.; Bunbury, A.; Tammara, A.; Mattson, M.P.; Taub, D.D. Age-associated alterations in the levels of cytotoxic lipid molecular species and oxidative stress in the murine thymus are reduced by growth hormone treatment. *Mech. Ageing Dev.* **2017**, *167*, 46–55. [CrossRef] [PubMed]

75. Langhi, L.G.; Andrade, L.R.; Shimabukuro, M.K.; van Ewijk, W.; Taub, D.D.; Borojevic, R.; de Mello Coelho, V. Lipid-laden multilocular cells in the aging thymus are phenotypically heterogeneous. *PLoS ONE* **2015**, *10*, e0141516. [CrossRef] [PubMed]

76. Taub, D.D.; Murphy, W.J.; Longo, D.L. Rejuvenation of the aging thymus: Growth hormone- and ghrelin-mediated signaling pathways. *Curr. Opin. Pharmacol.* **2010**, *10*, 408–424. [CrossRef] [PubMed]

77. De Mello-Coelho, V.; Bunbury, A.; Rangel, L.B.; Giri, B.; Weeraratna, A.; Morin, P.J.; Bernier, M.; Taub, D.D. Fat-storing multilocular cells expressing CCR5 increase in the thymus with advancing age: Potential role for CCR5 ligands on the differentiation and migration of preadipocytes. *Int. J. Med. Sci.* **2010**, *7*, 1–14. [CrossRef]

78. Gautam, N.; Das, S.; Mahapatra, S.K.; Chakraborti, S.P.; Kundu, P.K.; Roy, S. Age associated oxidative damage in lymphocytes. *Oxid. Med. Cell. Longev.* **2010**, *3*, 275–282. [CrossRef] [PubMed]

79. Ma, X. Epidemiology of myelodysplastic syndromes. *Am. J. Med.* **2012**, *125*, S2–S5. [CrossRef] [PubMed]

80. Shumilov, E.; Flach, J.; Kohlmann, A.; Banz, Y.; Bonadies, N.; Fiedler, M.; Pabst, T.; Bacher, U. Current status and trends in the diagnostics of AML and MDS. *Blood Rev.* **2018**. [CrossRef] [PubMed]

81. Malcovati, L.; Della Porta, M.G.; Pascutto, C.; Invernizzi, R.; Boni, M.; Travaglino, E.; Passamonti, F.; Arcaini, L.; Maffioli, M.; Bernasconi, P.; et al. Prognostic factors and life expectancy in myelodysplastic syndromes classified according to WHO criteria: A basis for clinical decision-making. *J. Clin. Oncol.* **2005**, *23*, 7594–7603. [CrossRef] [PubMed]

82. Farquhar, M.J.; Bowen, D.T. Oxidative stress and the myelodysplastic syndromes. *Int. J. Hematol.* **2003**, *77*, 342–350. [CrossRef] [PubMed]

83. De Souza, G.F.; Barbosa, M.C.; Santos, T.E.; Carvalho, T.M.; de Freitas, R.M.; Martins, M.R.; Gonçalves, R.P.; Pinheiro, R.F.; Magalhães, S.M. Increased parameters of oxidative stress and its relation to transfusion iron overload in patients with myelodysplastic syndromes. *J. Clin. Pathol.* **2013**, *66*, 996–998. [CrossRef] [PubMed]

84. Gan, Z.S.; Wang, Q.Q.; Li, J.H.; Wang, X.L.; Wang, Y.Z.; Du, H.H. Iron Reduces M1 Macrophage Polarization in RAW264.7 Macrophages Associated with Inhibition of STAT1. *Mediat. Inflamm.* **2017**, *8570818*. [CrossRef] [PubMed]

85. Angelucci, E.; Cianciulli, P.; Finelli, C.; Mecucci, C.; Voso, M.T.; Tura, S. Unraveling the mechanisms behind iron overload and ineffective hematopoiesis in myelodysplastic syndromes. *Leuk. Res.* **2017**, *62*, 108–115. [CrossRef] [PubMed]

86. Ivars, D.; Orero, M.T.; Javier, K.; Díaz-Vico, L.; García-Giménez, J.L.; Mena, S.; Tormos, C.; Egea, M.; Pérez, P.L.; Arrizabalaga, B.; et al. Oxidative imbalance in low/intermediate-1-risk myelodysplastic syndrome patients: The influence of iron overload. *Clin. Biochem.* **2017**, *50*, 911–917. [CrossRef] [PubMed]

87. Ghoti, H.; Amer, J.; Winder, A.; Rachmilewitz, E.; Fibach, E. Oxidative stress in red blood cells, platelets and polymorphonuclear leukocytes from patients with myelodysplastic syndromes. *Eur. J. Haematol.* **2007**, *79*, 463–467. [CrossRef] [PubMed]

88. Rassool, F.V.; Gaymes, T.J.; Omidvar, N.; Brady, N.; Beurlet, S.; Pla, M.; Reboul, M.; Lea, N.; Chomienne, C.; Thomas, N.S.; et al. Reactive oxygen species, DNA damage, and error-prone repair: A model for genomic instability with progression in myeloid leukemia? *Cancer Res.* **2007**, *67*, 8762–8771. [CrossRef] [PubMed]

89. Smith, C.; Gasparetto, M.; Jordan, C.; Pollyea, D.A.; Vasiliou, V. The effects of alcohol and aldehyde dehydrogenases on disorders of hematopoiesis. *Adv. Exp. Med. Biol.* **2015**, *815*, 349–359. [CrossRef] [PubMed]

90. Gasparetto, M.; Smith, C.A. ALDHs in normal and malignant hematopoietic cells: Potential new avenues for treatment of AML and other blood cancers. *Chem. Biol. Interact.* **2017**, *276*, 46–51. [CrossRef] [PubMed]

91. Semlitsch, T.; Tillian, H.M.; Zarkovic, N.; Borovic, S.; Purtscher, M.; Hohenwarter, O.; Schaur, R.J. Differential influence of the lipid peroxidation product 4-hydroxynonenal on the growth of human lymphatic leukaemia cells and human peripheral blood lymphocytes. *Anticancer Res.* **2002**, *22*, 1689–1697. [PubMed]

92. Rinaldi, M.; Barrera, G.; Spinsanti, P.; Pizzimenti, S.; Ciafrè, S.A.; Parella, P.; Farace, M.G.; Signori, E.; Dianzani, M.U.; Fazio, V.M. Growth inhibition and differentiation induction in murine erythroleukemia cells by 4-hydroxynonenal. *Free Radic. Res.* **2001**, *34*, 629–637. [CrossRef] [PubMed]

93. Gasparetto, M.; Pei, S.; Minhajuddin, M.; Khan, N.; Pollyea, D.A.; Myers, J.R.; Ashton, J.M.; Becker, M.W.; Vasiliou, V.; Humphries, K.R.; et al. Targeted therapy for a subset of acute myeloid leukemias that lack expression of aldehyde dehydrogenase1A1. *Haematologica* **2017**, *102*, 1054–1065. [CrossRef] [PubMed]

94. Koelling, T.M.; Yeager, A.M.; Hilton, J.; Haynie, D.T.; Wiley, J.M. Development and characterization of a cyclophosphamide-resistant subline of acute myeloid leukemia in the Lewis × Brown Norway hybrid rat. *Blood* **1990**, *76*, 1209–1213. [PubMed]

95. Ran, D.; Schubert, M.; Pietsch, L.; Taubert, I.; Wuchter, P.; Eckstein, V.; Bruckner, T.; Zoeller, M.; Ho, A.D. Aldehyde dehydrogenase activity among primary leukemia cells is associated with stem cell features and correlates with adverse clinical outcomes. *Exp. Hematol.* **2009**, *37*, 1423–1434. [CrossRef] [PubMed]

96. Hassane, D.C.; Guzman, M.L.; Corbett, C.; Li, X.; Abboud, R.; Young, F.; Liesveld, J.L.; Carroll, M.; Jordan, C.T. Discovery of agents that eradicate leukemia stem cells using an in silico screen of public gene expression data. *Blood* **2008**, *111*, 5654–5662. [CrossRef] [PubMed]

97. Pimková, K.; Chrastinová, L.; Suttnar, J.; Štikarová, J.; Kotlín, R.; Hermák, J.E. Plasma Levels of Aminothiols, Nitrite, Nitrate, and Malondialdehyde in Myelodysplastic Syndromes in the Context of Clinical Outcomes and as a Consequence of Iron Overload. *Oxid. Med. Cell. Longev.* **2014**, *2014*, 416028. [CrossRef] [PubMed]

98. Gradinaru, D.; Margina, D.; Ilie, M.; Borsa, C.; Ionescu, C.; Prada, G.I. Correlation between erythropoietin serum levels and erythrocyte susceptibility to lipid peroxidation in elderly with type 2 diabetes. *Acta Physiol. Hung.* **2015**, *102*, 400–408. [CrossRef] [PubMed]
99. Esfahani, A.; Ghoreishi, Z.; Nikanfar, A.; Sanaat, Z.; Ghorbanihaghjo, A. Influence of chemotherapy on the lipid peroxidation and antioxidant status in patients with acute myeloid leukemia. *Acta Med. Iran.* **2012**, *50*, 454–458. [PubMed]
100. Brioche, T.; Lemoine-Morel, S. Oxidative Stress, Sarcopenia, Antioxidant Strategies and Exercise: Molecular Aspects. *Curr. Pharm. Des.* **2016**, *22*, 2664–2678. [CrossRef] [PubMed]
101. Espino, J.; Pariente, J.A.; Rodríguez, A.B. Oxidative stress and immunosenescence: Therapeutic effects of melatonin. *Oxid. Med. Cell. Longev.* **2012**, *2012*, 670294. [CrossRef] [PubMed]

antioxidants

MDPI

Article

Unbiased Identification of Proteins Covalently Modified by Complex Mixtures of Peroxidized Lipids Using a Combination of Electrophoretic Mobility Band Shift with Mass Spectrometry

Bernd Gesslbauer * [iD], David Kuerzl, Niko Valpatic and Valery N. Bochkov [iD]

Institute of Pharmaceutical Sciences, Department of Pharmaceutical Chemistry, University of Graz, Humboldtstrasse 46, 8010 Graz, Austria; david.kuerzl@edu.uni-graz.at (D.K.); niko.valpatic@edu.uni-graz.at (N.V.); valery.bochkov@uni-graz.at (V.N.B.)
* Correspondence: bernd.gesslbauer@uni-graz.at

Received: 12 July 2018; Accepted: 29 August 2018; Published: 30 August 2018

Abstract: Covalent modification of functionally important cell proteins by lipid oxidation products (LOPs) is a known mechanism initiating pathological consequences of oxidative stress. Identification of new proteins covalently modified by electrophilic lipids can be performed by a combination of chemical, immunological, and mass spectrometry-based methods, but requires prior knowledge either on the exact molecular structure of LOPs (e.g., 4-hydroxynonenal) or candidate protein targets. However, under the conditions of oxidative stress in vivo, a complex mixture of proteins (e.g., cytosolic proteome) reacts with a complex mixture of LOPs. Here we describe a method for detection of lipid-modified proteins that does not require an a priori knowledge on the chemical structure of LOPs or identity of target proteins. The method is based on the change of electrophoretic mobility of lipid-modified proteins, which is induced by conformational changes and cross-linking with other proteins. Abnormally migrating proteins are detected by mass spectrometry-based protein peptide sequencing. We applied this method to study effects of oxidized palmitoyl-arachidonoyl-phosphatidylcholine (OxPAPC) on endothelial cells. Several known, but also many new, OxPAPC-binding proteins were identified. We expect that this technically relatively simple method can be widely applied for label-free analysis of lipid-protein interactions in complex protein samples treated with different LOPs.

Keywords: lipid oxidation; oxidized phospholipids; lipid-protein adducts; electrophoretic mobility shift assay; gel-shift; proteomics; mass spectrometry

---

1. Introduction

Lipid oxidation is a characteristic feature of chronic inflammation associated with a variety of pathologies including, e.g., cardiovascular and neurodegenerative disease [1–5]. Oxidation of free or esterified polyunsaturated fatty acids (PUFAs) can be induced by reactive oxygen and nitrogen species or by multiple enzymes, generating a variety of bioactive compounds. In particular, oxidation of sn-2 PUFA residues in phospholipids (PLs) generates a variety of full-length and truncated reactive oxidized PLs (OxPLs), which are increasingly recognized for their pleiotropic biological action including proinflammatory and prothrombotic effects, as well as stimulation of angiogenesis and protection of lung endothelial barrier [6]. OxPLs are known to be present in blood and tissues; their total concentrations in pathological conditions can reach micromolar range, while in atherosclerotic vessels total concentrations of several molecular species may be in a high micromolar range [1]. Lipid oxidation products (LOPs) containing electrophilic terminal aldehyde or α,β-unsaturated carbonyl groups can react with nucleophilic groups of biomacromolecules, such as thiol and amino groups of proteins, forming covalent Michael adducts or Schiff bases. Adduction of LOPs to proteins can alter protein structure and function, and consequently influence cellular signal transduction pathways [1,7,8]. This covalent modification of proteins may be quite selective, for example, modification of the β-subunit of IκB kinase prevents IκBα degradation and NF–κB activation [9]. Therefore, the knowledge of protein targets of LOPs is a prerequisite for better understanding of mechanisms of oxidative stress.

Several approaches to identify protein targets of LOPs that have been developed in the last years can be divided in two groups. The majority of studies identified proteins in crude preparations, e.g., cytosolic proteins, that were modified by a specific chemically-pure LOP, e.g., HNE (4-hydroxynonenal). Another group of publications analysed specific amino acid residues within a defined protein, which may be modified by one or different reactive LOPs. The most extensively investigated lipid-derived electrophile is HNE. Identification of HNE-protein adducts has been performed mainly by following three approaches: (i) tandem mass spectrometry (MS/MS and MS^3)-based detection of HNE-modified peptides [10,11]; (ii) immune detection of HNE-protein adducts by Western blotting followed by liquid chromatography tandem mass spectrometry (LC-MS/MS)-based identification of immunoreactive protein bands [12–14]; and (iii) attachment of biotin through click chemistry and enrichment of modified proteins or peptides using immobilized streptavidin. A variety of carbonyl-reactive probes have been developed to capture HNE-modified proteins and other carbonyl-containing compounds. These probes contain typically hydrazide or aminooxy functional groups, which are conjugated with biotin for streptavidin-based enrichment [15–19]. In addition to these post-labeling strategies, synthetic alkynyl- and azido-derivatives of HNE have been used to treat cells or isolated proteomes followed by biotin/streptavidin-based purification of modified proteins [20,21]. Site-specific characterization of cysteine modifications by different LOPs using thiol-specific probes was reported by Wang et al. [22]. With this chemoproteomic method, which is based on the quantification of unmodified cysteines (because LOP-modified cysteines are protected from thiol-specific labelling), >1000 reactive cysteines were quantified in isolated proteomes.

In contrast to the relatively well investigated protein targets of HNE, proteomic approaches to identify OxPL-protein adducts, are rather limited, because the enrichment methods that are reported for HNE-protein adducts cannot be directly applied to OxPLs.

Gugiu et al. used biotinylated 1-palmitoyl-2-arachidonoyl-*sn*-glycero-3-phosphoethanolamine (PAPE), which was oxidized on air to generate biotin-labeled OxPAPE (OxPAPE-N-biotin), to identify OxPAPE protein targets from human aortic endothelial cells [23]. Biological activity of OxPAPE-N-biotin was proven by upregulation of mRNA of inflammatory genes. Biotinylated OxPAPE-protein adducts were enriched with avidin beads, separated by sodium dodecyl sulfate polyacrylamide gel electrophoresis (SDS-PAGE) and detected by Western blotting using HRP-streptavidin (horseradish peroxidase-streptavidin). Protein bands containing biotinylated adducts detected by Western blot analysis were cut from an SYPRO Ruby stained gel and subjected to mass spectrometric identification. Overall, 29 mainly cytosolic and mitochondrial proteins could be identified from positively-stained protein bands, which were proposed as potential targets of OxPAPE in endothelial cells.

A similar approach was reported by Tallman et al. using a biotin-modified 1-palmitoyl-2-linoleoyl-*sn*-glycero-3-phosphatidylcholine (PLPC) probe to identify protein targets of OxPLs in human plasma [24,25]. In contrast to Gugiu et al., the eluate from streptavidin column was directly analysed by mass spectrometry (MS) resulting in the identification of 21 plasma proteins representing potential targets of this synthetic PC.

Fluorescently-labelled OxPL species represent a further method to detect OxPL-protein adducts. Stemmer et al. used synthetic BODIPY-labeled 1-palmitoyl-2-(5-oxovaleroyl)-*sn*-glycero-phosphoethanolamine to identify OxPL target proteins from mouse macrophages [26]. Proteins were separated by 1D and 2D gel electrophoresis followed by in-gel digestion and MS/MS based identification of proteins present in fluorescent protein spots/bands.

Multistep reversed phase C18-solid phase extraction enrichment in combination with aminolysis was applied by Gao et al. to enrich OxPL-modified peptides from unmodified peptides based on differences in hydrophobicity [27]. In this study, human platelets were incubated with liposomes containing 9-keto-12-oxo-10-dodecenoate-PC (KODA-PC) followed by cell lysis and tryptic in-solution digestion. Resulting peptides were first fractionated by C18-SPE into hydrophobic OxPL-modified peptides and hydrophilic peptides. Afterwards, hydrophobic OxPL-modified peptides were subjected to aminolysis, resulting in cleavage of the oxidized fatty acid at sn-2 position from the backbone. The resulting peptide-lipid electrophile adducts, which are much more hydrophilic than their precursors, were finally separated by C18-SPE from other hydrophobic compounds. MS/MS-based identification of peptide-lipid electrophile adducts resulted in the identification of 6 KODA-PC-modified proteins.

The abovementioned methods allowed the identification of a variety of potential targets of OxPLs. However, all these approaches have some drawbacks because they are either quite biased, since only a single LOP or a very small set of well-defined LOPs can be used, or require labelling or chemical modification of LOPs. Label-free identification of LOP-protein/peptide adducts using MS/MS requires knowledge of the exact structure of the LOP used. Furthermore, protein identification is often based on only one or two sequenced LOP-modified peptides. Antibody-based detection of unlabelled LOP-protein complexes from gels may result in false-positive hits, since not all proteins identified from a single spot/band are modified. Labelling or chemical modification for click chemistry may alter intrinsic properties of LOPs and, thus, may influence reactivity, selectivity and specificity of binding.

Here we report a novel straightforward approach to identify protein targets of OxPLs without labelling of the lipids or chemical reactions after the binding has occurred. The method is based on MS detection of changes in electrophoretic mobility of proteins covalently modified with electrophilic OxPLs as compared to intact proteins. The method is in principle independent of the general chemical nature of electrophilic LOPs (e.g., free or esterified), chemistry of covalent protein modification (e.g., Michael adduction, Schiff base formation), complexity of the LOP-sample (e.g., individual molecular species or complex mixtures such as OxPAPC, OxPAPE, oxidized palmitoyl-arachidonoyl-phosphatidylserine, etc.) or proteome (e.g., a pure protein, cell lysate, membrane-associated proteins, etc.). In other words, this is a technically straightforward, universal and unbiased method. Here we describe the application of this method for identification of OxPL-binding proteins in cytosolic and membrane fractions of endothelial cells.

2. Materials and Methods

Unless otherwise specified, all reagents were from Sigma-Aldrich (St. Louis, MO, USA).

2.1. Lipids and Cell Culture

1-Palmitoyl-2-arachidonoyl-*sn*-glycero-3-phosphocholine (PAPC) was obtained from Avanti Polar Lipids (Alabaster, AL, USA) and oxidized by prolonged exposure of pure lipid to air [28]. The extent of oxidation was monitored by positive ion electrospray MS.

Human umbilical vein endothelial cells (HUVEC) immortalized by telomerase overexpression were kindly provided by Dr. Hannes Stockinger (Medical University of Vienna, Austria). The cells were cultured in Nunclon Delta T75 flasks or six-well plates (Thermo Fisher Scientific, Waltham, MA, USA) in medium 199 (Gibco, Carlsbad, CA, USA) supplemented with 20% foetal calf serum, penicillin-streptomycin-amphotericin B (Lonza, Basel, Switzerland), and ECGS-heparin (PromoCell, Heidelberg, Germany) in the 95% humidified atmosphere with 5% CO_2 at 37 °C. For protein isolation, 90% confluent HUVECs in flasks were washed three times with 10 ml of phosphate buffered saline (PBS) followed by harvesting in PBS containing cOmplete™ Protease Inhibitor Cocktail (PBS-PI) using a scraper. Cells were pelleted by centrifugation at $450 \times g$ for 10 min.

For Western blotting, HUVECs were cultured until 80% confluence in six-well plates. Prior to OxPAPC treatment, cells were washed with serum-free medium. OxPAPC in PBS (or the same volume of PBS without OxPAPC) was added to a final concentration of 80 µmol/L. Cell medium was removed immediately or after 6 h followed by cell scraping in PBS-PI. Cells were pelleted by centrifugation at $450 \times g$ for 10 min and resulting cell pellets were resuspended in RIPA buffer (radioimmunoprecipitation assay buffer) containing protease inhibitors (RIPA-PI). Cell lysis was performed by sonication on ice.

2.2. Isolation of HUVEC Proteomes

Cells harvested from 10 T75 flasks were pooled, resuspended in 5 mL ice-cold PBS-PI and lysed by sonication on ice. After incubation of the suspension at 4 °C for 30 min with gentle shaking, PBS-insoluble material was pelleted by centrifugation at $17,000 \times g$ for 40 min. The supernatant containing "PBS-soluble proteins" was transferred to a new vial. The remaining pellet was washed five times with 5 mL of PBS-PI followed by centrifugation at $17,000 \times g$ for 40 min. The supernatant was discarded, the pellet redissolved in RIPA buffer and incubated at 4 °C for 30 min with gentle shaking. RIPA buffer enables efficient cell lysis and solubilization of membrane proteins while avoiding the loss of protein immunoreactivity. The final solution is composed of 150 mmol/L NaCl, 1.0% (*v/v*) IGEPAL® CA-630, 0.5% (*w/v*) sodium deoxycholate, and 0.1% (*w/v*) SDS (sodium dodecyl sulfate), 50 mmol/L Tris, pH 8.0. After centrifugation at $17,000 \times g$ for 40 min, the supernatant ("PBS-insoluble/RIPA-soluble proteins") was transferred to a new vial. Protein concentration was determined using the Pierce™ BCA Protein Assay Kit (Thermo Fisher Scientific).

2.3. Electrophoretic Mobility Band Shift Analysis (EMSA)

OxPAPC treatment of isolated proteomes was performed with 12 µg of protein in PBS-PI or RIPA-PI buffer with varying OxPAPC concentrations ranging from 40–100 µmol/L in a final volume of 30 µL. Samples were incubated at 37 °C with gentle shaking. Formation of OxPL-protein adducts was stopped by addition of 10 µL $4\times$ Laemmli buffer (250 mmol/L Tris-HCl, pH 6.8, 40% (*v/v*) glycerol, 8% (*w/v*) SDS, 0.02% (*w/v*) bromophenol blue, and 10% (*v/v*) 2-mercaptoethanol) after 0 or 90 min followed by heating for 5 min at 95 °C. Band shift analysis was performed by standard SDS-PAGE using commercial 4–20% TGX gels (Bio-Rad Laboratories, Hercules, CA, USA) or homemade 10% acrylamide Tris/glycine gels (composition of the separating gel: 10% (*v/v*) acrylamide/bis-acrylamide (37.5:1), 375 mmol/L Tris-HCl, pH 8.8, 0.1% (*w/v*) SDS, 0.033% (*w/v*) ammonium persulfate, and 0.066% (*v/v*) tetramethylethylenediamine; composition of the stacking gel: 3.9% (*v/v*) acrylamide/bis-acrylamide (37.5:1), 125 mmol/L Tris-HCl, pH 6.8, 0.1% (*w/v*) SDS, 0.05% (*w/v*) ammonium persulfate, and 0.1% (*v/v*) tetramethylethylenediamine). Electrophoresis was performed with the Mini-Protean Tetra System (Bio-Rad) using a Tris/glycine/SDS running buffer (25 mmol/L Tris, 192 mmol/L glycine, 0.1% (*w/v*) SDS, pH 8.3) and keeping the current constant at 15 mA per gel. A total amount of 5 µg of protein in Laemmli buffer was applied to each slot. Visualization of protein bands was done by the EMBL silver staining protocol [29]. Proteins from samples incubated for 90 min with and without OxPAPC were selected for proteomics-based identification. To this end, the whole gel (above a molecular weight of ~15 kDa) was cut into multiple pieces with special attention to cut gel pieces of treated and untreated samples at exactly the same position.

2.4. In-Gel Protein Digestion and Nano-LC-MS/MS Analysis

For in-gel digestion protein bands were excised, washed for 10 min each with 200 µL 50 mmol/L NH$_4$HCO$_3$ and NH$_4$HCO$_3$: acetonitrile 1:1, followed by dehydration with acetonitrile (HPLC grade, VWR, Radnor, PA, USA). Each sample was then subjected to reduction with 100 µL of a 50 mmol/L NH$_4$HCO$_3$ solution, containing 10 mmol/L dithiothreitol at 54 °C for 30 min, and to alkylation with 150 µL of a 50 mmol/L NH$_4$HCO$_3$ solution, containing 50 mmol/L iodoacetamide for 20 min at room temperature in the dark. The gel slices were then washed again with 50 mmol/L NH$_4$HCO$_3$ followed by washing with acetonitrile and drying in a vacuum centrifuge. For in-gel digestion, the gel slices were rehydrated in 15 µL of a trypsin solution (5 ng/µL; sequencing grade Trypsin, Roche, Basel, Switzerland) in 50 mmol/L NH$_4$HCO$_3$ and incubated overnight at 37 °C. Peptides were extracted from the gel matrix with 25 μL 50 mmol/L NH$_4$HCO$_3$ and subsequently two times with 25 μL of 5% (*v/v*) formic acid in an ultrasonic bath. All supernatants were combined and analysed by nano-LC-MS/MS.

Nano-LC separations were performed on an UltiMate 3000 RSLCnano system (Thermo Fisher Scientific, Waltham, MA, USA). The flow rate of the nano-HPLC system was set at 250 nL/min and the UV absorbance was detected at λ = 214 nm. The flow rate of the loading pump was set at 20 μL/min. The trap column dimensions were 300 μm id × 5 mm length, packed with PepMap C18 (Thermo Fisher Scientific, Waltham, MA, USA). After a sample loading time of 10 min, the trap column was switched in line with the nanocolumn. The sample was eluted in back flush mode. The mobile phases were: (A) 99.9% (*v/v*) water (HPLC grade, VWR) and 0.1% (*v/v*) formic acid; and (B) 80% (*v/v*) acetonitrile (HPLC grade, VWR) and 0.08% (*v/v*) formic acid. The mobile phase for the loading pump was water with 0.05% (*v/v*) trifluoroacetic acid. The HPLC gradient for separation was 4% B for 10 min, 4–40% B in 90 min, and 40–90% B in 5 min, 90% B for 5 min, 4% B for 20 min. For separation an Acclaim PepMap RSLC column (C18, 75 μm × 150 mm, 2 μm, 100 Å; Thermo Fisher Scientific, Waltham, MA, USA) was used. Eluted peptides were ionized via stainless steel emitters using a Nanospray Flex™ ion source (Thermo Fisher Scientific) and directly introduced into a LTQ XL mass spectrometer (Thermo Fisher Scientific).

The following electrospray ionization parameters were used: spray voltage, 1.7 kV; capillary temperature, 200 °C; capillary voltage, 30 V. The collision energy was set automatically depending on the mass of the parent ion. The data were collected in the centroid mode using one MS experiment followed by five MS/MS experiments of the most intensive ions (intensity at least 5 × 10^3). Dynamic exclusion was used for data acquisition with exclusion duration of 2 min and an exclusion mass width of ±1.5 Da.

For peptide identification, RAW-files were converted into MGF-files using ProteoWizard [30] and analysed with the MASCOT search engine (Matrix Science, London, UK). All MS/MS spectra were searched against the SwissProt protein sequence database. Following search parameters were used: carbamidomethylation on cysteine was set as a fixed modification; oxidation on methionine was set as a variable modification; trypsin was set as enzyme; the precursor mass tolerance was set to 3 Da; the fragment mass tolerance was set to 0.8 Da; the maximal number of missed cleavages was set to 2. The results were filtered to peptide scores ≥30 and to a 1% false discovery rate using Mascot. Representative MS/MS spectra of certain OxPAPC modified proteins can be found in Supplementary Material (Figures S2–S6). To calculate protein abundance of all proteins in a particular proteome fraction based on Exponentially Modified Protein Abundance Index (emPAI) score [31], MGF-files of all samples of a particular fraction were merged and analysed with the MASCOT search engine.

2.5. Data Analysis

Mascot search results from all single bands from treated and untreated samples were compared. A protein was regarded as modified by OxPAPC, if the corresponding protein was found in the primary band of both, OxPAPC treated and untreated samples, as well as in higher molecular weight regions of OxPAPC-treated samples compared to control samples. At least five unique peptides of each modified protein had to be identified in higher molecular weight regions.

2.6. Protein-Protein Interaction Network Construction and Functional Annotation

To analyse protein-protein interaction networks and to perform functional annotation of identified proteins, the Search Tool for the Retrieval of Interacting Genes (STRING) database was used.

2.7. Western Blotting

For Western blotting, samples from OxPAPC-treated live cells (see Section 2.1) and samples from gel band shift analysis (see Section 2.3) were used. SDS-PAGE was performed as described in Section 2.3. Proteins were transferred by semi-dry blotting to a BioTrace NT nitrocellulose membrane (Pall Corporation, Port Washington, NY, USA) using the Trans-Blot Turbo System (Bio-Rad) for 30 min keeping the voltage constant at 25 V. Afterwards the membrane was incubated in Tris-buffered saline, 0.05 % (*v/v*) Tween 20 (TBST) containing 3% (*w/v*) BSA for 60 min. Primary antibodies diluted in TBST containing 1% (*w/v*) BSA, were added to the membrane and incubated over night at 4 °C with gentle shaking. After three washing steps for 5 min with TBST, the membrane was incubated with horse radish peroxidase (HRP)-linked secondary antibodies for 60 min. After five washing steps for 5 min, the membrane was incubated with the Clarity ECL substrate (Bio-Rad) for 4 min. Chemiluminescence imaging was performed using the G:BOX Chemi XX6 (Syngene, Frederick, MD, USA). The following primary antibodies and corresponding dilutions were used: anti-HSP90β (anti-heat shock protein HSP 90-beta) (ADI-SPA-844-050; Enzo Life Sciences, Farmingdale, NY, USA), 1:1000 dilution; anti-β-Actin (A2103; Sigma-Aldrich, St. Louis, MO, USA) 1:1000 dilution; anti-BAP31 (B-cell receptor-associated protein 31), anti-Integrin α2, anti-CD13, anti-VDAC1 (sc-393810, sc-74466, sc-166105, sc-390996; Santa Cruz Biotechnology, Dallas, TX, USA) all diluted 1:500. The following secondary antibodies and corresponding dilutions were used: HRP-linked Anti-rabbit IgG (7074; Cell Signalling Technology, Danvers, MA, USA), 1:3000 dilution; or HRP-linked mouse-IgGκ binding protein (sc-390996; Santa Cruz Biotechnology, Dallas, TX, USA), 1:8000 dilution.

3. Results

3.1. Incubation with OxPLs Changes the Electrophoretic Mobility of Certain Proteins

In order to reduce the complexity of the proteome and to separate water-soluble, mainly cytosolic proteins, from membrane-associated proteins, we split the HUVEC proteome into a PBS-soluble part and a PBS-insoluble/RIPA-soluble part. Both samples were subjected to EMSA after treatment with varying concentrations of OxPAPC. If OxPAPC was not covalently bound to proteins (i.e., at time 0), the electrophoretic mobility of PBS-soluble proteins was not influenced (Figure 1, lane 2 compared with control lane 1). Incubation of the proteome at 37 °C for 90 min without OxPAPC, likewise did not influence the electrophoretic mobility of proteins (Figure 1, lane 4). However, if the sample was incubated with OxPAPC for 90 min, clear changes, including band shifts, band-smearing, and intensity changes can be observed (Figure 1, lane 3).

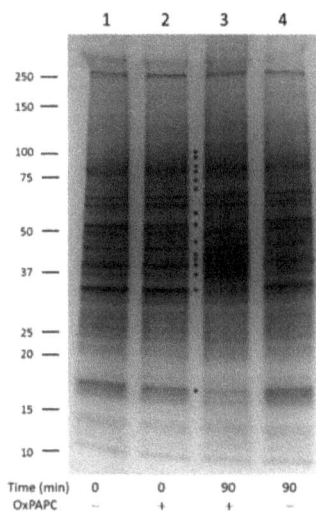

Figure 1. Gel band shift analysis of PBS-soluble human umbilical vein endothelial cell (HUVEC) proteins. Oxidized palmitoyl-arachidonoyl-phosphatidylcholine (OxPAPC) incubated with proteins for 90 min significantly changed the electrophoretic protein profile (lane 3). OxPAPC which is not covalently bound to proteins does not influence the electrophoretic mobility of proteins (lane 2). Lane 1: control sample time point 0; lane 4: control sample time point 90. The final OxPAPC concentration in treated samples was 100 μmol/L. Main visible changes are marked with asterisks.

These OxPAPC-induced changes of the electrophoretic protein profile were also observed within the PBS-insoluble/RIPA-soluble proteome (Figure 2). Similarly to the experiment with hydrophilic proteins, unbound OxPAPC did not influence the electrophoretic mobility of hydrophobic proteins (Figure 2, lane 2 compared with control lane 1). Treatment of proteins with two different concentrations of OxPAPC for 90 min induced significant changes such as band shifts, band smearing, and intensity changes (Figure 2, lanes 4 and 5).

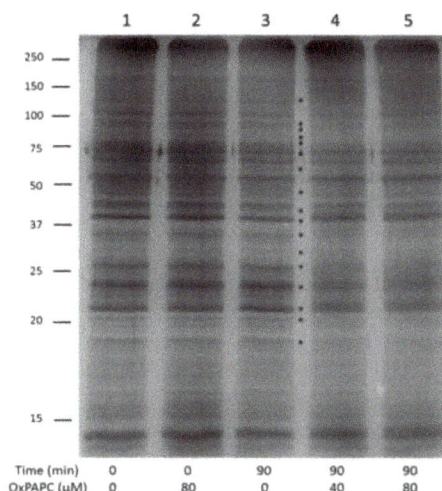

Figure 2. Gel band shift analysis of PBS-insoluble/RIPA-soluble HUVEC proteins. OxPAPC incubated with proteins for 90 min significantly changed the electrophoretic protein profile (lanes 4 and 5). OxPAPC which is not covalently bound to proteins does not influence the electrophoretic mobility of proteins (lane 2). Lane 1: control sample time point 0; lane 3: control sample time point 90. The final OxPAPC concentration in treated samples was 40 and 80 µmol/L. Main visible changes are marked with asterisks.

The band shift assay was performed using several independent protein preparations from HUVECs at different passages. In all cases we observed clearly visible and reproducible changes of the electrophoretic protein profile in silver-stained gels (data not shown). Since each visible band contained multiple proteins, it was impossible to reliably identify OxPAPC-modified proteins by targeted excision and sequencing of bands that became weaker or stronger upon OxPAPC treatment. Therefore, we applied an alternative approach, namely cut the whole gel into pieces and identified position of each protein in the gel as described in the next paragraph. The method successfully identified OxPAPC-modified proteins due to their abnormal migration in the gel.

3.2. Protein Targets of OxPAPC Identified by LC-MS/MS

In order to identify proteins that changed their electrophoretic mobility and thus represent targets of OxPAPC, protein bands from samples incubated with and without OxPAPC were analyzed by nano-LC-MS/MS. The whole gel (above a molecular weight of ~15 kDa) was cut into multiple pieces with special attention to cut gel pieces of treated and untreated samples at exactly the same position (Figure 3).

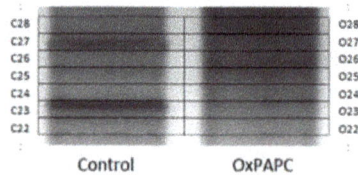

Figure 3. Representative region of a gel showing changes in pattern of protein silver staining. The grid shows how the bands were excised for in-gel trypsinisation and LC-MS/MS (liquid chromatography tandem mass spectrometry) analysis.

The bands were subjected to in-gel digestion and resulting peptides were identified by LC-MS/MS. Comparison of proteins identified from individual bands from treated and untreated samples clearly showed that in OxPAPC-treated samples several proteins shifted from their major band location to positions in higher molecular weight regions. A protein was regarded as modified by OxPAPC, if the corresponding protein was found in (i) the primary band of both samples and (ii) in higher molecular weight regions of OxPAPC-treated samples compared to control samples.

From the PBS-soluble fraction, 21 proteins were shifted to higher molecular weight regions in OxPAPC treated samples and thus regarded as modified by OxPAPC (Table 1). Since in total over 500 proteins were identified in both control and OxPAPC-treated samples, ~4% of the detected proteome was identified as modified.

Table 1. Potential OxPAPC target proteins identified from the PBS-soluble fraction.

Protein	Accession	Reported OxPL Target *
14-3-3 protein theta	1433T_HUMAN	
40S ribosomal protein SA	RSSA_HUMAN	
60 kDa heat shock protein, mitochondrial	CH60_HUMAN	
78 kDa glucose-regulated protein	GRP78_HUMAN	[23,26]
Actin, cytoplasmic 1	ACTB_HUMAN	[23]
Chloride intracellular channel protein 1	CLIC1_HUMAN	
Elongation factor 1-gamma	EF1G_HUMAN	
Elongation factor 2	EF2_HUMAN	
Endoplasmin	ENPL_HUMAN	[26]
Glyceraldehyde-3-phosphate dehydrogenase	G3P_HUMAN	
Heat shock protein HSP 90-beta	HS90B_HUMAN	
L-lactate dehydrogenase A chain	LDHA_HUMAN	
L-lactate dehydrogenase B chain	LDHB_HUMAN	
Myosin-9	MYH9_HUMAN	[23]
Protein-glutamine gamma-glutamyltransferase 2	TGM2_HUMAN	
Puromycin-sensitive aminopeptidase	PSA_HUMAN	
Pyruvate kinase PKM	KPYM_HUMAN	[26]
Spectrin alpha chain, non-erythrocytic 1	SPTN1_HUMAN	
Tubulin alpha-1A chain	TBA1A_HUMAN	
Tubulin beta chain	TBB5_HUMAN	[23]
Vimentin	VIME_HUMAN	[23,26]

* These proteins were identified to by targets of oxidized phospholipids (OxPLs) by proteomic based studies using comparable OxPLs and endothelial cells [23] or macrophages [26]. OxPAPC, oxidized palmitoyl-arachidonoyl-phosphatidylcholine; PBS, phosphate buffered saline.

In the PBS-insoluble/RIPA-soluble fraction 37 proteins were identified as potential targets of OxPAPC (Table 2). In this proteome, over 700 proteins could be detected both in control and OxPAPC-treated samples, thus ~5% were detected as modified.

Table 2. Potential OxPAPC target proteins identified from the PBS-insoluble/RIPA-soluble fraction.

Protein	Accession	Reported OxPL Target *
ADP/ATP Translocase 2	ADT2_HUMAN	
Aminopeptidase N	AMPN_HUMAN	
ATP synthase F(0) complex subunit B1	AT5F1_HUMAN	
ATP synthase subunit alpha, mitochondrial	ATPA_HUMAN	[23]
Brain acid soluble protein 1	BASP1_HUMAN	
Coiled-coil domain-containing protein 47	CCD47_HUMAN	
Cytoskeleton-associated protein 4	CKAP4_HUMAN	[23]
Endoplasmin	ENPL_HUMAN	[26]
Erlin 1	ERLN1_HUMAN	
Fatty aldehyde dehydrogenase	AL3A2_HUMAN	
Flotillin 1	FLOT1_HUMAN	
GTPase IMAP family member 1	GIMA1_HUMAN	
Guanine nucleotide-binding protein G (i) subunit alpha-2	GNAI2_HUMAN	[23]
Guanine nucleotide-binding protein G(I)/G(S)/G(T) subunit beta-1	GBB1_HUMAN	[26]
Guanine nucleotide-binding protein G(I)/G(S)/G(T) subunit beta-2	GBB2_HUMAN	
Inhibitor of nuclear factor kappa-B kinase-interacting protein	IKIP_HUMAN	
Integrin alpha-2	ITA2_HUMAN	
Leucine-rich repeat-containing protein 59	LRC59_HUMAN	
Myoferlin	MYOF_HUMAN	
NADH-cytochrome b5 reductase 3	NB5R3_HUMAN	
NADH dehydrogenase [ubiquinone] 1 alpha subcomplex subunit 9	NDUA9_HUMAN	
Perilipin 3	PLIN3_HUMAN	
Polypeptide N-acetylgalactosaminyltransferase 1	GALT1_HUMAN	
Polypyrimidine tract-binding protein 1	PTBP1_HUMAN	
Prohibitin	PHB_HUMAN	
Prohibitin 2	PHB2_HUMAN	[26]
RNA-binding protein Raly	RALY_HUMAN	
Serine/arginine-rich splicing factor 1	SRSF1_HUMAN	
Serine/arginine-rich splicing factor 9	SRSF9_HUMAN	
Sideroflexin-3	SFXN3_HUMAN	
Trifunctional enzyme subunit alpha, mitochondrial	ECHA_HUMAN	
Trifunctional enzyme subunit beta, mitochondria	ECHB_HUMAN	[23]
Vesicle-trafficking protein SEC22b	SC22B_HUMAN	
Voltage-dependent anion-selective channel protein 1	VDAC1_HUMAN	[26]
Voltage-dependent anion-selective channel protein 2	VDAC2_HUMAN	[26]
X-ray repair cross-complementing protein 6	XRCC6_HUMAN	
60S ribosomal protein L19	RL19_HUMAN	

* These proteins were identified to by targets of OxPLs by proteomic based studies using comparable OxPLs and endothelial cells [23] or macrophages [26].

The number of identified peptides of a modified protein was plotted against the estimated molecular weight region on the gel, where the protein was detected. The distribution of peptides corresponding to individual proteins in gel bands cut out in different gel regions is shown in Figure 4. The examples include voltage-dependent anion-selective channel protein 1 (VDAC-1), heat shock protein HSP 90-beta (HSP90β), and actin cytoplasmic 1 (β-actin). A clear increase of the number of peptides identified in higher molecular weight regions was observed in OxPAPC-treated samples.

To investigate if proteins modified by OxPAPC represent only highly abundant proteins in the particular proteome, the Exponentially Modified Protein Abundance Index (emPAI), which is automatically calculated by Mascot, was used. The MS/MS RAW-data of all protein bands of a particular proteome fraction were merged and searched against the SwissProt protein sequence database using Mascot. emPAI scores of the most abundant proteins, representing together ~85% of total proteins, are shown in Figure 5. The analysis was performed separately for soluble and membrane RIPA-soluble proteins. In the PBS-soluble fraction, several highly abundant proteins were found to be modified by OxPAPC (Figure 5A), while the most highly abundant proteins from the RIPA-soluble fraction formed no adducts with OxPAPC (Figure 5B). Nevertheless, the majority of highly abundant proteins was not modified in both fractions and vice versa. Additionally, low abundance proteins were modified in both proteomes. The data suggest that the property of proteins to be modified by OxPAPC is not a simple function of its abundance.

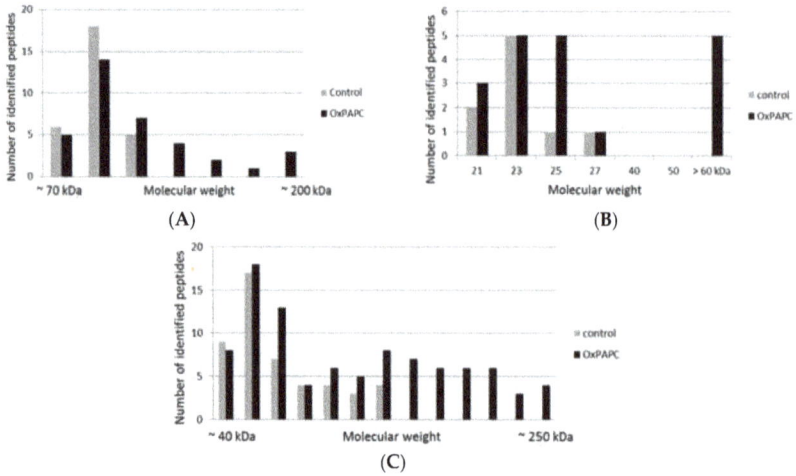

Figure 4. Peptide distribution of selected representative OxPAPC-modified proteins. A clear increase of peptides identified in higher molecular weight regions is observed after OxPAPC treatment. (**A**) Peptide distribution of HSP 90-beta; (**B**) Peptide distribution of VDAC-1; (**C**) Peptide distribution of Actin, cytoplasmic 1.

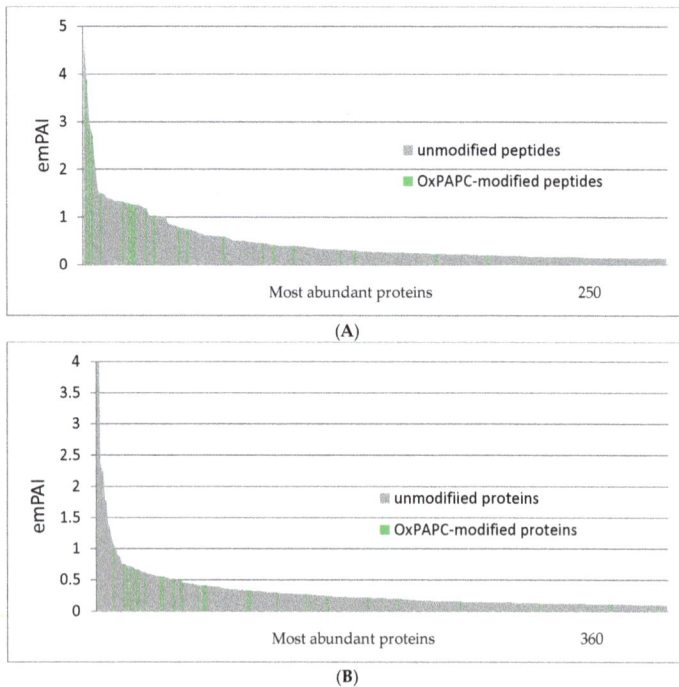

Figure 5. Protein abundance, based on exponentially modified protein abundance index (emPAI) score. Most abundant proteins, representing ~85% of total proteins, were plotted. The property of proteins to be modified by OxPAPC is not a simple function of their abundance. (**A**) PBS-soluble proteins; and (**B**) PBS-insoluble/RIPA-soluble proteins.

STRING analysis of OxPAPC-modified proteins from the RIPA-soluble fraction showed a clear enrichment of proteins located at membrane organelles in general (Figure 6A; green nodes) and proteins located at the mitochondrial membrane (Figure 6A; blue nodes) in particular. Furthermore, these mitochondrial proteins form a tight protein-protein interaction network. STRING analysis of the PBS-soluble fraction showed an enrichment of proteins related to protein folding (Figure 6B; red nodes) and proteins representing structural constituents of the cytoskeleton. Based on these results, mitochondrial (membrane-) proteins, proteins involved in protein folding and cytoskeletal proteins, seem to be major targets of OxPAPC.

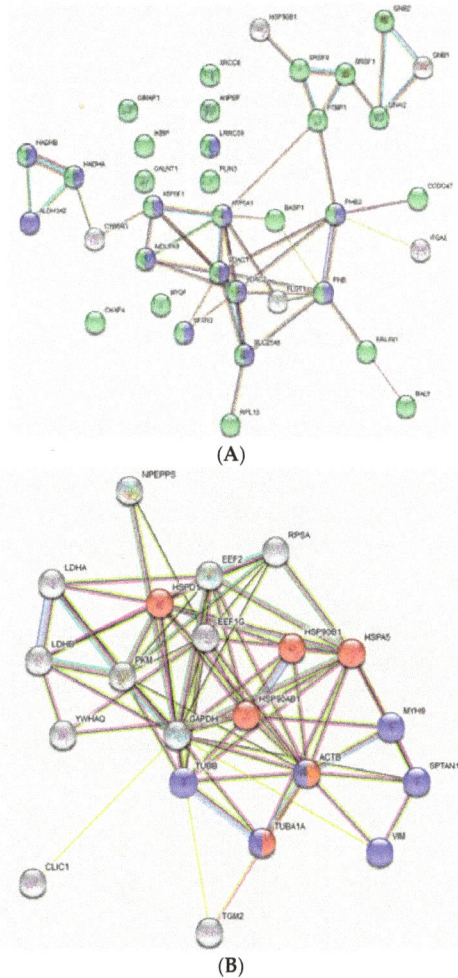

(A)

(B)

Figure 6. STRING analysis of proteins identified as targets of OxPAPC from the PBS-insoluble/RIPA-soluble fraction (**A**) and from the PBS-soluble fraction (**B**). The RIPA-soluble fraction of modified proteins contains mainly proteins located at membrane organelles (green nodes: membrane-bounded organelle; GO: 0043227). Furthermore, several modified proteins are located at the mitochondrial membrane (blue nodes: mitochondrial part; GO: 0044429). In the PBS-soluble fraction, proteins related to protein folding (red nodes: protein folding; GO: 0006457) and proteins related to the cytoskeleton (blue nodes: structural constituent of cytoskeleton; GO: 0005200) are enriched.

3.3. Confirmation of OxPAPC-Induced Protein Modifications by Western Blotting

To confirm that the apparent abnormal migration of OxPAPC-treated proteins in the gel was not an artifact of MS detection, we performed western blot analysis of selected proteins to detect these proteins by antibodies. Since β-actin is a rather highly abundant protein, which was identified to be modified by OxPAPC in the PBS-soluble fraction, we used a polyclonal anti-β-Actin antibody to analyse OxPAPC-actin adducts. We could observe a significant decrease of the amount of β-actin at its major band, as well as a slight band shift of actin in OxPAPC treated samples (Figure 7). Results from MS however, indicated that β-actin was also present in higher molecular weight regions (Figure 4). However, no β-actin was detected in higher molecular weight areas by this antibody.

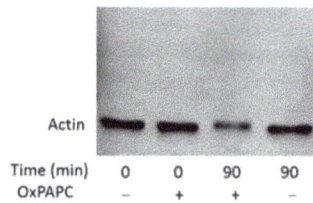

Actin				
Time (min)	0	0	90	90
OxPAPC	−	+	+	−

Figure 7. Western blot analysis of β-Actin from the OxPAPC-treated PBS-soluble fraction. The OxPAPC-treated sample incubated for 90 min, shows a decrease of actin immunostaining and a slightly slower band migration. The final OxPAPC concentration was 100 μmol/L.

Since two other polyclonal and monoclonal anti-β-actin antibodies were also not able to detect OxPAPC-actin adducts (data not shown), we concluded, that OxPAPC-induced modifications and cross-links of β-actin hampered its recognition by antibodies. Another possible explanation is that the amount of OxPAPC-modified β-actin in high molecular weight regions of gel/blot may be too low for antibody-based detection. To test this possibility, we cross-linked proteins using formaldehyde. Similarly to OxPAPC-treated samples, the antibody could not detect formaldehyde-modified β-actin (Figure 8). After 10 min incubation time the protein amount was decreased, after 30 min the signal was completely lost, but still no β-actin could be detected in higher molecular weight regions (Figure 8). Application of other anti-β-actin antibodies showed the same result (data not shown). We hypothesize that electrophilic molecules either induce irreversible denaturation of protein or modify immunodominant domains of β-actin and, thus, prevent its recognition by antibodies. Thus, although Western blotting cannot detect high molecular weight adducts of actin, the loss of the intact protein upon OxPAPC treatment clearly shows that the protein is modified by OxPAPC thus confirming the findings of MS analysis.

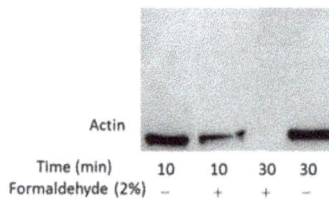

Actin				
Time (min)	10	10	30	30
Formaldehyde (2%)	−	+	+	−

Figure 8. Western blot analysis of β-Actin from the formaldehyde-treated samples. After formaldehyde-induced artificial cross-linking of proteins, actin could not be detected by antibodies any more.

Another protein identified to be modified by OxPAPC using the band shift assay was HSP90β. Using anti-HSP90β antibody we could detect a slight shift of the major band, as well as protein adducts in the higher molecular weight region upon treatment with OxPAPC (Figure 9). These findings are in agreement with the peptide distribution detected by LC-MS/MS.

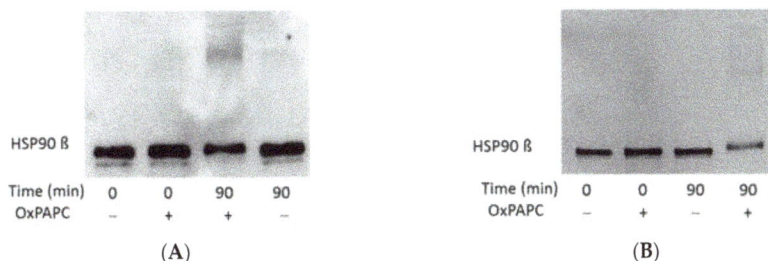

Figure 9. Western blot analysis of HSP90β from the OxPAPC-treated PBS-soluble fractions from two different preparations (**A,B**). OxPAPC induces a slight band shift of HSP90β and the generation of high molecular weight protein-OxPAPC complexes. Samples were treated for 0 or 90 min with a final OxPAPC concentration of 0 or 100 µmol/L.

3.4. OxPAPC Modifies Proteins in Live HUVECs

To investigate if identified protein targets of OxPAPC from isolated proteomes are also modified in cell culture, HUVECs were treated with OxPAPC in cell culture medium and subjected to Western blotting analysis. OxPAPC-protein adduct formation was analysed with three different proteins from the RIPA-soluble fraction (Figure 10). Proteins located in the plasma membrane and/or mitochondrial outer membrane, were selected for detection. Similarly to experiments with β-actin, antibodies were not able to detect band shifts in the higher molecular weight region, apparently due to a loss of protein immunoreactivity. However, similarly to β-actin, a decrease of signal intensity was observed for all three proteins after treatment with OxPAPC. These data confirm modification of proteins by OxPAPC and are in agreement with the results of MS analysis.

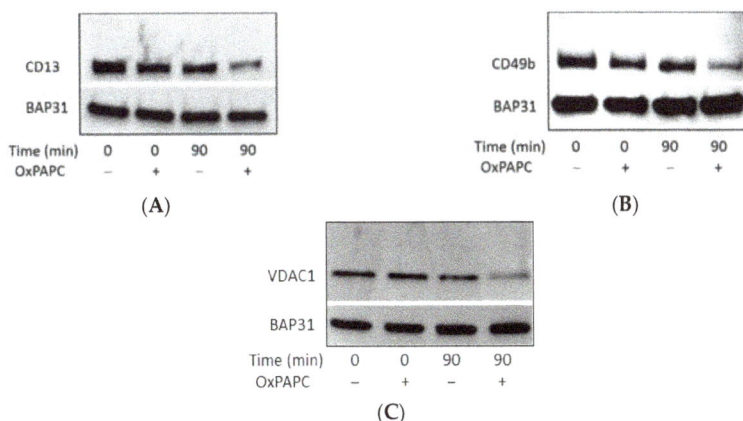

Figure 10. Western blot analysis of (**A**) Aminopeptidase N (CD13), (**B**) Integrin α-2 (CD49b) and (**C**) VDAC-1 after treatment of HUVECs with OxPAPC. Samples were treated for 0 or 90 min with a final OxPAPC concentration of 0 or 80 µmol/L. BAP31: B-cell receptor-associated protein 31 which was identified to not be modified by OxPAPC.

4. Discussion

OxPLs are increasingly recognized as signalling molecules inducing pleiotropic effects in different cell types [6]. Some of the effects (e.g., angiogenesis) depend on electrophilic properties of OxPLs [32] but cannot be explained by simple toxicity of electrophiles. Rather, sub-toxic concentrations of OxPLs seem to regulate specific signal transduction pathways, thus inducing important cellular reactions. OxPL-induced angiogenesis is just one example justifying the need for analysis of proteins that are covalently modified by OxPLs and may be involved in development of (patho-) physiological reactions to these lipids. Such an analysis is not a simple task because (i) oxidation of PUFA-PLs generates a complex mixture of molecular species containing electrophiles with different size, lipophilicity and reactivity; and (ii) cells contain thousands of proteins, some of which may be more prone to oxidation than other. Several methods for identification of OxPL-protein adducts have been published (see Introduction); all of them are based on identification of proteins by mass spectrometry. Although these methods produced important results on cellular targets of electrophilic OxPLs, each procedure has certain limitations. Three of these methods used OxPCs or OxPEs labelled at the head groups with biotin or fluorescent dye [23–26]. It is apparent that such chemical modification (i) may be more difficult to introduce into other classes of OxPLs and (ii) addition of a bulky group can change the binding characteristics of OxPLs. The method published by Gao et al. [27] is based on chromatographic separation of peptide adducts of OxPLs according to their hydrophobicity before and after cleavage of the fatty acid, which is a highly demanding technique that may be difficult to establish for other types of LOP.

Here we describe a simple method for identification of proteins covalently modified by OxPLs. The method is based on a combination of EMSA with LC-MS/MS (EMSA-MS). The procedure may be characterized as "double-unbiased" because no preliminary information on either oxidized lipids or proteins that are treated with these lipids is needed for the identification of the most reactive protein targets. In other words, this method is independent of the complexity of both the proteome and the LOPs, or from the type of covalent adducts. Furthermore, this approach does not require any kind of chemical tag- or antibody-assisted protein extraction or detection. We demonstrate application of this method for extracts from endothelial cells treated with OxPAPC, but it is to be expected that other types of LOPs and protein mixtures can be analysed using this universal procedure as well. In support of this possibility, we observed band shifts of HUVEC proteins treated with a chemically different LOP, 2-chlorohexadec-15-yn-1-al (Figure S1).

By applying the EMSA-MS to HUVEC lysates, we were able to identify several protein targets of OxPAPC both in soluble and membrane cell fractions (Tables 1 and 2). It was found that OxPAPC induced not only small shifts in protein band migration, but also the formation of SDS-resistant (supposedly covalent) complexes migrating in high molecular weight regions of the gel (Figure 4). Intermolecular cross-linking is the most probable mechanism generating significantly shifted bands. Since our EMSA analysis used isolated proteomes, any kind of OxPAPC-mediated biological protein modification or changes in protein expression levels as the cause of observed band shifts can be excluded. Comparison of semi-quantitative abundance (emPAI) scores of modified and unmodified proteins showed that the property of proteins to be modified by OxPAPC is not a simple function of their abundance (Figure 5). These results imply that OxPL-protein adducts are not randomly formed and that a certain selectivity of binding is existing. Using Western blot analysis we could prove that selected OxPAPC protein targets identified from isolated proteomes were also modified in live cells treated with OxPAPC (Figure 10).

We found that <5% of detectable proteins from HUVECs demonstrated a band shift after treatment with OxPAPC. In absolute numbers, we found 21 OxPL-modified proteins in the soluble fraction of HUVECs and 37 proteins in RIPA-solubilized membrane fraction. Gugiu et al. have analysed binding of OxPLs to proteins in human aortic endothelial cells using a different detection principle, namely biotin-labelled OxPAPE [23]. Twenty-eight modified proteins were identified, which is close to the number found in our work. Interestingly, nine proteins were also found in our samples (Tables 1 and 2). Stemmer et al. used fluorescently labelled OxPAPC components POVPC and PGPG and found >50 modified proteins in mouse macrophage cell line [26]. From these, nine proteins were also identified in our study (Tables 1 and 2). In summary, comparison of our results with two published studies on cells treated with OxPLs points to a similar sensitivity of the methods (dozens of modified proteins were identified in each cell type). Partial, but significant, overlap of our hits with those obtained using independent methods supports the reliability of results generated by EMSA-MS. Furthermore, using STRING analysis, we observed enrichment of proteins related to (i) cytoskeleton; (ii) protein folding; and (iii) mitochondria (Figure 6). All these are known as common themes in electrophile-modified proteome, thus further supporting the validity of the EMSA-MS method [11,15,20,33–40].

Similarly to other methods described above, our technique has limitations. Since the method does not identify exact structures of lipids modifying amino acids, false positive hits (i.e., proteins modified by chemically different LOPs) may arise due to transformation of added LOPs to their reactive derivatives, or due to generation of reactive oxygen species in the course of incubation. However, because the 'target' LOP is present in large molar excess as compared to its secondary products generated in the course of incubation, the secondary products can only have a limited impact on the whole electrophile-modified proteome and false positive results are unlikely to significantly bias the results. On the other hand, false negative results are also possible. Certain modifications by LOPs can have no effect on protein electrophoretic mobility and, therefore, some modified proteins can be falsely characterized as non-modified. Further studies are needed in order to understand how lipid adduction changes protein mobility. It seems likely that the method predominantly identifies multiply modified protein. This has certain advantages, because such proteins may be especially sensitive to LOPs and, as a consequence, play a mechanistic role in oxidative stress-induced damage. In other words, although the lipid specificity of EMSA-MS is not investigated, the method may be especially helpful for identification of proteins that are most sensitive to LOPs. Thus, the major application of the EMSA-MS technique is a search for protein targets of LOPs with the following mechanistic analysis of their role in oxidation-induced pathology.

An unexpected finding done in this work is that modification with OxPLs strongly inhibited recognition of proteins by standard commercial antibodies. OxPAPC-induced cross-linked high molecular weight complexes detected by MS could only be immunostained in case of HSP90β (Figure 9). Slight band shifts were observed on Western blots of modified β-actin and HSP90β (Figures 7 and 9). For all other analysed proteins, only a distinct decrease of signal intensity was observed by Western blotting after treatment with OxPAPC (Figure 10). Since artificial cross-linking using formaldehyde did not improve detection of high molecular weight adducts of actin, insufficient actin amount in shifted bands as the cause of the lack of signal in high molecular weight areas of gel can be excluded. We hypothesize that electrophilic molecules may induce irreversible denaturation of proteins that prevents partial refolding of SDS-treated proteins on blotting membranes and, thus, prevent binding of antibodies that recognize 3-D epitopes. Alternatively, for antibodies recognizing linear epitopes, immunodominant domains of proteins may be covalently modified by LOPs thus preventing their recognition by antibodies. Independently of the mechanism(s) of immunoreactivity loss, our data suggest that (i) Western blotting cannot be recommended for detection of band shifts induced by lipid adduction to proteins and (ii) β-actin should be used with caution as a housekeeping gene in experiments where the cells are exposed to oxidative stress.

In summary, here we describe a novel combination of electrophoretic mobility shift assay with MS-based protein sequencing, which allows identification of proteins modified by LOPs without an a priori knowledge of reactive lipid species or target proteins. We applied this method to endothelial cells and identified several known, but also many new, OxPAPC-binding proteins. We expect that this technically relatively simple method can be widely applied for label-free analysis of lipid-protein interaction in variable protein samples treated with different LOPs.

Supplementary Materials: The following are available online at http://www.mdpi.com/2076-3921/7/9/116/s1, Figure S1 Gel band shift analysis of PBS-soluble HUVEC proteins. Figure S2 Selected MS/MS spectra of VDAC-1.

Figure S3 Selected MS/MS spectra of Aminopeptidase N. Figure S4 Selected MS/MS spectra of HSP90β. Figure S5 Selected MS/MS spectra of Integrin alpha-2. Figure S6 Selected MS/MS spectra of ADP/ATP translocase 2.

Author Contributions: Conceptualization: B.G. and V.N.B.; investigation: B.G., D.K. and N.V.; supervision: B.G. and V.N.B.; writing—original draft: B.G.; writing—review and editing: V.N.B.

Funding: This work was supported by the Austrian Science Fund (grant No. P27682-B30).

Acknowledgments: The authors thank Wolfgang Sattler (Medical University of Graz, Austria) for providing 2-chlorohexadec-15-yn-1-al and Olga Oskolkova (University of Graz, Ausria) for providing OxPAPC.

Conflicts of Interest: The authors declare no conflict of interest.

Abbreviations

BAP31	B-cell receptor-associated protein 31
emPAI	Exponentially Modified Protein Abundance Index
EMSA	Electrophoretic mobility band shift analysis
HNE	4-hydroxynonenal
HRP	horseradish peroxidase
HSP90β	heat shock protein HSP 90-beta
HUVEC	human umbilical vein endothelial cell
KODA-PC	9-keto-12-oxo-10-dodecenoate-phosphatidylcholine
LOP	lipid oxidation product
MS	mass spectrometry
OxPAPC	oxidized palmitoyl-arachidonoyl-phosphatidylcholine
oxPL	oxidized phospholipid
PAPC	palmitoyl-arachidonoyl-phosphatidylcholine
PAPE	1-palmitoyl-2-arachidonoyl-*sn*-glycero-3-phosphoethanolamine
PAPS	1-palmitoyl-2-arachidonoyl-*sn*-glycero-3-phospho-l-serine
PBS	Phosphate-buffered saline
PC	phosphatidylcholine
PE	phosphoethanolamine
PBS-PI	phosphate buffered containing protease inhibitor cocktail
RIPA buffer,	Radioimmunoprecipitation assay buffer
PL	phospholipid
PLPC	1-palmitoyl-2-linoleoyl-*sn*-glycero-3-phosphatidylcholine
PUFA	polyunsaturated fatty acid
SDS	sodium dodecyl sulfate
VDAC-1	voltage-dependent anion-selective channel protein 1

References

1. Bochko, V.N.; Oskolkova, O.V.; Birukov, K.G.; Levonen, A.L.; Binder, C.J.; Stöckl, J. Generation and biological activities of oxidized phospholipids. *Antioxid. Redox Signal.* **2010**, *12*, 1009–1059. [CrossRef] [PubMed]
2. Lee, S.; Birukov, K.G.; Romanoski, C.E.; Springstead, J.R.; Lusis, A.J.; Berliner, J.A. Role of phospholipid oxidation products in atherosclerosis. *Circ. Res.* **2012**, *111*, 778–799. [CrossRef] [PubMed]
3. Negre-Salvayre, A.; Coatrieux, C.; Ingueneau, C.; Salvayre, R. Advanced lipid peroxidation end products in oxidative damage to proteins. Potential role in diseases and therapeutic prospects for the inhibitors. *Br. J. Pharmacol.* **2008**, *153*, 6–20. [CrossRef] [PubMed]
4. Adibhatla, R.M.; Hatcher, J.F. Lipid oxidation and peroxidation in CNS health and disease: From molecular mechanisms to therapeutic opportunities. *Antioxid. Redox Signal.* **2010**, *12*, 125–169. [CrossRef] [PubMed]
5. Bochkov, V.N. Inflammatory profile of oxidized phospholipids. *Thromb. Haemost.* **2007**, *97*, 348–354. [CrossRef] [PubMed]
6. Bochkov, V.; Gesslbauer, B.; Mauerhofer, C.; Philippova, M.; Erne, P.; Oskolkova, O.V. Pleiotropic effects of oxidized phospholipids. *Free Radic. Biol. Med.* **2017**, *111*, 6–24. [CrossRef] [PubMed]

7.	Ullery, J.C.; Marnett, L.J. Protein modification by oxidized phospholipids and hydrolytically released lipid electrophiles: Investigating cellular responses. *Biochim. Biophys. Acta* **2012**, *1818*, 2424–2435. [CrossRef] [PubMed]

8.	West, J.D.; Marnett, L.J. Endogenous reactive intermediates as modulators of cell signaling and cell death. *Chem. Res. Toxicol.* **2006**, *19*, 173–194. [CrossRef] [PubMed]

9.	Rossi, A.; Kapahi, P.; Natoli, G.; Takahashi, T.; Chen, Y.; Karin, M.; Santoro, M.G. Anti-inflammatory cyclopentenone prostaglandins are direct inhibitors of IkappaB kinase. *Nature* **2000**, *403*, 103–118. [CrossRef] [PubMed]

10.	Golizeh, M.; Geib, T.; Sleno, L. Identification of 4-hydroxynonenal protein targets in rat, mouse and human liver microsomes by two-dimensional liquid chromatography/tandem mass spectrometry. *Rapid Commun. Mass Spectrom.* **2016**, *30*, 1488–1494. [CrossRef] [PubMed]

11.	Rauniyar, N.; Stevens, S.M.; Prokai-Tatrai, K.; Prokai, L. Characterization of 4-hydroxy-2-nonenal-modified peptides by liquid chromatography-tandem mass spectrometry using data-dependent acquisition: Neutral loss-driven MS3 versus neutral loss-driven electron capture dissociation. *Anal. Chem.* **2009**, *81*, 782–789. [CrossRef] [PubMed]

12.	Andringa, K.K.; Udoh, U.S.; Landar, A.; Bailey, S.M. Proteomic analysis of 4-hydroxynonenal (4-HNE) modified proteins in liver mitochondria from chronic ethanol-fed rats. *Redox Biol.* **2014**, *2*, 1038–1047. [CrossRef] [PubMed]

13.	Mendez, D.; Hernaez, M.L.; Diez, A.; Puyet, A.; Bautista, J.M. Combined proteomic approaches for the identification of specific amino acid residues modified by 4-hydroxy-2-nonenal under physiological conditions. *J. Proteome Res.* **2010**, *9*, 5770–5781. [CrossRef] [PubMed]

14.	Reed, T.T.; Pierce, W.M.; Markesbery, W.R.; Butterfield, D.A. Proteomic identification of HNE-bound proteins in early Alzheimer disease: Insights into the role of lipid peroxidation in the progression of AD. *Brain Res.* **2009**, *1274*, 66–76. [CrossRef] [PubMed]

15.	Chen, Y.; Cong, Y.; Quan, B.; Lan, T.; Chu, X.; Ye, Z.; Hou, X.; Wang, C. Chemoproteomic profiling of targets of lipid-derived electrophiles by bioorthogonal aminooxy probe. *Redox Biol.* **2017**, *12*, 712–718. [CrossRef] [PubMed]

16.	Codreanu, S.G.; Zhang, B.; Sobecki, S.M.; Billheimer, D.D.; Liebler, D.C. Global analysis of protein damage by the lipid electrophile 4-hydroxy-2-nonenal. *Mol. Cell. Proteomics* **2009**, *8*, 670–680. [CrossRef] [PubMed]

17.	Roe, M.R.; Xie, H.; Bandhakavi, S.; Griffin, T.J. Proteomic mapping of 4-hydroxynonenal protein modification sites by solid-phase hydrazide chemistry and mass spectrometry. *Anal. Chem.* **2007**, *79*, 3747–3756. [CrossRef] [PubMed]

18.	Han, B.; Stevens, J.F.; Maier, C.S. Design, synthesis, and application of a hydrazide-functionalized isotope-coded affinity tag for the quantification of oxylipid-protein conjugates. *Anal. Chem.* **2007**, *79*, 3342–3354. [CrossRef] [PubMed]

19.	Chavez, J.; Wu, J.; Han, B.; Chung, W.G.; Maier, C.S. New role for an old probe: Affinity labeling of oxylipid protein conjugates by N'-aminooxymethylcarbonylhydrazino d-biotin. *Anal. Chem.* **2006**, *78*, 6847–6854. [CrossRef] [PubMed]

20.	Vila, A.; Tallman, K.A.; Jacobs, A.T.; Liebler, D.C.; Porter, N.A.; Marnett, L.J. Identification of protein targets of 4-hydroxynonenal using click chemistry for ex vivo biotinylation of azido and alkynyl derivatives. *Chem. Res. Toxicol.* **2008**, *21*, 432–444. [CrossRef] [PubMed]

21.	Codreanu, S.G.; Ullery, J.C.; Zhu, J.; Tallman, K.A.; Beavers, W.N.; Porter, N.A.; Marnett, L.J.; Zhang, B.; Liebler, D.C. Alkylation damage by lipid electrophiles targets functional protein systems. *Mol. Cell. Proteomics* **2014**, *13*, 849–859. [CrossRef] [PubMed]

22.	Wang, C.; Weerapana, E.; Blewett, M.M.; Cravatt, B.F. A chemoproteomic platform to quantitatively map targets of lipid-derived electrophiles. *Nat. Methods* **2014**, *11*, 79–85. [CrossRef] [PubMed]

23.	Gugiu, B.G.; Mouillesseaux, K.; Duong, V.; Herzog, T.; Hekimian, A.; Koroniak, L.; Vondriska, T.M.; Watson, A.D. Protein targets of oxidized phospholipids in endothelial cells. *J. Lipid Res.* **2008**, *49*, 510–520. [CrossRef] [PubMed]

24.	Tallman, K.A.; Kim, H.Y.; Ji, J.X.; Szapacs, M.E.; Yin, H.; McIntosh, T.J.; Liebler, D.C.; Porter, N.A. Phospholipid-protein adducts of lipid peroxidation: Synthesis and study of new biotinylated phosphatidylcholines. *Chem. Res. Toxicol.* **2007**, *20*, 227–234. [CrossRef] [PubMed]

25. Szapacs, M.E.; Kim, H.Y.; Porter, N.A.; Liebler, D.C. Identification of proteins adducted by lipid peroxidation products in plasma and modifications of apolipoprotein A1 with a novel biotinylated phospholipid probe. *J. Proteome Res.* **2008**, *7*, 4237–4246. [CrossRef] [PubMed]

26. Stemmer, U.; Ramprecht, C.; Zenzmaier, E.; Stojčić, B.; Rechberger, G.; Kollroser, M.; Hermetter, A. Uptake and protein targeting of fluorescent oxidized phospholipids in cultured RAW 264.7 macrophages. *Biochim. Biophys. Acta* **2012**, *1821*, 706–718. [CrossRef] [PubMed]

27. Gao, D.; Willard, B.; Podrez, E.A. Analysis of covalent modifications of proteins by oxidized phospholipids using a novel method of peptide enrichment. *Anal. Chem.* **2014**, *86*, 1254–1262. [CrossRef] [PubMed]

28. Oskolkova, O.V.; Afonyushkin, T.; Leitner, A.; Schlieffen, E.V.; Gargalovic, P.S.; Lusis, A.J.; Binder, B.R.; Bochkov, V.N. ATF4-dependent transcription is a key mechanism in VEGF up-regulation by oxidized phospholipids: critical role of oxidized sn-2 residues in activation of unfolded protein response. *Blood* **2008**, *112*, 330–339. [CrossRef] [PubMed]

29. Shevchenko, A.; Wilm, M.; Vorm, O.; Mann, M. Mass spectrometric sequencing of proteins silver-stained polyacrylamide gels. *Anal. Chem.* **1996**, *68*, 850–858. [CrossRef] [PubMed]

30. Chambers, M.C.; Maclean, B.; Burke, R.; Amodei, D.; Ruderman, D.L.; Neumann, S.; Gatto, L.; Fischer, B.; Pratt, B.; Egertson, J.; et al. A cross-platform toolkit for mass spectrometry and proteomics. *Nat. Biotechnol.* **2012**, *30*, 918–920. [CrossRef] [PubMed]

31. Ishihama, Y.; Oda, Y.; Tabata, T.; Sato, T.; Nagasu, T.; Rappsilber, J.; Mann, M. Exponentially modified protein abundance index (emPAI) for estimation of absolute protein amount in proteomics by the number of sequenced peptides per protein. *Mol. Cell. Proteomics* **2005**, *4*, 1265–1272. [CrossRef] [PubMed]

32. Afonyushkin, T.; Oskolkova, O.V.; Philippova, M.; Resink, T.J.; Erne, P.; Binder, B.R.; Bochkov, V.N. Oxidized phospholipids regulate expression of ATF4 and VEGF in endothelial cells via NRF2-dependent mechanism: novel point of convergence between electrophilic and unfolded protein stress pathways. *Arterioscler. Thromb. Vasc. Biol.* **2010**, *30*, 1007–1013. [CrossRef] [PubMed]

33. Pecorelli, A.; Cervellati, C.; Cortelazzo, A.; Cervellati, F.; Sticozzi, C.; Mirasole, C.; Guerranti, R.; Trentini, A.; Zolla, L.; Savelli, V.; et al. Proteomic analysis of 4-hydroxynonenal and nitrotyrosine modified proteins in RTT fibroblasts. *Int. J. Biochem. Cell Biol.* **2016**, *81*, 236–245. [CrossRef] [PubMed]

34. Usatyuk, P.V.; Natarajan, V. Hydroxyalkenals and oxidized phospholipids modulation of endothelial cytoskeleton, focal adhesion and adherens junction proteins in regulating endothelial barrier function. *Microvasc. Res.* **2012**, *83*, 45–55. [CrossRef] [PubMed]

35. Chavez, J.; Chung, W.G.; Miranda, C.L.; Singhal, M.; Stevens, J.F.; Maier, C.S. Site-specific protein adducts of 4-hydroxy-2(E)-nonenal in human THP-1 monocytic cells: protein carbonylation is diminished by ascorbic acid. *Chem. Res. Toxicol.* **2010**, *23*, 37–47. [CrossRef] [PubMed]

36. Ahmed, E.K.; Rogowska-Wrzesinska, A.; Roepstorff, P.; Bulteau, A.L.; Friguet, B. Protein modification and replicative senescence of WI-38 human embryonic fibroblasts. *Aging Cell.* **2010**, *9*, 252–272. [CrossRef] [PubMed]

37. Vladykovskaya, E.; Sithu, S.D.; Haberzettl, P.; Wickramasinghe, N.S.; Merchant, M.L.; Hill, B.G.; McCracken, J.; Agarwal, A.; Dougherty, S.; Gordon, S.A.; et al. Lipid peroxidation product 4-hydroxy-trans-2-nonenal causes endothelial activation by inducing endoplasmic reticulum stress. *J. Biol. Chem.* **2012**, *287*, 11398–11409. [CrossRef] [PubMed]

38. Beavers, W.N.; Rose, K.L.; Galligan, J.J.; Mitchener, M.M.; Rouzer, C.A.; Tallman, K.A.; Lamberson, C.R.; Wang, X.; Hill, S.; Ivanova, P.T.; et al. Protein Modification by Endogenously Generated Lipid Electrophiles: Mitochondria as the Source and Target. *ACS Chem. Biol.* **2017**, *12*, 2062–2069. [CrossRef] [PubMed]

39. Landar, A.; Zmijewski, J.W.; Dickinson, D.A.; Le, G.C.; Johnson, M.S.; Milne, G.L.; Zanoni, G.; Vidari, G.; Morrow, J.D.; Darley-Usmar, V.M. Interaction of electrophilic lipid oxidation products with mitochondria in endothelial cells and formation of reactive oxygen species. *Am. J. Physiol. Heart Circ. Physiol.* **2006**, *290*, H1777–H1787. [CrossRef] [PubMed]

40. Reed, T.T. Lipid peroxidation and neurodegenerative disease. *Free Radic. Biol. Med.* **2011**, *51*, 1302–1319. [CrossRef] [PubMed]

antioxidants

MDPI

Review

4-Hydroxynonenal in Redox Homeostasis of Gastrointestinal Mucosa: Implications for the Stomach in Health and Diseases

Andriy Cherkas [1] and **Neven Zarkovic [2],***

[1] Department of Internal Medicine #1, Danylo Halystkyi Lviv National Medical University, 79010 Lviv, Ukraine; cherkasandriy@yahoo.com
[2] Laboratory for Oxidative Stress (LabOS), Institute "Rudjer Boskovic", HR-10000 Zagreb, Croatia
* Correspondence: zarkovic@irb.hr

Received: 26 July 2018; Accepted: 30 August 2018; Published: 3 September 2018

Abstract: Maintenance of integrity and function of the gastric mucosa (GM) requires a high regeneration rate of epithelial cells during the whole life span. The health of the gastric epithelium highly depends on redox homeostasis, antioxidant defense, and activity of detoxifying systems within the cells, as well as robustness of blood supply. Bioactive products of lipid peroxidation, in particular, second messengers of free radicals, the bellwether of which is 4-hydroxynonenal (HNE), are important mediators in physiological adaptive reactions and signaling, but they are also thought to be implicated in the pathogenesis of numerous gastric diseases. Molecular mechanisms and consequences of increased production of HNE, and its protein adducts, in response to stressors during acute and chronic gastric injury, are well studied. However, several important issues related to the role of HNE in gastric carcinogenesis, tumor growth and progression, the condition of GM after eradication of *Helicobacter pylori*, or the relevance of antioxidants for HNE-related redox homeostasis in GM, still need more studies and new comprehensive approaches. In this regard, preclinical studies and clinical intervention trials are required, which should also include the use of state-of-the-art analytical techniques, such as HNE determination by immunohistochemistry and enzyme-linked immunosorbent assay (ELISA), as well as modern mass-spectroscopy methods.

Keywords: 4-hydroxynonenal; lipid peroxidation; redox balance; oxidative stress; stomach; peptic ulcer; gastritis; *Helicobacter pylori*; gastric cancer; non-steroid anti-inflammatory drugs-induced gastropathy

---

1. Introduction

The gastrointestinal tract (GIT) represents a highly specialized interface between the environment and an organism's internal medium, aimed primarily to digest food, and absorb nutrients and water. In addition, it fulfils a wide variety of other functions, including, but not limited to, immune defense, excretion of metabolic waste/detoxification, secretory and regulatory functions, and as a physical barrier. Last but not least, it is a vital niche for gut bacteria [1]. The GIT has to withstand harsh conditions, due to exposure to food/chyme, digestive enzymes, different, often very aggressive pH conditions, and numerous bacteria; therefore, high efficiency of protection and regeneration is required for its maintenance and function. This is particularly important in the case of the stomach, whose lumen contains aggressive hydrochloric acid, often reaching pH values of 1–2, and proteolytic enzymes, such as pepsin [2]. Toxins, which may be ingested together with food, as well as some drugs, may contribute to damaging the gastric mucosa (GM). Furthermore, in more than half of the human population worldwide, *Helicobacter pylori* (*H. pylori*) bacteria [3] persist in the GM, and may cause chronic gastritis and peptic ulcer, thus being a major contributor to the pathogenesis of gastric adenocarcinoma and mucosa-associated lymphoid tissue (MALT)-lymphoma [4].

The redox balance is a major homeostatic parameter and a regulatory factor for the metabolic functions of the whole organism and also the GIT [5]. Redox imbalance, often referred to as "oxidative stress", may be caused either by excessive exposure to oxidants, or by decreased activity of counter-regulatory enzymatic systems and a lack of antioxidants [6]. A certain degree of lipid peroxidation may take place in many cellular processes under physiological conditions, but redox imbalance that is observed in many diseases very often leads to excessive accumulation of oxidized lipids and their degradation products. Among such products of lipid peroxidation, 4-hydroxy-2-nonenal (HNE) is ubiquitous, and one of the most studied compounds, also considered as a "second messenger of free radicals" [7]. HNE is generated from omega-6 fatty acids. Along with its role in the pathogenesis of multiple diseases, it has been shown to be involved in various signaling pathways. It contributes to the regulation of energy metabolism, detoxification, cell proliferation and differentiation, maintenance of the cytoskeleton, and metabolic adaptations to redox derangements [8–10].

Considering the sophisticated functions of the mucous membrane and the wide variety of damaging exposures, the maintenance of the redox balance in GM is particularly challenging [5]. In order to sustain the lifelong function of the GIT, the cells of the mucosal epithelium have a high rate of proliferation, and an exceptional regenerative potential. However, this system is prone to derangements, which can result in gastritis, peptic ulcer, and gastric cancer. Gastrointestinal diseases cause severe health problems and overall socioeconomic damage [3]. Progress in understanding the roles of lipid peroxidation and its reaction product, HNE, in health and disease, stimulated studies focused on specific diseases of the GIT. This review is aimed at addressing important issues related to the role of HNE in normal functioning and in the development of diseases of the stomach.

2. Approaches to Determine HNE in Samples of Patients Suffering from Stomach Diseases

Along with conventional approaches to measure the concentrations of substances of interest in biological liquids like blood (serum, plasma, and whole blood), urine, cerebrospinal fluid, etc., with several other options available in the case of stomach diseases. First, the stomach is accessible to endoscopy, which is a routine clinical intervention. During endoscopy, it is possible to obtain biopsies of the mucous membrane from different parts of stomach for further morphological studies. Second, gastric juice can be obtained for chemical analysis. Third, a number of "breath-tests" (determination of metabolites of ingested reagents in exhaled air) are available for gastroenterological diagnostics. Finally, feces samples can be taken, for example, to test *H. pylori* bacterial contamination [3]. The researchers have to keep in mind that blood flows from stomach through the portal vein to the liver; many substances, such as xenobiotics, lipid peroxidation products, some hormones, and cytokines, may be degraded there and, thus, may be measured in the peripheral blood within normal concentration ranges, despite evidence of toxicity/inflammation [11].

HNE and other lipid peroxidation products, including acrolein, malonic dialdehyde, and many others, can be measured as biomarkers of redox imbalance [7]. However, their high reactivity and capacity for interactions with multiple functional groups of macromolecules, as well as their transfer to blood and/or urine from other compartments of the organism, may significantly lower steady-state concentrations of free lipid peroxidation products. Most of the detectable HNE are found to be conjugated to proteins or glutathione (GSH). Through a Michael-type reaction of nucleophilic addition, HNE binds covalently to cysteine, lysine, and histidine residues within proteins [12]. Development of specific antibodies against HNE–histidine adducts facilitated further research and enabled implementation of respective analytical methods [13,14].

In this regard, HNE-immunohistochemistry (qualitative/semiquantitative evaluation) is a widely used method of HNE determination, in order to map tissue or intracellular distribution of respective HNE-conjugates in human samples obtained by gastric biopsy [12]. A variety of HNE-ELISAs have been introduced that are applicable for quantitative evaluation of the levels of HNE-adducts in biological fluids, like blood serum, urine, or gastric juice [15]. Other antibody-based methods, which are often applied successfully, include immunofluorescence, immunogold electron microscopy, and immunoblotting [16]. However, since the use of antibodies for analytical purposes is often associated with technical problems, such as inaccessibility of some epitopes and/or their alterations, this may result in incomplete quantification [15]. Furthermore, higher degrees of protein modification can decrease the epitope recognition (non-linear dependence); therefore, the results of analyses based on antibody-dependent techniques should be interpreted very carefully. For clinical purposes, in particular, when histological samples are evaluated, the use of semiquantitative methods for HNE detection may be particularly reasonable [17–20]. Free HNE can be accurately determined by high performance liquid chromatography, and a number of modifications of mass spectroscopy-based methods [21]. However, due to the high reactivity of free HNE and its low steady-state levels, the determination of HNE conjugates reveals more biologically/clinically relevant information, and may have substantial advantages [12]. The formation of protein conjugates is proportional to the mean levels of free HNE; therefore, antibody-based methods of staining and quantitative determination of HNE are considered to be quite accurate and reliable, especially if HNE–histidine adducts are monitored [12]. The last generation mass spectrometric techniques and instrumentations, in combination with enrichment and separation techniques, have been successfully applied to the determination not only of HNE, but also of its adducts with amino acids in proteins [22–24].

3. HNE in the Stomach under Physiological Conditions

In the lumen of the stomach, ingested food is exposed to low pH (hydrochloric acid) and proteolytic enzymes, such as pepsin, contributing to denaturation and degradation of proteins. However, a highly acidic medium facilitates, also, a variety of chemical reactions between different food components [1,2]. Modeling of chemical processes taking place during gastric digestion reveals the possibility of iron- or metmyoglobin-catalyzed generation of substantial amounts of hydroperoxides and other lipid peroxidation products from components of common diets containing meat and unsaturated fats at low pH in the presence of water-dissolved oxygen. Notably, the ingestion of food rich in polyphenols dramatically lowers the generation of hydroperoxides, which may be, at least in part, responsible for the preventive effects of fruits and vegetables [2]. On the other hand, accumulation of lipid peroxidation products in the GM may be enhanced by consumption of large amounts of unsaturated fats that may be a part of many "healthy" diets or popular supplements containing polyunsaturated fatty acids (PUFAs) [25]. Therefore, food products containing significant quantities of PUFAs should be carefully processed and properly stored, in order to prevent their oxidation. Steady-state levels of HNE in the GM result from the rates of their generation/absorption and utilization [26]. The acidity of the chyme may also influence the stability of hydroperoxides and the likelihood of Michael addition within the gastric lumen, whereas the cells of the gastric epithelium are well protected from the acidic content by mucus. Noteworthy, *H. pylori* bacteria produce ammonia to provide protection from hydrochloric acid [27,28] and, therefore, create an alkaline local microenvironment at infection sites, that is more favorable for Michael reactions (Figure 1).

Formation of HNE conjugates with glutathione and adducts with proteins may have heterogeneous consequences for the cells, depending on the role of respective residues. Depletion of reduced glutathione may increase vulnerability of the cells to oxidants and shift the redox balance to the pro-oxidant side. Addition of HNE to cysteine residues may alter function of proteins, and may have significant regulatory consequences, whereas binding to other sites (for example, histidine or lysine residues) may have not much effect on function, and can rather reflect the degree of HNE accumulation and possible oxidative damage [12,29,30] (Figure 2).

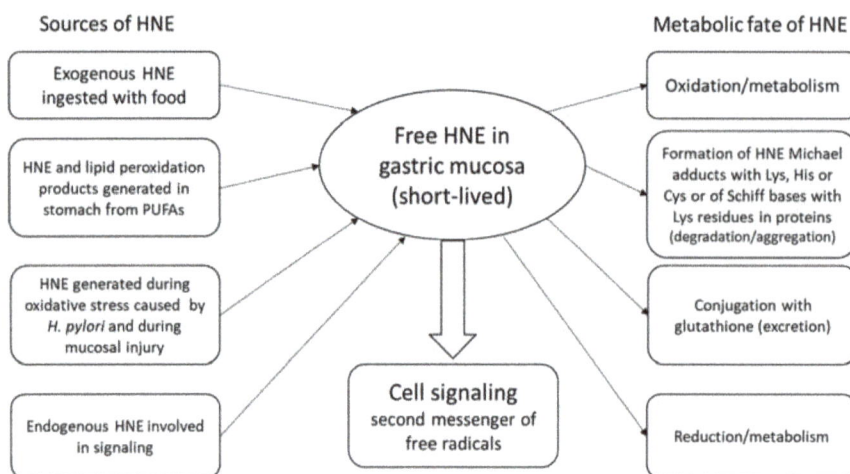

Figure 1. Schematic presentation of major sources of 4-hydroxynonenal (HNE) in gastric mucosa and the ways of its further transformations. Free HNE is a highly reactive molecule, capable of reacting with numerous targets within cells. HNE interfering with redox-sensitive pathways (for example, by binding to cysteine residues) may affect the function of redox-sensitive proteins. Conjugation of HNE with histidine or lysine residues of peptides and proteins are thought to be less important for signaling. However, even in these cases, HNE may bind enzymes, cytokines, and receptors, so they may have important regulatory roles. Hence, such aldehyde-protein adducts can represent a source of HNE and cause secondary oxidative stress, while they can also be used as biomarkers for immunochemical detection of HNE, denoted as advanced lipoxidation end products (ALEs).

The epithelium of the GIT is highly proliferating and, depending on the location, it is completely renewed every 3–10 days. Therefore, the immunohistochemical pattern of HNE adducts mainly reflects the metabolic conditions within the mucosa (e.g., oxidative stress, exposure to xenobiotics) during the last few days before taking the sample. Different HNE levels may occur rather as a result of recent alterations, than due to accumulation (for example with age), and are likely to depend on both the renewal rate of epithelial cells and the rate of lipid peroxidation.

A certain degree of accumulation of HNE–histidine adducts in the mucosa of gastric corpus and antrum was demonstrated for the majority of healthy volunteers [19]. Notably, almost all the samples, obtained from asymptomatic apparently healthy subjects, regardless of whether the patients have been *H. pylori*-positive or not, have shown mild to moderate HNE-immunopositivity in the cytoplasm of the gastric glandular epithelium, with only a few HNE-negative samples [19]. A likely explanation of these findings suggests that HNE may play a role in normal signaling and regulation of cellular functions in the GM under physiological conditions. HNE levels appear to be strictly maintained within a homeostatic range, providing adaptations to adverse factors, like metabolic or emotional stress, exogenous toxins that are occasionally ingested with food, or latent *H. pylori* infection. Only excessive and/or prolonged oxidative stress may cause GM injury and inflammation, as discussed below (Figure 2).

Interestingly, most of the *H. pylori*-positive subjects never experience clinically overt forms of gastritis, peptic ulcer, or gastric cancer [31]. This observation is in line with observations that apparently healthy *H. pylori*-positive subjects show no difference in HNE–histidine conjugates in GM compared to controls, despite occasional presence of inflammatory cells in the samples [19]. It is likely that asymptomatic subjects have sufficient compensatory power to cope with the negative influence of the pathogen. Only excessive virulence of certain *H. pylori* strains or lowered resistance of the host may result in clinically significant manifestations. In this regard, it is known that sedentary lifestyle may cause deleterious metabolic changes associated with activation of sympathetic tone (with subsequent parasympathetic impairment) [32]. Genetic defects, psychoemotional stress, and a number of other factors, may also contribute to autonomic imbalance that may lead to increased vulnerability of the GM [33–35]. It has been reported that patients with chronic peptic ulcer disease show altered autonomic function, as measured by Holter electrocardiogram monitoring [36]. The relationships of heart rate variability alterations with endothelial dysfunction [37], as well as oxidative stress [38], were earlier noticed. Hence, not clearly intuitive, the relationships of autonomic function and redox balance attract growing attention. For example, an anti-inflammatory action of cholinergic (parasympathetic) signaling [39,40], adrenergic pathways' interference with H_2O_2-mediated insulin signaling [41] and thermogenesis in adipose tissue [42], were demonstrated. Moreover, the link between redox balance and autonomic function was hypothesized [43], and is further confirmed by a recent observation that selective Nrf2 deletion in the rostral ventrolateral medulla in mice evokes hypertension and "sympatho-excitation" [44].

Numerous epidemiological observations associate *H. pylori*-positivity with so called extra-gastric manifestations that include, but are not limited to, atherosclerosis, insulin resistance/diabetes type 2, diseases of liver and pancreas, and others [45–49]. Proposed pathogenesis mechanisms include initial damage of GM caused by *H. pylori* and its virulence factors, oxidative stress and lipid peroxidation, local inflammation, release of pro-inflammatory cytokines and other bioactive mediators to the blood circulation, causing systemic effects and metabolic derangements [4,50–52]. Indeed, in *H. pylori*-positive healthy male subjects with a sedentary lifestyle, higher levels of fasting insulin and elevated homeostatic model assessment index (HOMA-index) were observed, compared to *H. pylori*-negative matches [47]. Another study showed significantly increased heart rate and sympathetic tone in *H. pylori*-positive asymptomatic volunteers. However, levels of the water-soluble HNE derivative 1,4-dihydroxynonane mercapturic acid (DHN-MA), iso-PGF2, pro- and anti-inflammatory cytokines, C-reactive protein, and a number of selected hormones, were not different between the groups, indicating that either the degree of local mucosal damage was not strong enough to cause marked elevation of studied parameters, or their mild/moderate elevation is obscured by the passage of blood through the liver [52].

4. HNE in Patients with *H. pylori*-Associated Gastritis and Peptic Ulcer

Despite its recent decline, the prevalence of *H. pylori* infection is still very high worldwide, ranging from rates between 20% and 40% in Western countries, to over 90% in many developing countries [3]. There is clear evidence that this microorganism is a causative factor for chronic gastritis type B and peptic ulcer. However, as mentioned above, most *H. pylori*-positive subjects are clinically healthy, and never develop gastritis or ulcer, suggesting that besides *H. pylori* and its virulence factors, conditions of the host organism play a crucial role in the outcome of this complex host–microbe interaction [4,53]. This idea fits well into the framework of the classical concept of balance of factors of "aggression" and "cytoprotection" in GM. On the cellular level, this paradigm is consistent with our current understanding of the principles of redox balance maintenance under stress conditions [6]. GM injury and subsequent inflammation may take place when the capacity of antioxidant mechanisms is not sufficient to protect the cells from the damaging factors and related oxidative stress [5].

Peptic ulcer and gastritis are, for a long time, known to be associated with redox imbalance and excessive lipid peroxidation [54], as confirmed in numerous studies and with different study models [55]. Clinical studies are less abundant, and only a few of them address the issue of oxidative stress and lipid peroxidation in GM. The use of gastric endoscopy enables obtaining of mucosal tissue samples for further histological examination. In the group of *H. pylori*-positive peptic ulcer patients, significantly higher accumulation of HNE–histidine adducts in GM compared to control group was clearly demonstrated [19]. In some cases, severe immunopositivity of nuclei and perinuclear spaces, along with diffuse accumulation of HNE–histidine conjugates in cytoplasm of the cells was observed, pointing to an impaired redox balance in the GM of these patients [18,19].

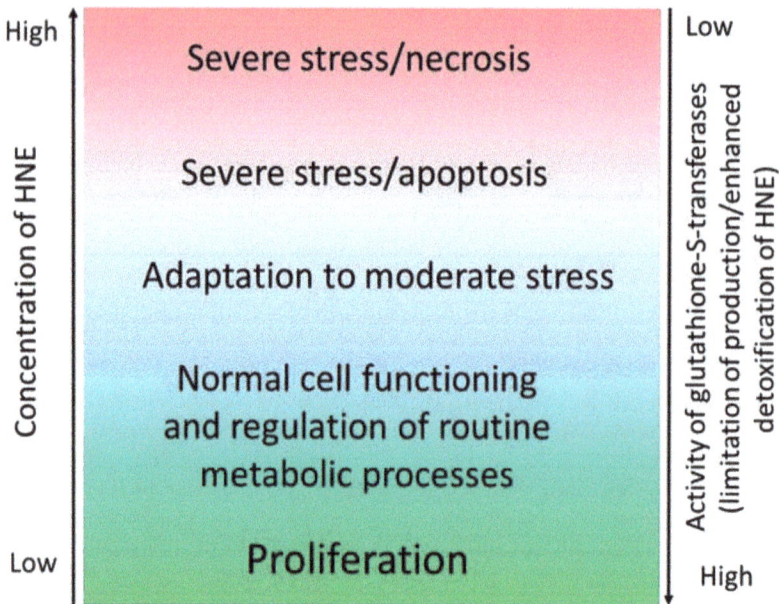

Figure 2. Physiological and pathophysiological effects of HNE on the gastric mucosa depend on the HNE concentration. Steady-state HNE levels inversely correlate with the cellular redox status, and are a function of the rate of its generation and metabolization. HNE content is regulated by the activities of alcohol and aldehyde dehydrogenases, and of glutathione S-transferases, depending mostly on the level of reduced glutathione and affinity to cellular proteins [56]. The overall pathophysiological consequences of HNE generation reflect the tissue/cellular redox (im)balance, and depend on the type of cells and the reaction of neighboring cells to the onset of lipid peroxidation. The cells often behave as individuals, not as a homogenous population, which is relevant for carcinogenic effects of HNE and for its involvement in (regulation of) host defense against cancer [57–60].

The pharmacological approach to treat chronic gastritis and peptic ulcer via eradication of *H. pylori* proved to be very successful from the clinical point of view, as it allows most of the patients to be cured of these diseases [61]. In addition, there are reasons to expect that eradication of this microorganism may be useful for prevention and/or treatment of other diseases associated with *H. pylori*, including metabolic syndrome, type 2 diabetes, non-alcoholic fatty liver disease, or atherosclerosis [62–64]. How an infection with *H. pylori* may result in systemic pathological effects, as well as the biochemical mechanisms that may contribute to metabolic deteriorations in *H. pylori*-positive patients, needs to be further elucidated.

Despite obvious clinical efficiency, there are reports indicating persistence of HNE–histidine adducts, hyperaccumulation in peptic ulcer patients, even after successful eradication of *H. pylori*, at least in the period of 4 weeks after completing antimicrobial treatment [18]. This is consistent with clinical observations that some patients still have symptoms (epigastric pain, nausea, reduced appetite, etc.) for several months after treatment [65]. It might be possible that metabolic dysfunction in these patients, as an integral part of ulcer disease, contributes to pathogenesis of gastric injury independent of persisting *H. pylori* occurrence. The combination of these two as well as any additional factors is known to increase the risk of ulcerations. In this regard, smoking, psychoemotional stress, unhealthy lifestyle, and suboptimal nutrition may be crucial for the outcome of host–microbial interaction [34,66]. Thus, it depends on the power of intrinsic cytoprotective mechanisms (genetics, sufficient blood microcirculation in stomach, effective autonomic regulation) and exogenous factors (*H. pylori*, ingestion of toxins, and products of PUFA peroxidation), and may vary from long-term asymptomatic carrying to chronic gastritis type B, with the periods of exacerbation and remission, peptic stomach ulcers, and/or duodenum or transformations in the form of MALT-lymphoma or gastric adenocarcinoma.

5. HNE in Gastric Carcinogenesis

The GM is exposed to different types of exogenous chemical agents, and reactive species are generated in the stomach during digestion. Some of them may be toxic and cause damage to the gastric epithelium, and some may also be carcinogenic [2]. Chronic inflammation and oxidative stress caused by *H. pylori* infection are also major contributors to malignant transformation of the cells of GM [50,67]. The idea to eradicate *H. pylori* in all carriers, even in asymptomatic ones, is gaining popularity, as some recently published trials showed positive results [68]. Moreover, eradication of *H. pylori* seems to be reasonable also in patients with early stages of gastric cancer undergoing endoscopic resection, since it decreases the rates of metachronous cancers compared to control group [69]. In this context, genotoxicity of supraphysiological levels of HNE and other lipid peroxidation products may be important for carcinogenesis, as well [70,71].

The role of HNE in malignant transformation and growth is ambiguous. On the one hand, HNE can diffuse from the site of generation into the nucleus and bind, covalently, to DNA molecules, causing mutations and supporting carcinogenesis [71], while, on the other hand, it is influencing pathways regulating proliferation, differentiation, and apoptosis of transformed cells. Depending on the activity of detoxifying systems in cancer cells, HNE may be toxic to them or can stimulate their growth and enforce resistance to cytostatic drugs [72].

While, in the case of acute and chronic GM injury caused by *H. pylori* and gastrotoxic agents, oxidative stress and increased lipid peroxidation is well documented, in case of gastric cancer, it is not. As it was shown by Ma et al. (2013), serum levels of major lipid peroxidation products, such as HNE, malonic dialdehyde, conjugated dienes, and 8-iso-prostaglandin F2α, were all decreased in cancer patients compared to control group [73]. Hence, though not statistically significant, lower levels of HNE were also observed in *H. pylori*-positive vs. *H. pylori*-negative patients, that may support the idea that moderate (or local) activation of lipid peroxidation may stimulate systemic activation of detoxification mechanisms through, for example, Nrf2-dependent mechanisms [72].

6. HNE in Alcohol- and Non-Steroid Anti-Inflammatory Drug (NSAID)-Induced Gastropathy

Alcohol and a rapidly growing use of NSAIDs are, jointly, the second most important cause of gastric injury after *H. pylori* [66]. Evidence from well-established animal models of GM injury suggests two principal mechanisms responsible for tissue damage. The first, a direct toxic effect on GM, and the second, limitation of gastric microcirculatory blood flow that is essential for a proper rate of proliferation, mucus secretion, etc., through decreased levels of gastroprotective prostaglandin E$_2$ with subsequent endothelial dysfunction and autonomic dysregulation, that may cause oxidative stress [74,75]. Both mechanisms contribute to the development of severe local oxidative stress, excessive lipid peroxidation, and accumulation of its products, including HNE, mostly covalently bound to proteins [54].

The important role of autonomic dysregulation is often ignored in case of diseases of stomach. It is known that an elevated sympathetic tone limits blood flow in the organs of gastrointestinal tract, and caused endothelial dysfunction, which is crucial for gastroprotection; therefore, autonomic imbalance may significantly potentiate the damaging effects of alcohol and NSAIDs [76,77].

7. Pharmacological and Non-Pharmacological Approaches to Reduce Redox Imbalance in GM

Considering multiple etiologic and pathogenic factors that may interact with each other and contribute to GM damage, there are a number of different approaches in order to prevent or treat gastric injuries (Table 1).

Eradication of *H. pylori* with a combination of two antibiotics and proton pump inhibitors has been proven to be effective in most of the *H. pylori*-positive patients suffering from gastritis and peptic ulcer [61]. However, in some of these patients, elimination of the microbial factor is not sufficient, and symptoms, as well as redox imbalance, may persist long after completion of the treatment [18,65]. Moreover, eradication of *H. pylori* does not significantly lower the risk of gastric cancer, at least within a few years after eradication, and the statistical difference becomes significant only after 8–10 years [78]. Therefore, other approaches are also needed in order to overcome these limitations and to address other aspects of GM injury pathogenesis.

Table 1. Selected pharmacological and non-pharmacological interventions and their effects on HNE production/utilization in gastric mucosa.

Intervention	Target Process/Pharmacological Effect	References
Proton pump inhibitors, H_2 histamine receptor inhibitors	Reduction of acidity, decreased proteolytic activity of gastric juice/decreased gastric injury (production of HNE)	[61,66]
Antibiotics	*H. pylori* eradication/decreased gastric injury (production of HNE)	[61,66]
NO, CO, H_2S-releasing NSAIDs	Release of CO, NO, and/or H_2S modulates redox signaling, improves endothelial function, and improves microcirculation/reduced production and improved utilization of HNE	[75,76]
Antioxidants/polyphenols present in food	Reduced lipid peroxidation of PUFAs in stomach/reduced absorption of exogenous HNE	[2,79]
Phytochemical and phytotoxins with moderate prooxidant action	Nrf-2 activators induce expression of antioxidant genes and increase detoxification of HNE	[20,80]
Interval hypoxic training	Improvement of autonomic control of microcirculation and function of internal organs	[81,82]
Exercise, intermittent fasting, caloric restriction	Activation of autophagy, reduction of systemic inflammatory response, improvement of protein quality control and autonomic regulation	[83]
Ulcer-healing drugs (actovegin, solcoseryl etc.)	Mechanism unknown, suggested influence on microcirculation and/or endothelial function	[84,85]

Since substantial amounts of gastrotoxic substances may be ingested with food or generated during digestion, the idea to use drugs, supplements, or certain types of food able to neutralize toxins or reduce the rate of lipid peroxidation was actively explored. Indeed, subjects consuming more fruits and vegetables show lower incidence of gastric diseases, especially gastric cancer [86]. Studies also show that polyphenols reduce the formation of hydroperoxides in stomach and in in vitro models of gastric digestion [2,79]. Pre- and probiotics [87], as well as a number of plant-derived traditional, medicines or extracts, were also shown to be protective against gastric and intestinal mucosal damage and may improve redox balance in mucous membranes in different parts of the GIT [20]. Thus, a number of natural compounds present in fruit and vegetables (e.g., phenolic flavonoids, lycopenes, carotenoids, glucosinolates) act as radical-trapping antioxidants, and they represent not a only useful and convenient beneficial health-promoting approach, due to their natural occurrence and abundance, but also a model for the development of novel drugs aimed to modulate redox balance [88].

The molecular mechanisms underlying protective effects of beneficial compounds are often not yet elucidated, but at least some of them may act via a hormetic response, when moderate prooxidant action causes the activation of defense mechanisms (for example, by induction of target genes of the Nrf-2 transcription factor) [72]. Alternatively, they may contribute to increased mucosal microcirculation through improvement of endothelial function or parasympathetic tone, as it has been shown for Actovegin, which has been used as an anti-ulcer drug for several decades [85]. Among non-pharmacological interventions that showed some efficiency in the case of peptic ulcer disease is also interval hypoxic training [81]. Exact gastroprotective mechanisms in this case are not clear as well, but it is likely that the mechanism includes improvements of autonomic balance and enhanced microcirculation [82].

Therapeutic use of NSAIDs is overwhelming, and in order to reduce their gastrotoxicity, a wide range of new formulations are introduced or are under development [89]. For example, a number of nitric oxide (NO)-, carbon monoxide (CO)-, or hydrogen sulfide (H_2S)-releasing derivatives of acetylsalicylic acid and other NSAIDs were shown to be as pharmacologically effective as traditional drugs, but have preventive effects against NSAID-induced gastrotoxicity via improvement of endothelial function, and anti-inflammatory and cytoprotective effects [75,76]. Protective actions of these drugs may be also closely related to HNE signaling pathways and maintenance of redox balance in GM.

8. Conclusions

The integrity, high functional activity, and sufficient regeneration rate of GM in harsh conditions is very challenging. The health of gastric epithelium highly depends on the efficiency of redox balance maintenance, antioxidant defense, and activity of detoxifying systems within the cells, as well as robustness of blood supply. The products of lipid peroxidation, in particular, of HNE and its protein/histidine adducts, are important mediators in physiological adaptive reactions, cell signaling, and are also implicated in pathogenesis of numerous gastric diseases. Hence, while the mechanisms and consequences of HNE generation in response to strong stressors during acute and chronic gastric injury are well studied, many other important issues related to gastric carcinogenesis, tumor growth and progression, the condition of GM after eradication of *H. pylori*, and many others, still need extensive studies and new comprehensive approaches.

Author Contributions: Both authors contributed to conceptualization, original draft preparation, review and editing, and preparation of the figures.

Funding: This work was supported by COST Actions B35 "LPO-lipid peroxidation associated disorders", CM1001 "Chemistry of non-enzymatic protein modification–modulation of protein structure and function", "BM1203", "EU-ROS", "New concepts and views in redox biology and oxidative stress research", CA16112 "Personalized Nutrition in aging society: redox control of major age-related diseases", A.C. was supported by the Georg Forster (HERMES) Scholarship from Alexander Von Humboldt Foundation (Bonn, Germany).

Acknowledgments: The authors are grateful to all the colleagues and collaborators that contributed to the research that has been done at the Department of Internal Medicine #1 and other departments of Danylo Halytskyi Lviv National Medical University (Lviv, Ukraine), Rudjer Boskovic Institute and School of Medicine of the University of Zagreb (Zagreb, Croatia). Authors express special gratefulness to Dr. Holger Steinbrenner (Department of Nutrigenomics, Institute of Nutrition, Friedrich Schiller University Jena, Jena, Germany) for valuable comments regarding manuscript.

Conflicts of Interest: The authors declare no conflict of interest.

References

1. Sherwood, L. *Human Physiology: From Cells to Systems*; Cengage Learning: Boston, MA, USA, 2015.
2. Kanner, J.; Lapidot, T. The stomach as a bioreactor: Dietary lipid peroxidation in the gastric fluid and the effects of plant-derived antioxidants. *Free Radic. Biol. Med.* **2001**, *31*, 1388–1395. [CrossRef]
3. Mentis, A.; Lehours, P.; Mégraud, F. Epidemiology and Diagnosis of Helicobacter pylori infection. *Helicobacter* **2015**, *20* (Suppl. 1), 1–7. [CrossRef]
4. Sgouras, D.N.; Trang, T.T.H.; Yamaoka, Y. Pathogenesis of Helicobacter pylori Infection. *Helicobacter* **2015**, *20* (Suppl. 1), 8–16. [CrossRef] [PubMed]
5. Pérez, S.; Taléns-Visconti, R.; Rius-Pérez, S.; Finamor, I.; Sastre, J. Redox signaling in the gastrointestinal tract. *Free Radic. Biol. Med.* **2017**, *104*, 75–103. [CrossRef] [PubMed]
6. Sies, H.; Berndt, C.; Jones, D.P. Oxidative Stress. *Annu. Rev. Biochem.* **2017**, *86*, 715–748. [CrossRef] [PubMed]

7. Zarkovic, N. 4-Hydroxynonenal as a bioactive marker of pathophysiological processes. *Mol. Asp. Med.* **2003**, *24*, 281–291. [CrossRef]

8. Zarkovic, N.; Cipak Gasparovic, A.; Cindrić, M.; Waeg, G.; Borović Šunjić, S.; Mrakovčić, L.; Jaganjac, M.; Kolenc, D.; Andrišić, L.; Gverić Ahmetašević, S.; et al. 4-Hydroxynonenal-protein adducts as biomarkers of oxidative stress, lipid peroxidation and oxidative homeostasis. In Proceedings of the Free Radicals Health and Lifestyle, Rome, Italy, 26–29 August 2009; pp. 37–44.

9. Žarković, N.; Žarković, K.; Schaur, R.J.; Štolc, S.; Schlag, G.; Redl, H.; Waeg, G.; Borović, S.; Lončarić, I.; Jurić, G.; et al. 4-Hydroxynonenal as a second messenger of free radicals and growth modifying factor. *Life Sci.* **1999**, *65*, 1901–1904. [CrossRef]

10. Schaur, R.; Siems, W.; Bresgen, N.; Eckl, P. 4-Hydroxy-nonenal—A Bioactive Lipid Peroxidation Product. *Biomolecules* **2015**, *5*, 2247–2337. [CrossRef] [PubMed]

11. Kirpich, I.A.; Marsano, L.S.; McClain, C.J. Gut-Liver Axis, Nutrition, and Non Alcoholic Fatty Liver Disease. *Clin. Biochem.* **2015**, *48*, 923–930. [CrossRef] [PubMed]

12. Zarkovic, K.; Jakovcevic, A.; Zarkovic, N. Contribution of the HNE-immunohistochemistry to modern pathological concepts of major human diseases. *Free Radic. Biol. Med.* **2017**, *111*, 110–126. [CrossRef] [PubMed]

13. Toyokuni, S.; Miyake, N.; Hiai, H.; Hagiwara, M.; Kawakishi, S.; Osawa, T.; Uchida, K. The monoclonal antibody specific for the 4-hydroxy-2-nonenal histidine adduct. *FEBS Lett.* **1995**, *359*, 189–191. [CrossRef]

14. Waeg, G.; Dimsity, G.; Esterbauer, H. Monoclonal antibodies for detection of 4-hydroxynonenal modified proteins. *Free Radic. Res.* **1996**, *25*, 149–159. [CrossRef] [PubMed]

15. Weber, D.; Milkovic, L.; Bennett, S.J.; Griffiths, H.R.; Zarkovic, N.; Grune, T. Measurement of HNE-protein adducts in human plasma and serum by ELISA-Comparison of two primary antibodies. *Redox Biol.* **2013**, *1*, 226–233. [CrossRef] [PubMed]

16. Sousa, B.C.; Pitt, A.R.; Spickett, C.M. Chemistry and analysis of HNE and other prominent carbonyl-containing lipid oxidation compounds. *Free Radic. Biol. Med.* **2017**, *111*, 294–308. [CrossRef] [PubMed]

17. Zarkovic, K.; Uchida, K.; Kolenc, D.; Hlupic, L.; Zarkovic, N. Tissue distribution of lipid peroxidation product acrolein in human colon carcinogenesis. *Free Radic. Res.* **2006**, *40*, 543–552. [CrossRef] [PubMed]

18. Cherkas, A.; Yelisyeyeva, O.; Semen, K.; Zarković, K.; Kaminskyy, D.; Gasparović, A.C.; Jaganjac, M.; Lutsyk, A.; Waeg, G.; Zarkovic, N. Persistent accumulation of 4-hydroxynonenal-protein adducts in gastric mucosa after Helicobacter pylori eradication. *Coll. Antropol.* **2009**, *33*, 815–821. [PubMed]

19. Yelisyeyeva, O.; Cherkas, A.; Zarkovic, K.; Semen, K.; Kaminskyy, D.; Waeg, G.; Zarkovic, N. The distribution of 4-hydroxynonenal-modified proteins in gastric mucosa of duodenal peptic ulcer patients. *Free Radic. Res.* **2008**, *42*, 205–211. [CrossRef] [PubMed]

20. Cherkas, A.; Zarkovic, K.; Cipak Gasparovic, A.; Jaganjac, M.; Milkovic, L.; Abrahamovych, O.; Yatskevych, O.; Waeg, G.; Yelisyeyeva, O.; Zarkovic, N. Amaranth oil reduces accumulation of 4-hydroxynonenal-histidine adducts in gastric mucosa and improves heart rate variability in duodenal peptic ulcer patients undergoing Helicobacter pylori eradication. *Free Radic. Res.* **2018**, *52*, 135–149. [CrossRef] [PubMed]

21. Spickett, C.M.; Wiswedel, I.; Siems, W.; Zarkovic, K.; Zarkovic, N. Advances in methods for the determination of biologically relevant lipid peroxidation products. *Free Radic. Res.* **2010**, *44*, 1172–1202. [CrossRef] [PubMed]

22. Spickett, C.M. The lipid peroxidation product 4-hydroxy-2-nonenal: Advances in chemistry and analysis. *Redox Biol.* **2013**, *1*, 145–152. [CrossRef] [PubMed]

23. Aslebagh, R.; Pfeffer, B.A.; Fliesler, S.J.; Darie, C.C. Mass spectrometry-based proteomics of oxidative stress: Identification of 4-hydroxy-2-nonenal (HNE) adducts of amino acids using lysozyme and bovine serum albumin as model proteins. *Electrophoresis* **2016**, *37*, 2615–2623. [CrossRef] [PubMed]

24. Aldini, G.; Gamberoni, L.; Orioli, M.; Beretta, G.; Regazzoni, L.; Facino, R.M.; Carini, M. Mass spectrometric characterization of covalent modification of human serum albumin by 4-hydroxy-trans-2-nonenal. *J. Mass Spectrom.* **2006**, *41*, 1149–1161. [CrossRef] [PubMed]

25. Tirosh, O.; Shpaizer, A.; Kanner, J. Lipid Peroxidation in a Stomach Medium Is Affected by Dietary Oils (Olive/Fish) and Antioxidants: The Mediterranean versus Western Diet. *J. Agric. Food Chem.* **2015**, *63*, 7016–7023. [CrossRef] [PubMed]

26. Negre-Salvayre, A.; Auge, N.; Ayala, V.; Basaga, H.; Boada, J.; Brenke, R.; Chapple, S.; Cohen, G.; Feher, J.; Grune, T.; et al. Pathological aspects of lipid peroxidation. *Free Radic. Res.* **2010**, *44*, 1125–1171. [CrossRef] [PubMed]

27. Abadi, A.T.B. Strategies used by helicobacter pylori to establish persistent infection. *World J. Gastroenterol.* **2017**, *23*, 2870–2882. [CrossRef] [PubMed]

28. Backert, S.; Schmidt, T.P.; Harrer, A.; Wessler, S. Exploiting the gastric epithelial barrier: Helicobacter pylori's attack on tight and adherens junctions. In *Current Topics in Microbiology and Immunology*; Springer: Berlin, Germany, 2017; ISBN 9783319505206.

29. Barrera, G.; Pizzimenti, S.; Ciamporcero, E.S.; Daga, M.; Ullio, C.; Arcaro, A.; Cetrangolo, G.P.; Ferretti, C.; Dianzani, C.; Lepore, A.; et al. Role of 4-Hydroxynonenal-Protein Adducts in Human Diseases. *Antioxid. Redox Signal.* **2015**, *22*, 1681–1702. [CrossRef] [PubMed]

30. Mol, M.; Regazzoni, L.; Altomare, A.; Degani, G.; Carini, M.; Vistoli, G.; Aldini, G. Enzymatic and non-enzymatic detoxification of 4-hydroxynonenal: Methodological aspects and biological consequences. *Free Radic. Biol. Med.* **2017**, *111*, 328–344. [CrossRef] [PubMed]

31. Cid, T.P.; Fernández, M.C.; Benito Martínez, S.; Jones, N.L. Pathogenesis of Helicobacter pylori infection. *Helicobacter* **2013**, *18* (Suppl. 1), 12–17. [CrossRef]

32. Cherkas, A.; Abrahamovych, O.; Golota, S.; Nersesyan, A.; Pichler, C.; Serhiyenko, V.; Knasmüller, S.; Zarkovic, N.; Eckl, P. The correlations of glycated hemoglobin and carbohydrate metabolism parameters with heart rate variability in apparently healthy sedentary young male subjects. *Redox Biol.* **2015**, *5*, 301–307. [CrossRef] [PubMed]

33. Cherkas, A.; Zhuraev, R. A marked decrease in heart rate variability in marfan syndrome patients with confirmed FBN1 mutations. *Cardiol. J.* **2016**, *23*, 23–33. [CrossRef] [PubMed]

34. Levenstein, S.; Rosenstock, S.; Jacobsen, R.K.; Jorgensen, T. Psychological stress increases risk for peptic ulcer, regardless of Helicobacter pylori infection or use of nonsteroidal anti-inflammatory drugs. *Clin. Gastroenterol. Hepatol.* **2015**, *13*, 495–506. [CrossRef] [PubMed]

35. Haensel, A.; Mills, P.J.; Nelesen, R.A.; Ziegler, M.G.; Dimsdale, J.E. The relationship between heart rate variability and inflammatory markers in cardiovascular diseases. *Psychoneuroendocrinology* **2008**, *33*, 1305–1312. [CrossRef] [PubMed]

36. Nomura, M.; Yukinaka, M.; Miyajima, H.; Nada, T.; Kondo, Y.; Okahisa, T.; Shibata, H.; Okamura, S.; Honda, H.; Shimizu, I.; et al. Is autonomic dysfunction a necessary condition for chronic peptic ulcer formation? *Aliment. Pharmacol. Ther.* **2000**, *14* (Suppl. 1), 82–86. [CrossRef] [PubMed]

37. Wehrens, S.M.T.; Hampton, S.M.; Skene, D.J. Heart rate variability and endothelial function after sleep deprivation and recovery sleep among male shift and non-shift workers. *Scand. J. Work Environ. Health* **2012**, *38*, 171–181. [CrossRef] [PubMed]

38. Yelisyeyeva, O.; Cherkas, A.; Semen, K.; Kaminskyy, D.; Lutsyk, A. Study of aerobic metabolism parameters and heart rate variability and their correlations in elite athletes: A modulatory effect of amaranth oil. *Clin. Exp. Med. J.* **2009**, *3*, 293–307. [CrossRef]

39. Matteoli, G.; Boeckxstaens, G.E. The vagal innervation of the gut and immune homeostasis. *Gut* **2013**, *62*, 1214–1222. [CrossRef] [PubMed]

40. Hoover, D.B. Cholinergic modulation of the immune system presents new approaches for treating inflammation. *Pharmacol. Ther.* **2017**, *179*, 1–16. [CrossRef] [PubMed]

41. Steinhorn, B.; Sartoretto, J.L.; Sorrentino, A.; Romero, N.; Kalwa, H.; Abel, E.D.; Michel, T. Insulin-dependent metabolic and inotropic responses in the heart are modulated by hydrogen peroxide from NADPH-oxidase isoforms NOX2 and NOX4. *Free Radic. Biol. Med.* **2017**, *113*, 16–25. [CrossRef] [PubMed]

42. Contreras, C.; Nogueiras, R.; Diéguez, C.; Rahmouni, K.; López, M. Traveling from the hypothalamus to the adipose tissue: The thermogenic pathway. *Redox Biol.* **2017**, *12*, 854–863. [CrossRef] [PubMed]

43. Cherkas, A.; Yatskevych, O. The amplitude of heart rate oscillations is dependent on metabolic status of sinoatrial node cells. *OA Med. Hypothesis* **2014**, *2*, 1–8.

44. Gao, L.; Zimmerman, M.C.; Biswal, S.; Zucker, I.H. Selective NRF2 Gene Deletion in the Rostral Ventrolateral Medulla Evokes Hypertension and Sympathoexcitation in Mice. *Hypertension* **2017**, *69*, 1198–1206. [CrossRef] [PubMed]

45. Polyzos, S.A.; Kountouras, J. Novel Advances in the Association Between Helicobacter pylori Infection, Metabolic Syndrome, and Related Morbidity. *Helicobacter* **2015**, *20*, 405–409. [CrossRef] [PubMed]

46. Chen, L.-W.; Chien, C.-Y.; Yang, K.-J.; Kuo, S.-F.; Chen, C.-H.; Chien, R.-N. Helicobacter pylori Infection Increases Insulin Resistance and Metabolic Syndrome in Residents Younger than 50 Years Old: A Community-Based Study. *PLoS ONE* **2015**, *10*, e0128671. [CrossRef] [PubMed]

47. Cherkas, A.; Eckl, P.; Gueraud, F.; Abrahamovych, O.; Serhiyenko, V.; Yatskevych, O.; Pliatsko, M.; Golota, S. Helicobacter pylori in sedentary men is linked to higher heart rate, sympathetic activity, and insulin resistance but not inflammation or oxidative stress. *Croat. Med. J.* **2016**, *57*, 141–149. [CrossRef] [PubMed]

48. Cardaropoli, S.; Rolfo, A.; Todros, T. Helicobacter pylori and pregnancy-related disorders. *World J. Gastroenterol.* **2014**, *20*, 654–664. [CrossRef] [PubMed]

49. Franceschi, F.; Gasbarrini, A.; Polyzos, S.A.; Kountouras, J. Extragastric Diseases and Helicobacter pylori. *Helicobacter* **2015**, *20* (Suppl. 1), 40–46. [CrossRef]

50. Hardbower, D.M.; de Sablet, T.; Chaturvedi, R.; Wilson, K.T. Chronic inflammation and oxidative stress: The smoking gun for Helicobacter pylori-induced gastric cancer? *Gut Microbes* **2013**, *4*, 475–481. [CrossRef] [PubMed]

51. Handa, O.; Naito, Y.; Yoshikawa, T. Helicobacter pylori: A ROS-inducing bacterial species in the stomach. *Inflamm. Res.* **2010**, *59*, 997–1003. [CrossRef] [PubMed]

52. Cherkas, A.; Golota, S.; Guéraud, F.; Abrahamovych, O.; Pichler, C.; Nersesyan, A.; Krupak, V.; Bugiichyk, V.; Yatskevych, O.; Pliatsko, M.; et al. A Helicobacter pylori-associated insulin resistance in asymptomatic sedentary young men does not correlate with inflammatory markers and urine levels of 8-iso-PGF2α or 1,4-dihydroxynonane mercapturic acid. *Arch. Physiol. Biochem.* **2018**, *124*, 275–285. [CrossRef] [PubMed]

53. Yeniova, A.O.; Uzman, M.; Kefeli, A.; Basyigit, S.; Ata, N.; Dal, K.; Guresci, S.; Nazligul, Y. Serum 8 Hydroxydeoxyguanosine and Cytotoxin Associated Gene A as Markers for Helicobacter pylori Infection. *Asian Pac. J. Cancer Prev.* **2015**, *16*, 5199–5203. [CrossRef] [PubMed]

54. Bhattacharyya, A.; Chattopadhyay, R.; Mitra, S.; Crowe, S.E. Oxidative Stress: An Essential Factor in the Pathogenesis of Gastrointestinal Mucosal Diseases. *Physiol. Rev.* **2014**, *94*, 329–354. [CrossRef] [PubMed]

55. Kwiecien, S.; Jasnos, K.; Magierowski, M.; Sliwowski, Z.; Pajdo, R.; Brzozowski, B.; Mach, T.; Wojcik, D.; Brzozowski, T. Lipid peroxidation, reactive oxygen species and antioxidative factors in the pathogenesis of gastric mucosal lesions and mechanism of protection against oxidative stress—Induced gastric injury. *J. Physiol. Pharmacol.* **2014**, *65*, 613–622. [PubMed]

56. Awasthi, Y.C.; Ramana, K.V.; Chaudhary, P.; Srivastava, S.K.; Awasthi, S. Regulatory roles of glutathione-S-transferases and 4-hydroxynonenal in stress-mediated signaling and toxicity. *Free Radic. Biol. Med.* **2017**, *111*, 235–243. [CrossRef] [PubMed]

57. Jaganjac, M.; Cacev, T.; Cipak, A.; Kapitanović, S.; Gall Troselj, K.; Zarković, N. Even stressed cells are individuals: Second messengers of free radicals in pathophysiology of cancer. *Croat. Med. J.* **2012**, *53*, 304–309. [CrossRef] [PubMed]

58. Bauer, G.; Zarkovic, N. Revealing mechanisms of selective, concentration-dependent potentials of 4-hydroxy-2-nonenal to induce apoptosis in cancer cells through inactivation of membrane-associated catalase. *Free Radic. Biol. Med.* **2015**, *81*, 128–144. [CrossRef] [PubMed]

59. Zhong, H.; Xiao, M.; Zarkovic, K.; Zhu, M.; Sa, R.; Lu, J.; Tao, Y.; Chen, Q.; Xia, L.; Cheng, S.; et al. Mitochondrial control of apoptosis through modulation of cardiolipin oxidation in hepatocellular carcinoma: A novel link between oxidative stress and cancer. *Free Radic. Biol. Med.* **2017**, *102*, 67–76. [CrossRef] [PubMed]

60. Živković, N.P.; Petrovečki, M.; Lončarić, Č.T.; Nikolić, I.; Waeg, G.; Jaganjac, M.; Žarković, K.; Žarković, N. Positron emission tomography-computed tomography and 4-hydroxynonenal-histidine immunohistochemistry reveal differential onset of lipid peroxidation in primary lung cancer and in pulmonary metastasis of remote malignancies. *Redox Biol.* **2017**, *11*, 600–605. [CrossRef] [PubMed]

61. Zagari, R.M.; Rabitti, S.; Eusebi, L.H.; Bazzoli, F. Treatment of Helicobacter pylori infection: A clinical practice update. *Eur. J. Clin. Investig.* **2018**, *48*. [CrossRef] [PubMed]

62. Buzás, G.M.; Gastroenterologist, C. Metabolic consequences of Helicobacter pylori infection and eradication. *World J. Gastroenterol.* **2014**, *20*, 5226–5234. [CrossRef] [PubMed]

63. Dogan, Z.; Sarikaya, M.; Ergul, B.; Filik, L. The effect of Helicobacter pylori eradication on insulin resistance and HbA1c level in people with normal glucose levels: A prospective study. *Biomed. Pap. Med. Fac. Univ. Palacký Olomouc Czechoslov.* **2015**, *159*, 242–245. [CrossRef] [PubMed]

64. Longo-Mbenza, B.; Nkondi Nsenga, J.; Vangu Ngoma, D. Prevention of the metabolic syndrome insulin resistance and the atherosclerotic diseases in Africans infected by Helicobacter pylori infection and treated by antibiotics. *Int. J. Cardiol.* **2007**, *121*, 229–238. [CrossRef] [PubMed]

65. Kumar, D.; Anjan, D.; Dattagupta, S.; Ahuja, V.; Mathur, M.; Sharma, M.P. Pre and post eradication gastric inflammation in Helicobacter pylori-associated duodenal ulcer. *Indian J. Gastroenterol.* **2002**, *21*, 7–10. [PubMed]

66. Lanas, A.; Chan, F.K.L. Peptic ulcer disease. *Lancet* **2017**, *390*, 613–624. [CrossRef]

67. Hardbower, D.M.; Peek, R.M.; Wilson, K.T. At the Bench: Helicobacter pylori, dysregulated host responses, DNA damage, and gastric cancer. *J. Leukoc. Biol.* **2014**, *96*, 201–212. [CrossRef] [PubMed]

68. Bae, S.E.; Choi, K.D.; Choe, J.; Kim, S.O.; Na, H.K.; Choi, J.Y.; Ahn, J.Y.; Jung, K.W.; Lee, J.; Kim, D.H.; et al. The effect of eradication of *Helicobacter pylori* on gastric cancer prevention in healthy asymptomatic populations. *Helicobacter* **2018**, *23*, e12464. [CrossRef] [PubMed]

69. Choi, I.J.; Kook, M.-C.; Kim, Y.-I.; Cho, S.-J.; Lee, J.Y.; Kim, C.G.; Park, B.; Nam, B.-H. *Helicobacter pylori* Therapy for the Prevention of Metachronous Gastric Cancer. *N. Engl. J. Med.* **2018**, *88*, 475–485. [CrossRef] [PubMed]

70. Eckl, P.M. Genotoxicity of HNE. *Mol. Asp. Med.* **2003**, *24*, 161–165. [CrossRef]

71. Gentile, F.; Arcaro, A.; Pizzimenti, S.; Daga, M.; Paolo Cetrangolo, G.; Dianzani, C.; Lepore, A.; Graf, M.; Ames, P.R.J.; Barrera, G. DNA damage by lipid peroxidation products: Implications in cancer, inflammation and autoimmunity. *AIMS Genet.* **2017**, *4*, 103–137. [CrossRef]

72. Gasparovic, A.C.; Milkovic, L.; Sunjic, S.B.; Zarkovic, N. Cancer growth regulation by 4-hydroxynonenal. *Free Radic. Biol. Med.* **2017**, *111*, 226–234. [CrossRef] [PubMed]

73. Ma, Y.; Zhang, L.; Rong, S.; Qu, H.; Zhang, Y.; Chang, D.; Pan, H.; Wang, W. Relation between gastric cancer and protein oxidation, DNA damage, and lipid peroxidation. *Oxid. Med. Cell. Longev.* **2013**, *2013*, 543760. [CrossRef] [PubMed]

74. Kwiecień, S.; Brzozowski, T.; Konturek, S.J. Effects of reactive oxygen species action on gastric mucosa in various models of mucosal injury. *J. Physiol. Pharmacol.* **2002**, *53*, 39–50. [PubMed]

75. Wallace, J.; Pshyk-Titko, I.; Muscara, M.N.; Bula, N.; Pavlovsky, Y.; Gavriluk, E.; Zayachkivska, O. Influence of hydrogen sulphide-releasing Aspirin on mucosal integrity of esophageal and gastric mucosa. *Proc. Shevchenko Sci. Soc.* **2015**, *43*, 63–74.

76. Kwiecien, S.; Magierowska, K.; Magierowski, M.; Surmiak, M.; Hubalewska-Mazgaj, M.; Pajdo, R.; Sliwowski, Z.; Chmura, A.; Wojcik, D.; Brzozowski, T. Role of sensory afferent nerves, lipid peroxidation and antioxidative enzymes in the carbon monoxide-induced gastroprotection against stress ulcerogensis. *J. Physiol. Pharmacol.* **2016**, *67*, 717–729. [PubMed]

77. Abrahamovych, O.; Cherkas, A.; Abrahamovych, U.; Abrahamovych, M.; Serhiyenko, V. *Heart Rate Variability: Physiological Bases, Clinical Importance and Peculiarities in Patients with Peptic Ulcer and after Resection of Stomach*; Abrahamovych, O., Ed.; Danylo Halytskyi Lviv National Medical University: Lviv, Ukraine, 2014; ISBN 978-966-7175-729.

78. O'Connor, A.; O'Morain, C.A.; Ford, A.C. Population screening and treatment of Helicobacter pylori infection. *Nat. Rev. Gastroenterol. Hepatol.* **2017**, *14*, 230–240. [CrossRef] [PubMed]

79. Kim, H.-G.; Bae, J.-H.; Jastrzebski, Z.; Cherkas, A.; Heo, B.-G.; Gorinstein, S.; Ku, Y.-G. Binding, Antioxidant and Anti-proliferative Properties of Bioactive Compounds of Sweet Paprika (*Capsicum annuum* L.). *Plant Foods Hum. Nutr.* **2016**, *71*, 129–136. [CrossRef] [PubMed]

80. Lushchak, V.I. Glutathione Homeostasis and Functions: Potential Targets for Medical Interventions. *J. Amino Acids* **2012**, *2012*, 736837. [CrossRef] [PubMed]

81. Semen, K.O.; Yelisyeyeva, O.P.; Kaminskyy, D.V.; Cherkas, A.P.; Zarkovic, K.; Lutsyk, O.; Cipak, A.; Jaganjac, M.; Zarkovic, N. Interval hypoxic training in complex treatment of Helicobacter pylori-associated peptic ulcer disease. *Acta Biochim. Pol.* **2010**, *57*, 199–208. [PubMed]

82. Ielisieieva, O.P.; Semen, K.O.; Cherkas, A.P.; Kaminsky, D.V.; Kurkevych, A.K. Specific mechanisms of individually adjusted interval hypoxic hypercapnic effects on heart rate variability in athletes. *Fiziol. Zhurnal* **2007**, *53*, 78–86.

83. Cherkas, A.; Golota, S. An intermittent exhaustion of the pool of glycogen in the human organism as a simple universal health promoting mechanism. *Med. Hypotheses* **2014**, *82*, 387–389. [CrossRef] [PubMed]

84. Gulevsky, A.K.; Abakumova, Y.S.; Moiseyeva, N.N.; Ivanov, Y.G. Influence of Cord Blood Fraction (below 5 kDa) on Reparative Processes during Subchronic Ulcerative Gastropathy. *Ulcers* **2011**, *2011*, 214124. [CrossRef] [PubMed]

85. Stelmakh, A.; Abrahamovych, O.; Cherkas, A. Highly purified calf hemodialysate (Actovegin®) may improve endothelial function by activation of proteasomes: A hypothesis explaining the possible mechanisms of action. *Med. Hypotheses* **2016**, *95*, 77–81. [CrossRef] [PubMed]

86. Wang, T.; Cai, H.; Sasazuki, S.; Tsugane, S.; Zheng, W.; Cho, E.R.; Jee, S.H.; Michel, A.; Pawlita, M.; Xiang, Y.B.; et al. Fruit and vegetable consumption, Helicobacter pylori antibodies, and gastric cancer risk: A pooled analysis of prospective studies in China, Japan, and Korea. *Int. J. Cancer* **2017**, *140*, 591–599. [CrossRef] [PubMed]

87. Shafaghi, A.; Pourkazemi, A.; Khosravani, M.; Fakhrie Asl, S.; Amir Maafi, A.; Atrkar Roshan, Z.; Abaspour Rahimabad, J. The Effect of Probiotic Plus Prebiotic Supplementation on the Tolerance and Efficacy of Helicobacter Pylori Eradication Quadruple Therapy: A Randomized Prospective Double Blind Controlled Trial. *Middle East J. Dig. Dis.* **2016**, *8*, 179–188. [CrossRef] [PubMed]

88. Ingold, K.U.; Pratt, D.A. Advances in radical-trapping antioxidant chemistry in the 21st century: A kinetics and mechanisms perspective. *Chem. Rev.* **2014**, *114*, 9022–9046. [CrossRef] [PubMed]

89. Gargallo, C.J.; Sostres, C.; Lanas, A. Prevention and Treatment of NSAID Gastropathy. *Curr. Treat. Opt. Gastroenterol.* **2014**, *12*, 398–413. [CrossRef] [PubMed]

antioxidants

MDPI

Review

Modulation of the Oxidative Stress and Lipid Peroxidation by Endocannabinoids and Their Lipid Analogues

Cristina Anna Gallelli [1,†] , Silvio Calcagnini [1,†] , Adele Romano [1], Justyna Barbara Koczwara [1] , Marialuisa de Ceglia [1] , Donatella Dante [1], Rosanna Villani [2], Anna Maria Giudetti [3] , Tommaso Cassano [4,*] and Silvana Gaetani [1]

[1] Department of Physiology and Pharmacology "V. Erspamer", Sapienza University of Rome, Piazzale Aldo Moro 5, 00185 Rome, Italy; cristinaanna.gallelli@uniroma1.it (C.A.G); silvio.calcagnini@uniroma1.it (S.C.); adele.romano@uniroma1.it (A.R.); justynabarbara.koczwara@uniroma1.it (J.B.K); marialuisa.deceglia@uniroma1.it (M.d.C.); donatella.dante@uniroma1.it (D.D.); silvana.gaetani@uniroma1.it (S.G.)

[2] C.U.R.E. University Centre for Liver Disease Research and Treatment, Department of Medical and Surgical Sciences, Institute of Internal Medicine, University of Foggia, 71122 Foggia, Italy; rosanna.villani@unifg.it

[3] Department of Biological and Environmental Sciences and Technologies, University of Salento, Via Monteroni, 73100 Lecce, Italy; anna.giudetti@unisalento.it

[4] Department of Clinical and Experimental Medicine, University of Foggia, Via Luigi Pinto, c/o Ospedali Riuniti, 71122 Foggia, Italy

* Correspondence: tommaso.cassano@unifg.it; Tel.: +39-0881-588-029

† These authors contributed equally to this study.

Received: 28 June 2018; Accepted: 13 July 2018; Published: 18 July 2018

Abstract: Growing evidence supports the pivotal role played by oxidative stress in tissue injury development, thus resulting in several pathologies including cardiovascular, renal, neuropsychiatric, and neurodegenerative disorders, all characterized by an altered oxidative status. Reactive oxygen and nitrogen species and lipid peroxidation-derived reactive aldehydes including acrolein, malondialdehyde, and 4-hydroxy-2-nonenal, among others, are the main responsible for cellular and tissue damages occurring in redox-dependent processes. In this scenario, a link between the endocannabinoid system (ECS) and redox homeostasis impairment appears to be crucial. Anandamide and 2-arachidonoylglycerol, the best characterized endocannabinoids, are able to modulate the activity of several antioxidant enzymes through targeting the cannabinoid receptors type 1 and 2 as well as additional receptors such as the transient receptor potential vanilloid 1, the peroxisome proliferator-activated receptor alpha, and the orphan G protein-coupled receptors 18 and 55. Moreover, the endocannabinoids lipid analogues N-acylethanolamines showed to protect cell damage and death from reactive aldehydes-induced oxidative stress by restoring the intracellular oxidants-antioxidants balance. In this review, we will provide a better understanding of the main mechanisms triggered by the cross-talk between the oxidative stress and the ECS, focusing also on the enzymatic and non-enzymatic antioxidants as scavengers of reactive aldehydes and their toxic bioactive adducts.

Keywords: oxidative stress; lipid peroxidation; reactive aldehydes; reactive oxygen and nitrogen species; free radicals; endocannabinoids; cannabinoid receptors; peroxisome proliferator-activated receptors; transient receptor potential vanilloid; G protein-coupled receptors

1. Introduction

Oxidative stress and lipid peroxidation are the consequences of a deregulated redox homeostasis that results in the accumulation of highly reactive molecules and cellular injury, especially in those tissues with a high oxygen consumption, such as heart, kidney, and brain, thus leading to cardiovascular [1,2], renal [3], and neurodegenerative diseases [4–6], just to mention a few. Examples of the possible repercussions of free radical damage are provided in this review with special emphasis on lipid peroxidation-derived reactive aldehydes including acrolein (ACR), malondialdehyde (MDA), and 4-hydroxy-2-nonenal (4-HNE), among others [7].

To get a deeper insight into the cellular pathways that regulate reactive oxygen and nitrogen species (ROS/RNS) as well as reactive aldehydes formation, there is a growing interest in identifying free radical scavenging molecules that can prevent cell death following oxidative stress-induced damage of cellular membranes. In this perspective, over the last few years, the endocannabinoid system (ECS) has attracted significant attention because of the existing cross-talk between endocannabinoids (ECs) as well as their lipid analogues and various redox-dependent processes. Therefore, the pathways by which the ECs and their lipid-related mediators contribute to the modulation of oxidative stress and lipid peroxidation represent a significant research area that will yield novel pharmaceutical strategies for the treatment of diseases characterized by a redox imbalance.

The cannabinoid receptors type 1 (CB1) and 2 (CB2), together with additional ECs receptor targets, take part in the complex ECS and, because of their wide distribution, they may play a role in mediating the antioxidant properties of ECs [8–10]. However, the great diversity of results in this field discloses the requirement of a better understanding on the pathways by which these receptors are involved in regulating oxidative stress and lipid peroxidation processes.

In this review, we will provide an overview of the role of the ECS in pathological conditions related to a redox status imbalance, leading to a better comprehension of the intricate routes that are associated to the antioxidant properties exerted by the ECs, thus enhancing the research in finding a therapeutic benefit for cannabinoid-based drugs in various redox-dependent disorders.

2. Oxidative Stress and Lipid Peroxidation

Oxidative stress can be described as an imbalance between the production of oxidant species and the antioxidant defenses, which may affect cellular redox homeostasis leading to molecular alterations and thus resulting in cell and tissue damage [11]. The term "oxidants" is a general term used to identify several groups of reactive molecules among which ROS and RNS are considered the most interesting from a biological point of view. ROS/RNS are natural byproducts of aerobic metabolism and are produced by all living multicellular organisms. ROS include free oxygen radicals and non-radical molecules, such as superoxide anion ($O_2\bullet^-$), hydroxyl ($\bullet OH$), peroxyl, alkyl, and alkoxyl radicals, as well as singlet oxygen ($^1O_2$), hydrogen peroxide (H_2O_2), ozone (O_3), and hypochlorous acid (HClO), while RNS include nitrogen compounds such as nitric oxide ($\bullet NO$), nitrogen dioxide ($NO_2\bullet$), nitrate (NO_3^-), nitrite (NO_2^-), and peroxynitrite ($ONOO^-$) [12,13].

In mammals, the main cellular sources of ROS/RNS are the mitochondrial and microsomal electron transport chains [14], the NADPH oxidase enzymes (NOXs), which consist of seven isoforms with various tissue distributions and mechanisms of activation [15,16], the flavoenzyme endoplasmic reticulum oxireductin 1 [17], nitric oxide synthase (NOS) [18], cytochrome P450 enzymes [19], cyclooxygenases (COXs), lipoxygenases (LOXs) [20], xanthine oxidase [21], diamine oxidase [22], and prostaglandin synthase [23]. In addition to these endogenous sources, the ionizing radiation, ultraviolet rays, pathogens, xenobiotics (e.g., drugs, herbicides, fungicides, trace metals, etc.), and environmental pollutants (e.g., smog, cigarette smoke, smoke from wood combustion, etc.) are identified as exogenous sources of ROS/RNS [24], which may seriously alter the fundamental oxidants-antioxidants balance.

To date, growing evidence confirms that ROS/RNS are produced by healthy cells in a highly regulated fashion in order to maintain the intracellular redox homeostasis. Moreover, ROS/RNS regulate several cellular functions ranging from immune defense to gene expression regulation, thus acting as reactive molecules secreted against circulating pathogens [25] or as second messengers of specific signaling pathways [26]. The crucial role played by ROS/RNS in immune defense was demonstrated by the discovery of the chronic granulomatous disorder (CGD), a hereditary disease characterized by NOX type 2 (NOX2)-defective phagocytes [27] which are unable to produce ROS/RNS. This genetic defect leads CGD patients in developing a primary immunodeficiency due to the inability of host innate defense to kill and digest ingested pathogens such as bacterial and fungal cells [28–31]. Moreover, ROS/RNS play also an important role in the cardiovascular system because of their ability to regulate blood pressure. In particular, the endothelial NOX2 isoform regulates the release of •NO, the endothelium-derived relaxing factor, which modulates the caliber of blood vessels, through the production of $O_2•^-$. In hypertension and other vascular pathologies, NOX2 seems to be up-regulated leading to a reduced •NO bioavailability and to the consequent oxidants-antioxidants imbalance in the endothelium, further worsening the oxidative state [32–34]. Moreover, in vivo studies of single nephron function and in vitro studies performed on perfused juxtaglomerular apparatus preparation demonstrated that also the normal renal functions are modulated by ROS/RNS. In particular, $O_2•^-$ and •NO, which are generated by NOX type 3 (NOX3) and NOS type 1 (NOS1) enzymes, respectively, modulate afferent arteriolar tone and control Na^+ reabsorption and renal oxygenation by regulating the tubuloglomerular feedback response [35–37]. Furthermore, in the loop of Henle, ROS/RNS increase the absorption of NaCl by modulating the activity of the Na^+/H^+ exchanger [38,39]. In airway and pulmonary artery smooth muscle cells of the lung, NOX2-generated ROS/RNS act as signaling intermediates, which regulate the proliferation and differentiation by the activation of the nuclear factor-κB (NF-κB) and NOS2, and they further show an important role in O_2 sensing [39–41].

Moreover, ROS/RNS formation by mucosal cells of the colon seems to modulate the serotonin production by enterochromaffin cells through a NOXs-dependent system, thus contributing to the regulation of serotonin secretion as well as intestinal motility [42]. ROS/RNS have also a fundamental role in the central nervous system (CNS), in particular in central autonomic neurons. To this regard, ROS/RNS produced by NOX2 in the nucleus of the solitary tract, in the hypothalamic paraventricular nucleus, and in the subfornical organ modulate angiotensin II signaling, thus contributing to the regulation of cardiovascular homeostasis [43,44]. Moreover, in microglia but not in astrocytes, H_2O_2 formation by NOX2 enzyme is involved in the regulation of cell proliferation [45].

Beyond the role as signaling molecules, it has been shown that the aberrant ROS/RNS formation is the leading cause of cell and tissue oxidative stress-induced damage. Indeed, it is well known that excessive levels of ROS/RNS may directly damage lipids containing carbon-carbon double bounds such as cholesterol, glycolipids, phospholipids, and polyunsatured fatty acids (PUFAs), which are abundant within cellular membranes. To this regard, free radical–mediated lipid peroxidation of PUFAs is one of the main mechanisms by which ROS/RNS induce the generation of reactive aldehydes [46]. Due to their abundance of reactive hydrogens, PUFAs are more oxidation-prone lipids compared to monounsaturated fatty acids. PUFAs include the ω-3 (e.g., linolenic acid, eicosapentaenoic acid and docosahexaenoic acid) and ω-6 (e.g., linoleic acid and arachidonic acids) fatty acids.

Lipid peroxidation is a chain reaction, which, once started, proceeds through three main steps referred to initiation, propagation and termination [47]. Moreover, lipid peroxidation may occur by several mechanisms: (1) free radical-mediated oxidation [47], (2) enzymatic oxidation, and (3) spontaneous oxidation [48]. In this review, we will focus mainly on the free radical–mediated mechanisms that lead to the formation of reactive aldehydes from PUFAs. In particular, the free radical-mediated oxidation of PUFAs occurs through the following reactions: (1) During the initiation phase, ROS/RNS free radicals attack PUFAs ripping off one hydrogen atom, leading to lipid radicals formation. (2) During the propagation phase, lipid radicals react with oxygen molecules, thus producing peroxyl radicals, which, in turn, react with nearby lipids resulting in the formation of new lipid radicals and lipid hydroperoxides. Due to their high instability, lipid hydroperoxides are further degraded into reactive secondary products, such as ACR, MDA, 4-HNE and other reactive aldehydes [7]. (3) During the termination phase, peroxyl radicals may react with other radicals thus generating less reactive compounds, which block the propagation phase (Figure 1) [49].

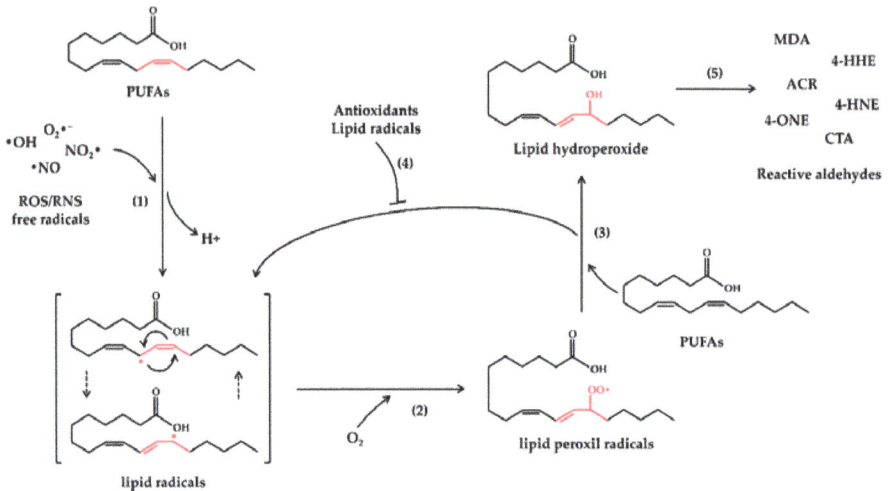

Figure 1. Schematic diagram of the free radicals-mediated peroxidation of polyunsatured fatty acids (PUFAs). ROS/RNS: reactive oxygen and nitrogen species; ACR: acrolein; MDA: malondialdehyde; CTA: crotonaldehyde; 4-HNE: 4-hydroxy-2-nonenal; 4-HHE: 4-hydroxy-hexanal; 4-ONE: 4-oxo-nonenal. During the initiation phase (1), ROS/RNS free radicals react with PUFAs and rip off an allylic hydrogen thus forming lipid radicals. Generally, lipid radicals tend to be stabilized by a molecular rearrangement. (2) In the propagation phase, lipid radicals react with oxygen to form lipid peroxyl radicals, which in turn react with PUFAs or other nearby lipids resulting in the formation of new lipid radicals and lipid hydroperoxides (3). During the termination phase (4), antioxidants or lipid radicals block the propagation phase by donating a hydrogen atom to lipid peroxyl radicals resulting in the formation of non-radical products. Nevertheless, lipid hydroperoxides are highly unstable therefore they are further degraded into reactive secondary products such as ACR, MDA, 4-HNE, and other reactive aldehydes (5).

Today, it is well accepted that oxidative stress and lipid peroxidation are key features in the pathogenesis of several disorders. Indeed, it has been reported that lipid peroxidation products may interfere in vivo with several biological processes, such as substrate-receptor interaction, signal transduction, gene expression, and homeostatic responses to intracellular and environmental stimuli [50–53]. Currently, the main objective of research focused on oxidative stress, lipid peroxidation, and reactive aldehydes is the characterization of the pathogenic mechanisms in several disorders as well as the identification of specific biomarkers for diseases.

Among the reactive aldehydes, the most frequently studied are ACR, MDA, 4-HNE, 4-hydroxy-hexanal (4-HHE), 4-oxo-nonenal (4-ONE), and crotonaldehyde (CTA) (Figure 2).

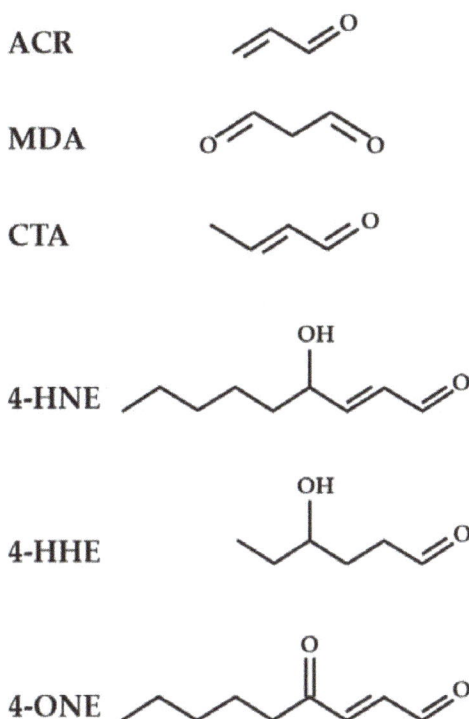

Figure 2. Chemical structures of the main reactive aldehydes produced by lipid peroxidation. ACR: acrolein; MDA: malondialdehyde; CTA: crotonaldehyde; 4-HNE: 4-hydroxy-2-nonenal; 4-HHE: 4-hydroxy-hexanal; 4-ONE: 4-oxo-nonenal.

Some of these compounds are known to contribute to the pathogenesis of several diseases, such as atherosclerosis, rheumatoid arthritis, neuropsychiatric disorders, heart disease, cellular reperfusion injury, cancer, and metabolic disorders such as diabetes and hepatic diseases [4,5,7,12,54]. Reactive aldehydes are a group of electrophilic molecules with different features: some of them are very unstable, characterized by a short half-life, while others are long-lived and highly reactive. In the past years, the endogenous formation of reactive aldehydes has drawn great interest. The ability of aldehydes to easily diffuse across biological membranes [55], and to form adducts with macromolecules such as phospholipids, nucleic acids and proteins [7,46,56–58], is of particular concern. Adducts consist of covalent modifications, which involve the formation of Schiff bases or Michael addition reactions. To this regard, the reactive aldehydes toxicity against peptides and proteins is due to their ability to alter their structure and/or function through the formation of cross-links between different amino acid chains, thus potentially leading to the production of aberrant protein aggregates (Figure 3) [59]. Concerning the toxicity of reactive aldehydes against DNA, it has been shown that these compounds may react against nucleobases, among which the most affected is guanine, due to its chemical structure prone to oxidative modifications. The most studied DNA modifications caused by reactive aldehydes are the exocyclic adducts (Figure 4) [57,58,60].

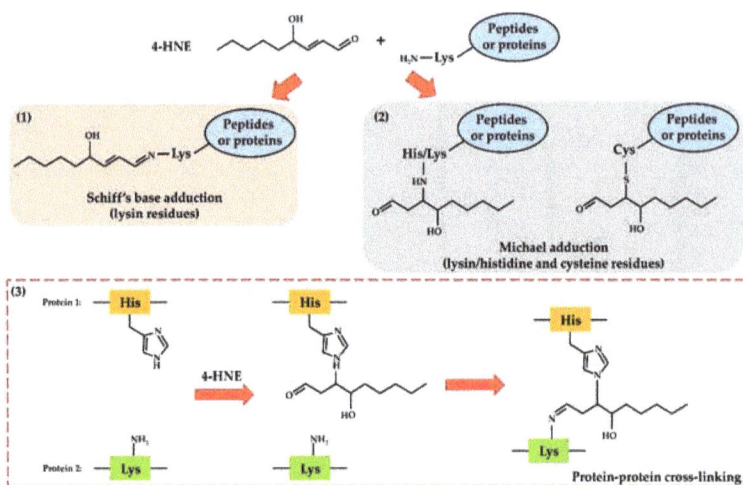

Figure 3. Schematic representation of protein adducts formation and protein-protein cross-linking by 4-HNE. Reactive aldehydes are able to modify peptides/proteins by the formation of toxic adducts which may alter the structure and/or the function of targeted peptides/proteins. These adducts consist of covalent modifications which occur through the formation of Schiff bases or through Michael addition reactions: (1) Schiff base formation on primary amine (lysine residue) through the reaction between peptides/proteins and 4-HNE, (2) Michael addition of 4-HNE on amino groups (lysine/histidine residues) or thiols (cysteine residue) through the reaction between peptides/proteins and 4-HNE, and (3) Protein-protein cross-linking through the reaction between 4-HNE with histidine and lysine residues from different peptides/proteins.

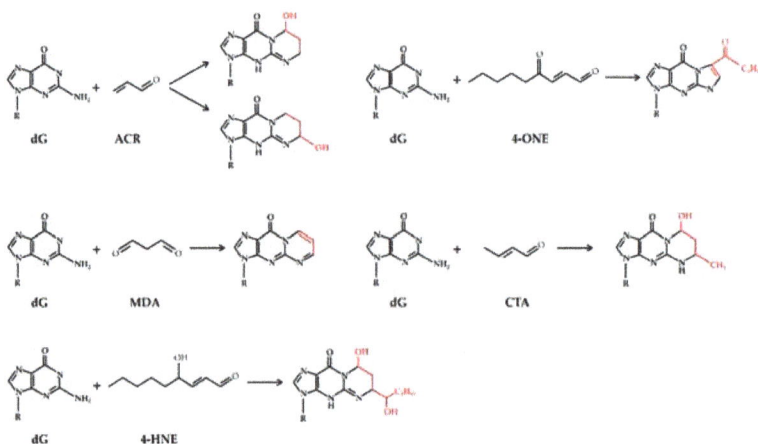

Figure 4. Hypothetical DNA adducts produced by reactive aldehydes.By reacting with DNA, in particular with the deoxyguanosine nucleobases, several reactive aldehydes such as ACR, MDA, 4-HNE, 4-ONE and CTA produce DNA modifications named exocyclic adducts that alter the DNA structure and, if not correctly repaired, may produce carcinogenic effects.

4-HNE and 4-ONE are generated from lipid peroxidation of ω-6 PUFAs (e.g., arachidonic acid and linoleic acid) [61]. Among reactive aldehydes, 4-HNE is the most studied, and its toxic effects can be explained by its ability to form protein adducts by reacting with thiols and amino groups of cysteine, histidine, and lysine amino acid residues [62]. For a detailed explanation of the main 4-HNE-modified proteins, see the following publication [63]. 4-ONE is an electrophilic compound that reacts both in vitro and in vivo with nucleobases, in particular with 2′-deoxyadenosine and 2′-deoxycytidine [64–67].

Unlike 4-HNE and 4-ONE, 4-HHE is generated from ω-3 PUFAs (e.g., docoshexaenoic acid, eicosapentaenoic acid and linolenic acid) and, because of its chemical structure, it is considered a soft electrophil with a lower reactivity compared to 4-HNE [7].

MDA, which is widely used as a marker of lipid peroxidation [68], contains at least two unsaturations [7] and is generally produced by PUFAs. Regarding its toxicity, MDA modifies target proteins through the formation of Schiff base complexes, which occur on the amino groups of lysine, histidine, arginine, glutamine, and asparagine amino acid residues as well as on the N-terminal of peptide chains [69]. For a detailed explanation of the main MDA-modified proteins, see the following publication [63]. Moreover, in vitro mutagenicity of MDA has been observed by several authors using the Salmonella tiphimurium assay [70–72]. Several studies showed the presence of both MDA and MDA-protein adducts in rheumatoid arthritis patients compared to healthy controls [73–76]. Moreover, high levels of circulating autoantibodies against MDA-modified epitopes have been detected in serum or plasma of patients affected by rheumatoid arthritis [77–79]. CTA or 2-butenal is a carcinogenic aldehyde formed by lipid peroxidation, which is also commonly found in air pollution, in cigarette smoke and in other combustion processes. CTA is able to form adducts with DNA [80–83] and proteins. In accordance with in vitro mutagenesis assay with Salmonella typhimorium, CTA is a mutagenic compound [84] able to induce hepatocellular carcinoma in rats [85]. About protein modifications, CTA reacts preferentially with lysine and histidine amino acid residues, thus forming β-substituted butanal adducts [86].

ACR or propenal is a metabolite of PUFAs lipid peroxidation, but it is also a ubiquitous environmental pollutant by-product derived by incomplete combustion of organic matter and plastic, cigarette smoke, overheated cooking oils, as well as by anticancer treatment with cyclophosphamide [87,88]. Among reactive aldehydes, ACR is the strongest electrophile, which shows a high reactivity with cysteine, histidine, and lysine amino acid residues. Moreover, ACR forms cyclic adducts with nucleosides in vitro, and is recognized as a potent mutagen [89].

Despite their harmful properties, growing evidence has also demonstrated the hormetic effects of reactive aldehydes [51,63,90–93]. The term "hormesis" refers to a highly conserved and dose-dependent response of biological systems in which low doses of noxious stimuli activate an adaptive response that increases the functionality and/or resistance of the systems to more severe stress. Conversely, high doses of noxious stimuli cause inhibition or detrimental effects [94]. To this regard, low levels of reactive aldehydes may modulate cell signaling, cellular proliferation and many other processes [7,61,89,95]. A typical example is represented by 4-HNE, which may also act as a signaling molecule by modulating the activity of different stress-related transcription factors, such as nuclear factor-erythroid 2-related factor 2 (Nrf2), activating protein-1, NF-κB, and peroxisome proliferator-activated receptors (PPARs) [96–100]. Moreover, low levels of 4-HNE may stimulate the activity of protein kinase C (PKC), may increase cell proliferation, and the expression of cyclooxygenase type 2 (COX-2) and prostaglandin E2 (PGE2) [51].

3. The Endocannabinoid System: Endocannabinoids, Their Lipid Analogues, and the Receptors

Over the last years, the ECS has attracted considerable attention as a signaling system because of its emerging regulatory functions in health and disease.

Several components jointly make up the ECS, and they specifically consist of (1) the ECs, endogenous bioactive lipid mediators generated in the brain and in several peripheral tissues; (2) two membrane G-protein-coupled receptors (GPCRs) referred to as CB1 and CB2, and others, not yet identified, receptors; and (3) several proteins implicated in the biosynthesis, release, transport, and degradation of these lipid mediators [101].

N-arachidonoyl-ethanolamine or anandamide (AEA) and 2-arachidonoyl-glycerol (2-AG), both derived from the arachidonic acid, are the best characterized members of the main families of ECs (*N*-acylethanolamines (NAEs) and monoacylglycerols (MAG), respectively) and exert their biological effects by interacting with CB1 and/or CB2 receptors [102]. AEA, an endogenous eicosanoid derivative isolated from pig brain in 1992, was the first EC to be identified [103], and it is well known to modulate several physiological functions being present in the autonomic and in the CNS as well as in the gastrointestinal tract and in the cardiovascular, immune and reproductive systems [104,105].

The second EC ligand to be discovered was 2-AG [106], which has been identified in brain and reproductive tissues in higher concentrations compared to AEA [107–109]. Moreover, 2-AG has also been found in the heart, endothelial cells and circulating cells such as macrophages and platelets [104].

Even though AEA and 2-AG interact with both CB1 and CB2 [110], they show different affinity and efficacy. In particular, depending of on the specific tissue, AEA can be either a partial or a full agonist of CB1, whereas it shows a low overall efficacy for CB2, for which it is a relatively weak ligand [111]. On the contrary, 2-AG appears to be a full agonist of both receptors [112] showing higher CB1 and CB2 efficacy than AEA.

Unlike what has been thought for many years, CB1 expression is not restricted to the brain, where it represents the most abundant of all GPCRs [113,114], but it has been also identified, albeit at much lower concentrations, in various peripheral tissues and cell types including adipose tissue, liver, skeletal muscle, kidney, bone, pancreas, myocardium, human coronary artery endothelial and smooth muscle cells and inflammatory cells (macrophages, lymphocytes) [104,115,116].

In the brain, CB1 is widely present in cerebral cortex, hippocampus, caudate-putamen, substantia nigra pars reticulata, globus pallidus, entopeduncular nucleus, and cerebellum [117]. Interestingly, accumulating evidence supports a new mechanism of action of CB1 signalling in the brain, since it has been found in mitochondria, where it probably modulates neuronal energy homeostasis [118]. On the other hand, the CB2, also known as the "immune cannabinoid receptor", is primarily expressed in immune and hematopoietic cells. However, its presence has also been established at lower, although functionally relevant, levels in the brain, liver, gut, exocrine and endocrine pancreas, reproductive cells, bone, myocardium, human coronary endothelial and smooth muscle cells, and inflammatory cells (e.g., lymphocytes, macrophages, neutrophils) [104,115,119].

CB1 and CB2 are seven-transmembrane-domain proteins both coupled with $G_{\alpha i/o}$ proteins, which inhibit adenylyl cyclase (AC) leading to a reduced protein kinase A (PKA) and PKC activity and to the consequent inhibition of voltage-gated Ca^{2+} channels and activation of inwardly rectifying K^+ currents [120]. Furthermore, through a common pathway mediated by $G_{\alpha o}$ proteins, CB1 and CB2 are also able to modulate Ras-related protein (Rap) (a member of the Ras small G protein family) and, in particular, it has been postulated that the activation of $G_{\alpha o}$ would release Rap1 guanosine triphosphatase (GTPase) activating protein (Rap1 GAP), which then would be free to inhibit the activity of Rap [121]. Moreover, several observations demonstrated that, depending on the CB1 agonist, this receptor could also interact with $G_{\alpha s}$ proteins [122,123].

On the basis of the cell type, the signaling of CB1 and CB2 may also involve G protein independent mechanisms, leading to the activation of mitogen-activated protein kinases (MAPKs) including p38- and p44/42-MAPKs, c-Jun *N*-terminal kinase (JNK), PKA and PKC, COX-2, and ceramide signaling [124–126].

However, beyond binding the CB1 and CB2 there is increasing pharmacological evidence for additional receptor targets for ECs [127], such as the transient receptor potential vanilloid 1 (TRPV1) [127–129], the PPARs family [130,131] and the orphan G protein-coupled receptors 119 (GPR119), 55 (GPR55) and 18 (GPR18) [132]. TRPV1 is a member of the vanilloid transient receptor potential cation channel subfamily, abundantly expressed in the cardiovascular system, peripheral nervous system, CNS and in epithelial cells of the bladder and the gastrointestinal tract. It is known to act by activating PKA and the endothelial nitric oxide synthase (eNOS), thus stimulating the production of •NO and the release of calcitonin gene-related peptide and substance P [133,134], which, in turn, lead to the altered ion permeability [135].

The finding that some pharmacological actions of AEA can be mediated by the activation of TRPV1 suggests the capability of this endogenous lipid compound to act as an "endovanilloid" [136,137], although AEA induces typical TRPV1-mediated effects with a lower affinity compared to CB1 [127].

PPARs are a family of transcription factors constituted by three different isoforms (α, β/δ, and γ), widely expressed in tissues with a higher oxidative capacity such as the cardiovascular system and, in particular, cardiomyocytes, endothelial cells, and vascular smooth muscle cells [104], but also in several brain areas and in peripheral tissues such as kidney and liver [138].

After being activated by a ligand, PPARs stimulate gene expression by creating heterodimers with the retinoid X receptor (RXR), thereby binding to specific peroxisome proliferator response elements (PPREs) in the promotor region of target genes [139]. They are involved in different biological processes, such as energy homeostasis, lipid and lipoprotein metabolism, cell proliferation and inflammation, blood pressure control and hypertensive-related complications, such as stroke and renal damage [140,141]. Furthermore, among the different members of the PPARs family, PPAR-α is recently attracting great attention for its anti-oxidative properties [142].

Moreover, AEA has been shown to exert anti-inflammatory and analgesic actions, and to control feeding behavior by activating the isoform α and γ of PPARs receptors [130,143,144]. Unlike AEA, 2-AG has no affinity for TRPV1 and is only able to activate PPARs [144,145].

As above mentioned, additional GPCRs were suggested to participate in non-CB1/CB2-mediated actions of ECs including the GPR18, GPR119 and GPR55 [146].

The GPR18, widely expressed in the cardiovascular system, CNS, spleen, and testis, is coupled with $G_{\alpha i/o}$ proteins whose activation results in the AC inhibition and in the modulation of the PI3K/Akt and extracellular signal-related kinases (ERK 1/2) pathways [104]. The $G_{\alpha s}$ coupled-GPR119, primarily expressed in human and rodent pancreas, foetal liver, gastrointestinal tract and in rodent brain, stimulates AC leading to increased intracellular adenosine 3′,5′-cyclic monophosphate (cAMP) levels, thus regulating incretin and insulin hormone secretion [147].

Finally, the GPR55, which is expressed in human brain and liver, but also in rat spleen, vasculature, intestine, foetal tissues, decidua, and placenta, is coupled with $G_{\alpha 12/13}$ proteins and increases intracellular Ca^{2+} via the activation of RhoGTPase nucleotide exchange factors (RhoGEFs) [148].

Different from the classical neurotransmitters, the ECs are not stored in intracellular vesicles but are synthesized "on demand" from membrane phospholipid precursors in response to stimuli that trigger an increase in intracellular Ca^{2+} levels [131], and then released from postsynaptic neurons to act on presynaptic CB1/CB2 through a retrograde mechanism [149,150]. However, recent findings suggested that AEA could be stored inside the cell into adiposomes, which are thought to connect plasma membrane to internal organelles along the metabolic route of this EC [151].

Although 2-AG and AEA are both derived from arachidonic acid, they do not share the same anabolic and catabolic enzymes [126]. Depending on the available precursors and the distinct physiological or pathological conditions [131], AEA can be synthesized by multiple routes. The main pathway for AEA biosynthesis consists of the enzymatic cleavage of the precursor N-acyl-phosphatidylethanolamine (NAPE), which is mediated by the NAPE-phospholipase D (NAPE-PLD) [152], whereas the biosynthesis of 2-AG begins with the hydrolysis of 2-arachidonoyl-phosphatidylinositol that occurs through the activity of diacylglycerol lipase (DAGL) and phospholipase Cβ [153].

ECs have a short duration of action, being rapidly metabolized by intracellular enzymes such as fatty acid amide hydrolase (FAAH), the main enzyme responsible for AEA degradation [154–156], and monoacylglycerol lipase (MAGL), which favors 2-AG catabolism [157].

Additional oxidative enzymes, including COX-2, LOXs and cytochrome P450 may also play a role in the metabolism of both AEA and 2-AG by transforming them in bioactive eicosanoids [158,159], which may activate cannabinoid receptor-independent mechanisms [160].

Beyond the ECs, several other endogenous mediators have attracted considerable attention, despite some of them showed poor affinity for CB1 and CB2 [126]. Among them, palmitoylethanolamide (PEA), stearoylethanolamide (SEA), and oleoylethanolamide (OEA), belonging to the family of NAEs, are the best characterized. However, other lipid analogues have recently been discovered and include *N*-arachidonoyldopamine (NADA), Cis-9,10-octadecanoamide (oleamide or ODA), and *N*-arachidonoylglycine (NAGly) [161], commonly referred to as endovanilloids because of their ability to activate TRPV1. Additionally, 2-arachidonoylglyceryl ether (noladin ether, 2-AGE), O-rachidonoylethanolamine (virodhamine), and arachidonoyl-L-serine (ARA-S) have also been identified [105].

Although still debated, NAEs are generally thought to be cannabinoid-receptor inactive, and they appeared to be responsible for enhancing AEA activity through the so-called "entourage effect", which consists in the inhibition of FAAH leading to an increase of AEA tissue levels [162].

PEA and OEA, shorter and fully saturated analogues of AEA, are well-documented high affinity PPAR-α and TRPV1 endogenous ligands and have been shown to exert roles in many physiological and pathological conditions such as satiety, inflammation, pain and memory consolidation [163–168]. Furthermore, due to their high expression in the CNS, growing evidence established their protective effects in neurodegenerative and neuropsychiatric disorders [169–172]. Moreover, PEA is also an endogenous agonist of GPR55, while OEA can bind GPR119.

As already mentioned, NADA belongs to the endovanilloid class of ECs and is an endogenous ligand of CB1, TRPV1 and PPAR-γ [105]. Since this compound is widely distributed in the brain, particularly in the striatum, hippocampus, cerebellum, and dorsal root ganglia, it has been shown to exert a role in neuronal pain and inflammation [105]. Interestingly, NADA also showed antioxidative and anti-inflammatory effects on glial cells [105].

2-AGE is an endogenous analogue of 2-AG, able to bind to CB1, PPAR-α and very weakly to CB2 [143,173]. Moreover, thanks to its chemical structure, 2-AGE is more stable compared to AEA and 2-AG, which are rapidly hydrolysed in vivo [102].

Virodhamine is the ester of arachidonic acid and ethanolamine and is more expressed in the periphery compared to the brain, where it is rapidly converted to AEA, due to its chemical instability. Virodhamine has been shown to act as a full agonist of CB2 and a partial agonist of CB1, whereas at higher concentrations it can be also a CB1 antagonist [174]. Furthermore, it appeared to activate also PPAR-α [143] and GPR55 [175].

NAGly is an efficacious ligand of the orphan GPR18, with no CB1, CB2, or TRPV1 activity, and shows analgesic, anti-inflammatory, and vasorelaxant properties [176].

AraS is another ECs-like compound structurally similar to AEA, which was demonstrated to produce endothelium-dependent arterial vasodilatation and to activate p44/42 MAPKs in cultured endothelial cells, effects also observed after ECs treatment [105]. To date, AraS has been shown to be a low efficacy agonist to GPR18 without binding CB1/CB2 or additional ECs receptors [105].

Lastly, ODA is a full agonist of cannabinoid receptors with selectivity for the CB1, whose activation is the primary responsible for ODA effects [105].

As suggested by the wide range and distribution of the cannabinoid receptors and by the several compounds that take part in the ECS, the latter is now considered as a complex signaling system that may play a key role in physiological and pathological conditions. Thus, targeting these intricate pathways can represent a challenge in finding a therapeutic benefit for cannabinoid-based drugs in various disorders.

4. Modulation of Oxidative Stress and Lipid Peroxidation through Cannabinoid Receptors by Endocannabinoids and Their Lipid Analogues

It is well documented that there is an important cross-talk between the ECS and various redox-dependent processes. Indeed, the ECS has been reported as a novel therapeutic target against free radical-induced lipid peroxidation. In fact, it has been shown that ECS is implicated in the development of a growing number of diseases linked with redox homeostasis deregulation, including those associated with metabolic disorders, such as type 2 diabetes and obesity, cardiovascular diseases, as well as various neuropsychiatric and neurodegenerative disorders, ischemia/reperfusion (I/R) injury, and renal diseases [2,4,5,54,177].

In the past decade, various and complex pathways have been studied to clarify the role of ECs in the modulation of redox imbalance, whose knowledge is the specific aim of this review.

There is accumulating evidence that shows the ability of ECs to alter the expression and/or the activity of enzymes implicated in the generation of these reactive small molecules (such as NOX2 and NOX4), and to modulate the production of cellular ROS/RNS by controlling mitochondrial-derived ROS/RNS generation [177].

Alternatively, ECs and their lipid analogues may modulate oxidative stress and lipid peroxidation either by conveying beneficial free radical scavenging effects or through targeting CB1 and CB2 [8–10]. Furthermore, CB1 and CB2 are differentially involved in oxidative stress modulation. In fact, several studies highlight that the activation of CB1 results in a redox imbalance enhancement, whereas CB2 stimulation is responsible for lowering ROS/RNS formation [9]. The beneficial or detrimental effects of ECs may be cell- and injury-type-specific and may depend on the stage of the disease progression as well [8].

This aspect was further investigated by Han and colleagues, who demonstrated a different role of CB1 and CB2 in regulating macrophage activity, and, in particular, the former appeared to be directly involved in the induction of intracellular ROS/RNS formation with consequent pro-inflammatory macrophage response, while the latter, after being activated by AEA, was able to negatively regulate CB1-stimulated ROS/RNS generation, through a pathway involving the small G protein, Rap1 [9]. The authors further showed that blocking CB1 while selectively activating CB2 might suppress pro-inflammatory responses of macrophages.

These data are consistent with other studies using cisplatin-induced renal dysfunction [178–181], in which it was observed that blocking the CB1 [179], or activating the CB2 [180,181], led to the attenuation of the cisplatin-induced increase of renal 4-HNE and ROS/RNS-generating enzymes (NOX2 and NOX4) expression, thus protecting against tubular damage.

Other examples of the opposite effects of CB1 and CB2 come from studies conducted in animal models of obesity and type 1 and 2 diabetes mellitus, where an increase of oxidative stress is observed [182–184]. In fact, in these models, increased levels of ECs in various renal cells contribute to the development of oxidative stress, as a result of renal CB1 activation, whereas inhibition of CB1 or activation of CB2 are able to ameliorate such effects (Figure 5) [185].

Overall, the over activation of the ECS that occurs in many type of tissue injury may induce oxidative stress, inflammatory cell infiltration, and the consequent cell death through CB1 activation [8,179], while it may also serve as an endogenous compensatory mechanism to limit early inflammatory response and interrelated oxidative stress-cell death through the activation of CB2 [186].

Interestingly, a cross-talk between redox homeostasis and ECS is particularly involved in the regulation of the cardiovascular system and metabolic tissues (i.e., liver, skeletal muscle and adipose tissue) [187,188], where CB1 and CB2 are widely distributed. Furthermore, previous studies have suggested increased ECs levels in many cardiovascular disorders, such as cardiomyopathies, atherosclerosis, and hypertension [189].

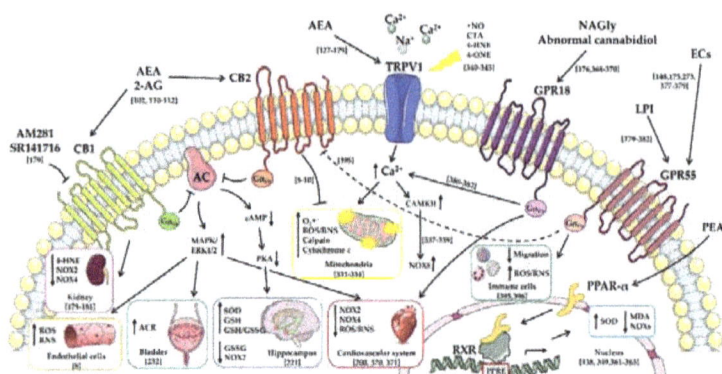

Figure 5. Role of endocannabinoids (ECs) and their lipid analogues in modulating reactive oxygen and nitrogen species (ROS/RNS) and reactive aldehydes formation. AM281: 1-(2,4-dichlorophenyl)-5-(4-iodophenyl)-4-methyl-N-4-morpholinyl-1H-pyrazole-3-carboxamide; SR141716: rimonabant; CB: can nabinoid receptors; AEA: anandamide; 2-AG: 2-arachidonoyl-glycerol; TRPV: transient receptor potential vanilloid; CTA: crotonaldehyde; NAGly: *N*-arachidonoylglycine; GPR18: G protein-coupled receptor 18; GPR55: G protein-coupled receptor 55; LPI: L-α-lysophosphatidylinositol; ECs: endocannabinoids; PEA: palmitoylethanolamide; PPARs: peroxisome proliferator-activated receptors; SOD: Cu^{2+}/Zn^{2+}-superoxide dismutase; MDA: malondialdehyde; PPRE: peroxisome proliferator response element; RXR: retinoid X receptor; NOX: NADPH oxidase enzyme; GSH: glutathione; GSSG: oxidized glutathione; ACR: acrolein; MAPK/ERK1/2: mitogen-activated protein kinases/extracellular signal-regulated kinases; PKA: protein kinase A; cAMP: adenosine 3′,5′-cyclic monophosphate; CAMKII: Ca^{2+}/calmodulin-dependent protein kinase; AC: adenylyl cyclase.

It is well known that cardiovascular diseases are associated with oxidative stress, which leads to the accumulation of lipid peroxidation-derived reactive aldehydes and may consequently cause an increase in the formation of ROS/RNS and/or a decrease in the antioxidant defense [2].

In this regard, it has been demonstrated that, after being activated by AEA, CB1 expressed in endothelial cells [190] and in cardiomyocytes in a murine model of doxorubicin-induced cardiomyopathy [8], induce the activation of the p38-JNK-MAPK pathway and increase the generation of ROS/RNS. These effects lead to cell death and resulted to be partially attenuated by the pharmacological inhibition of CB1 [9].

In contrast to CB1, the activation of CB2 appeared to exert cardioprotective effects by reducing $O_2\bullet^-$ production and decreasing endothelial cell activation. These findings are in agreement with recent studies showing that CB2 activation, by ECs and their analogue lipid mediators, protects against oxidative stress-induced tissue damage in experimental models of I/R injury [191–195], cardiovascular inflammation, and/or atherosclerosis [191,196,197].

Among the cardiovascular diseases, atherosclerosis is due to altered homeostatic redox processes with progressive ROS/RNS over production, which leads to the generation and deposition of toxic oxidized low-density lipoproteins (oxLDL) in the vessel wall. It has been clearly demonstrated that OxLDL promote the activation of NOXs and the synthesis of $O_2\bullet^-$ by a cluster of differentiation 36 (CD36) scavenger receptor-mediated method, effects that can be counteracted by several compensatory mechanisms including the involvement of the ECS [198].

Support for this comes from the observation that increased production of $O_2\bullet^-$ and enhanced NOXs activation in atherosclerosis correlated with increased rates of 2-AG biosynthesis in the vessel wall, which may be a compensatory response to oxidative stress via CB2 signaling [199].

In agreement with these results, it has been observed that the genetic disruption of CB2 in Apolipoprotein E-deficient mice (ApoE$^{-/-}$), a murine model of atherosclerosis, is the cause of boosted $O_2\bullet^-$ generation, whereas its stimulation reduced vascular $O_2\bullet^-$ release, resulting in the suppression of ROS/RNS generation and a subsequent reduction in the size of atherosclerotic lesions (Figure 5) [200].

Further evidence of the protective effects of ECs in atherosclerosis comes from the demonstration that CB1 inhibition in ApoE$^{-/-}$ mice is able to promote the down-regulation of vascular angiotensin II type 1 receptor (AT1), which is responsible for NOXs activation when stimulated by angiotensin II [201]. Consequently, the decreased expression of AT1, mediated by CB1 inhibition, leads to the reduction of NOXs activity and oxidative stress, thereby improving endothelial function and exerting beneficial direct vascular effects [201].

Since the discovery that the levels of NAEs are higher in several pathological conditions linked with redox homeostasis impairment, these compounds are attracting great attention as a survival response toward oxidative damage [202].

Indeed, it has been clearly shown that NAEs, particularly 16:0 and 18:0, exert protective effects in many diseases by the inhibition of free radical-induced lipid peroxidation [203], which is considered one of the main causes of cell damage and death [204].

In particular, previous findings discovered an involvement of two long-chain NAEs, PEA and SEA, in the inhibition of lipid peroxidation in liver mitochondria membranes of acute hypoxic hypoxia animal model [203], a pathological condition associated with an increase in partially reduced oxygen products, which represent the main cause of lipid oxidation-induced formation of reactive aldehydes [205]. The authors suggested that the inhibitory effect of NAEs on lipid peroxidation depends on the length of acyl chain and is related to their ability to protect membranes [206].

These results are in good agreement with other data showing that OEA treatment of rat heart mitochondria is able to reduce the production of MDA, which is one of the end products of lipid peroxidation in cell membrane [203].

Among NAEs, OEA, PEA, and AEA appeared to inhibit Cu^{2+}-induced in vitro lipid peroxidation in plasma lipoproteins [202] and cardiac mitochondria [207], consequently showing antioxidant properties in the pathogenesis of atherosclerosis. Moreover, Zolese and collaborators demonstrated that, depending on its concentration of incubation, PEA exerts both anti-oxidative and pro-oxidative effects on radical-induced oxidation of plasma LDL [208]. The authors showed that higher PEA concentrations could be responsible for its pro-oxidant effect, whereas PEA at lower levels is able to suppress reactive aldehydes, generated by lipid peroxidation, and to decrease the consumption rate of LDL endogenous anti-oxidants, thereby showing anti-oxidant properties [208].

In the context of cardiovascular diseases is also interesting to mention hypertension, which is characterized by (1) deregulation of ECS with increased activity of FAAH and MAGL, (2) increased levels of AEA, 2-AG, and NADA, and (3) increased expression of CB1 [209], effects that are accompanied by an imbalance of redox homeostasis (decreased activities of glutathione peroxidase (GPx), glutathione reductase (GR) and the antioxidant enzymes Cu^{2+}/Zn^{2+}-superoxide dismutase (SOD) and catalase (CAT)).

It has been demonstrated that increased levels of AEA, following chronic administration of the FAAH inhibitor URB597 in a rat model of hypertension [210], significantly enhanced the expression of the CB1, thus preventing the hypertension-induced decrease of SOD, glutathione (GSH) and glutathione transferase (GT) activities and consequently lowering ROS generation and inducing hypotension. However, it has been postulated that the enhanced AEA levels are responsible for the perturbation of membrane phospholipid metabolism resulting in PUFAs chain cyclization or fragmentation. This causes an increase in the formation of α,β-unsaturated reactive aldehydes such as 4-HNE, MDA, and 4-ONE in the liver of hypertensive rats [209].

It is well documented that ECS and oxidative stress may also play a role in the pathophysiology of liver diseases [188,211]. For instance, DeLeve and collaborators [212] reported that CB1 activation is responsible for liver inflammation and, therefore, induces non-alcoholic liver disease, whereas the CB2 stimulation appeared to have protective effects in liver damage through reducing liver oxidative stress [213].

Accumulating evidence supports the involvement of ECS as a therapeutic potential in many neurodegenerative pathologies such as Alzheimer's and Parkinson's diseases, in which oxidative stress has been recognized as one of the hallmarks of the pathology [4,171,172,214–217].

Indeed, the brain is a tissue with a high oxygen consumption whose cell membranes are particularly rich in PUFA side-chains and, therefore, highly sensitive to lipid peroxidation and oxidative damage [54,183,218].

NOXs enzymes have been shown to be significant sources of ROS/RNS during tissue injury and, in particular, it has been observed that the activation of NOX2 contributes to oxidative imbalance–induced CNS damage [219], while its inhibition is able to ameliorate cerebral oxidative stress injury [220].

A recent study conducted by Jia and collaborators defined AEA as a promising candidate for the treatment of oxidative stress–related neurological disorders [221]. In particular, AEA has been found to protect a mouse hippocampal neuron cell line from H_2O_2-induced redox imbalance by increasing SOD and GSH intracellular levels, reducing oxidized glutathione (GSSG), increasing the GSH/GSSG ratio, and lowering NOX2 expression. All of these effects were completely abolished by both CB1 antagonist administration and CB1-siRNA, suggesting that the ability of AEA to ameliorate oxidative stress in hippocampal neurons may be mediated by CB1 activation (Figure 5) [221].

Similarly, it has been also reported that the stimulation of CB1 is able to reduce intracellular ROS/RNS generation and NOX2 expression thus enhancing nigrostriatal dopaminergic neurons survival in a mouse model of Parkinson's disease [222].

These findings supporting the beneficial effects of CB1 activation against ROS/RNS formation in the brain seem to be controversial in comparison to what above mentioned for the cardiovascular and renal tissues. An explanation for this argument comes from growing evidence suggesting that the pathways underlying the interplay between cannabinoid receptors and oxidative stress modulation may be cell type–specific [177].

Notably, as well as responses mediated by CB1, further data showed that the modulation of CB2 signaling, either by using specific CB2 agonists [223–225] or by inhibiting 2-AG degrading enzyme MAGL [226], can ameliorate the morphological changes induced by oxidative stress and attenuate cerebral β-amyloid plaque accumulation in a mouse model of Alzheimer's disease carrying mutated human APPswe and PS1dE9 genes [227,228].

Interestingly, in vitro studies revealed that a selective CB1 agonist, arachidonyl-2-chloroethylamide, decreased the Fe^{2+}-induced lipid peroxidation in the brain, through a metal-chelating mechanism, as well as the •OH radicals generated by the Fenton system [229].

Moreover, the activation of the recently discovered mitochondrial CB1 by arachidonyl-2-chloroethylamide has been demonstrated to reduce oxidative stress, thereby exerting neuroprotective effects in I/R injury [227]. To this regard, CB2 activation also appeared to have a role in attenuating I/R damage through lowering ROS/RNS production and lipid peroxidation [227].

The involvement of CB2 in I/R injury has also been investigated in a context of propofol cardioprotection in an in vivo model of myocardial I/R injury, in which it has been observed that CB2 inactivation reverses propofol cardioprotective and anti-oxidative effects [230]. These findings imply that the enhancement of ECs release and the subsequent activation of CB2 signaling are responsible for the reduced oxidative stress mediated by propofol cardioprotection in myocardial I/R injury [230].

Furthermore, CB2 are expressed in the bladder [231] and are involved in the treatment of hemorrhagic cystitis, a common side effect of Cyclophosphamide, an antineoplastic alkylating agent usually metabolized by the liver to ACR, which is accumulated in urine and therefore is considered to be the main responsible for Cyclophosphamide-induced cystitis [232]. The findings of this study revealed that, following stimulation, CB2 attenuated ACR-induced cystitis through modulating ERK1/2 MAPK pathways (Figure 5) [232].

AEA and 2-AG are also involved in the progression of cancer, where they were shown to exert protective effects against increased ROS/RNS production–induced tumor [233], leading to apoptosis in normal and cancer cells by modulating ERK and ROS/RNS pathways [234].

5. Modulation of Oxidative Stress and Lipid Peroxidation through the Transient Receptor Potential Vanilloid Channels by Endocannabinoids and Their Lipid Analogues

The transient receptor potential (TRP) channels superfamily is a wide group of tetrameric channels formed by six transmembrane domains and a cation-selective pore. On the basis of its amino acid sequence homology, TRP superfamily, in mammals, is organized into six subfamilies, which include TRP canonical, TRP melastatin, TRP ankyrin, TRP mucipilin, TRP vanilloid, and TRP polycistin channels. TRP channels are ubiquitously expressed in most mammalian cells [235,236] and they depolarize cells by altering membrane potential or intracellular Ca^{2+} concentration. With the exception of some TRP channels, most of them are non-selective and weakly voltage-sensitive [237]. TRP channels are fundamental players of sensory physiology as they respond to environmental stimuli such as taste, light, sound, smell, touch, temperature, and osmolarity [238]. Today, only a few endogenous ligands are known to activate TRP channels, and it is not yet clear how they are activated in vivo [237]. However, several experiments performed on knockout mice are revealing the complexity and the different functions of TRP channels [238–240].

In this review, we will focus mainly on the vanilloid TRP (TRPV) channels subfamily and how they respond to oxidative stress and lipid peroxidation-induced cell damage. Currently, six TRPV channels (TRPV1-6) have been identified and divided into two subgroups: TRPV1-4 and TRPV5-6, based on their amino acid sequence, functions, and cation selectivity. A detailed review on TRPV channels pharmacology has been provided by Vriens and colleagues [241]. Briefly, TRPV1 is expressed in primary sensory neurons, in few brain regions (hypothalamus, intrafascicular, supramammillary and rostral raphe nuclei, entorhinal cortex, hippocampus, and periaqueductal gray), as well as in smooth muscle cells of several thermoregulatory tissues (skin, dura, tongue, trachea, cremaster muscle, and ear) [242]. TRPV1 seem to be activated by heat above 43 °C, by low pH [243–245], by vanilloid compounds (e.g., capsaicin and capsinate) [243,246], by ethanol [247,248], as well as by several endogenous compounds such as AEA [127], OEA [249], NADA [250], *N*-oleoyldopamine (OLDA) [251], and arachidonic acid-derived metabolites released by LOXs [252]. Moreover, TRPV1 activity is modulated by various intracellular molecules and signals including calmodulin [253,254], ATP [255], phosphatidylinositol 4,5-bisphosphate (PIP2) and phosphatidylinositol 3,4,5-trisphosphate (PIP3) [256], PKC [257], PKA [258], as well as protein phosphatase calcineurin [259].

Among the main functions, in addition to acting as a thermoreceptor, TRPV1 regulates the normal functioning of urinary bladder [260], controls the gut afferent sensitivity to distension and acids [261] and it also allows the taste perception of sodium chloride [262]. From a physiopathological point of view, TRPV1 has a direct role in the behavioral response to ethanol [247,248,263], as well as in inflammatory airway diseases [264]. Moreover, TRPV1 is also involved in vascular dementia as well as in Huntington's disease, where its activation promotes neuroprotection, increase learning and memory, and reduce oxidative stress [265–267].

Differently, TRPV2 is a weakly Ca^{2+}-selective channel, which seems to be activated by thermal stimuli above 53 °C but not by low pH or vanilloid compounds [268]. TRPV2 is expressed in different tissues including brain, spinal cord, spleen, and intestine, as well as in vas deferens, bladder, heart, kidney [269], and immune cells such as monocytes and dendritic cells [270]. It is noteworthy that TRPV2 signaling plays an important role in the endosomal pathway, where TRPV2 modulates the fusion between endosomal membranes by releasing Ca^{2+} from early endosomes [271,272] as well as in phagocytosis [273,274].

TRPV3 is a non-selective cation channel activated by temperatures of 33–39 °C, which showed a marked sensitization following repeated heat stimuli [275,276]. Moreover, TRPV3 could be activated by several vegetable-derived molecules, such as eugenol, thymol, camphor and carvacrol [277,278]. Furthermore, other agents such as PIP2/PIP3, calmodulin, ATP, and inflammatory mediators like histamine, bradykinin, and PGE2 are able to sensitize TRPV3 function [278–281]. Moreover, it was hypothesized that, in rodent skin cells, heat-induced TRPV3 signaling could mediate an autonomous response to heat stimulation, thus acting as thermoreceptors in keratinocytes [275,282]. In support of this evidence, TRPV3 knock-out mice showed strong deficits in response to heat stimulation [277]. Likewise, TRPV4 is also activated by heat, in particular by temperatures of 27–34 °C, as well as by osmotic and mechanical stimuli [283,284]. Among putative endogenous ligands, it was observed that AEA, 2-AG, and arachidonic acid indirectly activate TRPV4 by epoxyeicosatrienoic acids released from cytochrome P450 epoxygenases [285,286]. As for TRPV1 and TRPV3, TRPV4 activity is modulated by PIP2/PIP3, calmodulin and ATP [287–289] and by several protein kinases, such as PKA, PKC, Src family kinases (SFKs), and serum glucocorticoid-induced protein kinase-1 (SGK1) [290–293]. TRPV4 channels are widely expressed in epithelial cells of the renal convoluted tubule, trachea, submucosal glands, as well as in neutrophils, in autonomic nerve fibers, in peripheral sensory ganglia, in hair cells of the inner ear, and brain structures such as vascular organ of the lamina terminalis and the hypothalamic median preoptic region [283,294,295]. Due to its widespread expression, TRPV4 is involved in several physiological functions. In particular, it mediates temperature sensation in skin keratinocytes, anterior hypothalamus, and sensory ganglia [275,283,284]. TRPV4 is also involved in mechanosensation [296] and contribute to the normal functioning of the urinary bladder [297,298] and pulmonary alveoli [299,300] and to the development of mechanical hyperalgesia in inflammatory states [301].

TRPV5 and TRPV6 share a high sequence homology (74% of identity) and form highly Ca^{2+}-selective channels, which are not activated by heat [302–304]. As for the other TRPV family members, the activity of TRPV5 and TRPV6 is modulated by a variety of second messengers, including Ca^{2+}, Mg^{2+}, ATP, PIP2, calmodulin, and PKC [302,303,305–313]. TRPV5 is expressed in several tissues but is mostly abundant in renal tubules, where it regulates transcellular transport and reabsorption of Ca^{2+} [314]. Furthermore, TRPV5 is also involved in bone remodeling [315,316]. TRPV6 is widely expressed [305,317,318] but is mostly distributed in the intestine, kidney, and placenta, where it respectively modulates the Ca^{2+} transcellular entry, reabsorption, and transfer to fetus [319–322].

Among endogenous ligands of TRPV, or endovanilloids, there are leukotriene B4 and 12-hydroperoxyeicosatetraenoic acid that belong to the eicosanoid family, produced by lipoxygenase-mediated oxidation of PUFAs (especially arachidonic acid), which are potent activators of TRPV1 [252,323]. Other lipid-derived mediators of TRPV are epoxyeicosatrienoic acids, such as 5′,6′-epoxyeicosatrienoic acid, which are synthesized from arachidonic acid by cytochrome P450 epoxygenases and may activate TRPV1 and TRPV4 [286,324].

As AEA is structurally similar to arachidonic acid as well as to PUFAs, it can be metabolized by COX-2 and LOXs. In particular, COX-2 converts AEA into prostaglandin-ethanolamides, which are endoperoxide molecules also known as prostamides [325,326]. On the other hand, LOXs convert AEA into hydroperoxy fatty acids, such as 12- and 15-hydroperoxyeicosatetraenoylethanolamide, which are, respectively, synthesized by 12-LOX and 15-LOX [327,328]. In guinea-pig bronchi, these oxidized lipid mediators seem to act as TRPV1 agonists and are also responsible, at least partially, for the contractile action of AEA [329].

Growing evidence supports a key role for TRPV, especially TRPV1, in the modulation of oxidative stress and lipid peroxidation mediated by endocannabinoids, their lipid analogues, and other lipid-related mediators. As known, AEA is considered an endovanilloid because of its ability to activate TRPV1 [127,136,330]: several in vitro analyses performed on human and rat cell lines have shown that AEA induces apoptotic effects via a TRPV1-mediated mechanism, which induces and increase in intracellular Ca^{2+} levels, mitochondrial uncoupling, oxidative stress due to increased $O_2 \bullet^-$ formation, cytochrome c release as well as calpain and caspase-3 activation [331–333]. Similarly, another in vitro study performed on human bladder cancer T24 cells showed that TRPV1 activation by capsaicin was correlated in a dose-dependent manner with an increase of cytosolic Ca^{2+} levels, with mithocondrial membrane depolarization and a marked ROS/RNS generation, which reduced T24 cells viability (Figure 5) [334].

Other studies showed that AEA was able to increase ROS/RNS production by targeting TRPV1, [335,336], which lead to the activation of the Ca^{2+}/calmodulin-dependent protein kinase II (CAMKII), and to the upregulation of NOX5 [337–339].

Moreover, it was observed, in the human esophageal epithelial cell line Het1A, that acid- or capsaicine-induced activation of TRPV1 leads to an increased production of intracellular ROS/RNS levels as well as to increased ROS/RNS- or HNE-modified proteins. In the same study, immunoprecipitation analyses of 4-HNE-stimulated Het1A cells revealed, also, that TRPV1 was modified by 4-HNE [340]. In addition to 4-HNE, TRPV1 is directly activated by •NO, oxidants and other chemical agents through the modification of cysteine free sulfhydryl groups [341]. Moreover, functional assays with mutated TRPV showed that cysteine residues 553 and 558, between the fifth and sixth transmembrane domains, are essential for •NO-induced activation of TRPV1, TRPV3, and TRPV4 and thus are potential targets of nitrosylation [342]. In addition, TRPV1 nitrosylation by •NO increased the intracellular Ca^{2+} levels and thus enhanced the channel sensitivity to H^+ and heat. These sensitizing effects induced by nitrosylation of cysteine residues were further supported by the use of oxidizing agents such as diamide and chloramine-T [343]. Furthermore, several studies reported that TRPV1 is also responsive to other electrophilic compounds generated during oxidative stress. To this regard, in TRPV1 channel-expressing human embryonic kidney (HEK) cells, a modest TRPV1 activation was observed following 4-ONE treatment (100 µM) [344]. Another TRPV1 activator is CTA. In particular, an in vitro study performed on murine cardiomyocytes incubated with CTA showed an increase in TRPV1 and NOXs levels, in ROS/RNS formation, in apoptotic events, and a decrease in the activity of mithocondrial proteins such as aconytase, uncoupling protein 2, and peroxisome proliferator-activated receptor-gamma coactivator-1alpha [345].

6. Modulation of Oxidative Stress and Lipid Peroxidation through the Peroxisome Proliferator-Activated Receptors-Alpha by Endocannabinoids and Their Lipid Analogues

Because of the high expression of PPAR-α in kidney, liver, heart, and brain, it is well documented that the activation of these transcription factors exerts protective roles in cardiovascular as well as renal, hepatic, and neurodegenerative diseases [138,346–349].

There is rising acknowledgment that the beneficial effects of PPAR-α stimulation could be explained by its ability to dampen oxidative stress in several pathological conditions linked to the redox impairment. A number of reports point to the involvement of various mechanisms through which PPAR-α agonists can modulate antioxidants.

In particular, the identification of PPREs elements in promoter regions of CAT and SOD genes in rat [347] additionally supported the involvement of these nuclear receptors in lowering ROS/RNS formation and lipid peroxidation products.

Nevertheless, PPAR-α is not only involved in suppressing ROS/RNS generation, but it can also play a role in modulating enzymes involved in ROS/RNS synthesis and/or scavenging. Consistently, the decrease in striatal SOD expression, which resulted in the 6-hydroxydopamine (6-OHDA)-induced Parkinson disease mouse model, was completely counteracted by PPAR-α agonists confirming the ability of this nuclear receptor to regulate the transcription of antioxidant enzymes (Figure 5) [138,346,350].

For instance, Diep and colleagues reported that the PPAR-α -induced suppression of oxidative stress in cardiovascular diseases is mediated by the ability of PPAR-α activators to inhibit angiotensin II-induced activation of NOXs in the vascular wall [348] and to increase scavenging enzymes as well.

Among the PPAR-α ligands, ECs and their lipids analogues have been shown to play a prominent role in affecting redox homeostasis in several oxidative stress-related pathologies, through a PPAR-α dependent mechanism. Consistently, it has been shown that PPAR-α stimulation by PEA lowers blood pressure and prevents hypertension-induced renal damage in hypertensive rats by inhibiting the subunit p47phox of NOXs (a key regulatory subunit essential for NOXs functioning) [349], and by significantly reducing the hypertension-induced increased levels of MDA in urine and renal tissues (Figure 5) [348].

Moreover, through PPAR-α activation, PEA appeared to simultaneously enhance the antioxidant defense by increasing SOD expression in the kidney [348], thus protecting from renal damage. In agreement with these results, other studies further support the potential beneficial effects of PEA activated-PPAR-α on kidney diseases [351]. For instance, it has been demonstrated that PEA, by targeting PPAR-α, is able to prevent kidney damage induced by I/R injury through dampening the lipid peroxidation products in the kidney, thereby leading to a reduction of neutrophil recruitment [352].

Moreover, because of the high expression of PPAR-α and its endogenous lipid agonists in the CNS, it has been demonstrated that PPAR-α activation can exert neuroprotective properties in several neuropathological conditions, especially in neurodegenerative disorders [169], by modulating the redox balance that resulted altered in these situations.

Further support for this comes from the observation that the brain areas that display the highest PPAR-α expression exhibit an overlapping expression pattern with key enzymes involved in ROS/RNS synthesis and/or scavenging including CAT, SOD1 and acyl-CoA oxidase 1 (ACOX1) [353–355], whose genes are known to be under the control of PPAR-α [356,357].

Thanks to its anti-oxidative properties, PPAR-α protects against normal brain aging and regulates the onset and progression of neurodegenerative disorders [358,359]. Interestingly, evidence suggests that in conditions of neurodegeneration, oxidative stress itself is responsible for the induction of PPAR-α expression. As a matter of fact, in hippocampal CA1 pyramidal cells of a transgenic mouse model of Alzheimer's disease, an increase in the levels of PPAR-α simultaneously with the production of ACR and 8-hydroxy(de)oxyguanosine, which represent markers of oxidative imbalance, was observed [360]. Such increase in hippocampal PPAR-α expression could trigger the induction of its target genes encoding for peroxisomal membrane protein-70 (PMP70) and ACOX1, which are involved in fatty acyl-CoA transport across peroxisomal membranes and peroxisomal β-oxidation respectively, by evoking a compensatory response to Aβ-mediated mitochondrial insult that occurs in early stage of Alzheimer's disease [4–6,360].

In this context, PEA was demonstrated to protect neurons and glia from oxidative stress by reducing MDA formation, thereby restoring a proper cellular redox state, and this effect appeared to be PPAR-α-dependent [171,172,361,362]. It has also been established that PEA neuroprotective effects are mediated, at least in part, through the de novo synthesis of neurosteroids (particularly allopregnanolone), which is triggered by PPAR-α activation [362].

The abovementioned findings, coupled with a recent report demonstrating that PEA treatment (through binding PPAR-α) is able to induce SOD and dampen ROS/RNS-induced oxidative damage in 6-OHDA-induced mouse model of Parkinson disease, additionally suggest the neuroprotective scavenging effects of this lipid compound (Figure 5) [363]. Beyond the ECs, several other synthetic ligands of PPAR-α have been shown to exert antioxidative properties. For instance, Wy14643 through binding PPAR-α is able to protect rabbit hearts from I/R injury by increasing the expression of the oxidative stress-inducible isoform of heme oxygenase and to preserve hippocampal neurons from H_2O_2 challenge by modulating mitochondrial fusion and fission events [360].

Moreover, it should be noted that the production of PPAR-α endogenous ligands, PEA and OEA as the mostly characterized, could be differently affected by physiological and pathological oxidative stress-related conditions. For instance, the ROS/RNS metabolism imbalance, which is responsible for oxidative stress-induced brain aging and neurodegeneration, can quantitatively and qualitatively modify the production of PPAR-α agonists and thus differently modulate PPAR-α-mediated pathways in neuronal and astroglial cells [169].

Additionally, the interplay between PPAR-α and oxidative-stress-induced lipid peroxidation comes also from the observation that NOXs activated-4-HNE is able to act as an endogenous PPAR-α activator leading to the discovery of the so called "lipid peroxidation products–PPARs–NOXs axis" [364]. The regulation of this axis, which represents an alternative pathway mediating ROS/RNS production, could ensure additional strategies to counteract oxidative-stress-related disorders.

7. Modulation of Oxidative Stress and Lipid Peroxidation through Other Receptors by Endocannabinoids and Their Lipid Analogues

Recently, in addition to PPAR-α and TRPV1, the orphan receptors GPR18, GPR55 and GPR119 were assessed as novel cannabinoid-related receptors [365]. Structurally, GPRs are GPCRs and, among them, GPR18, GPR55 and GPR119 share a limited primary sequence homology with CB1 and CB2.

GPR18 was discovered for the first time in 1997 by Gantz and colleagues [366]. GPR18 is widely expressed in testis and spleen, and in lesser extent in several other tissues such as thymus, lymph nodes, peripheral blood leukocytes, small intestine, and appendix, thus suggesting a regulatory role for GPR18 in the immune system [366]. Moreover, GPR18 was also found in several brain regions such as hypothalamus, brainstem, cerebellum, and striatum as well as in lung, thyroid and ovary [367]. Several studies reported that NAGly is the endogenous ligand of GPR18 that induces an elevation of intracellular Ca^{2+} levels [176]. The same authors demonstrated also that GPR18 activation was pertussis toxin-sensitive, suggesting the involvement of a $G_{\alpha i/o}$ protein in this response [176]. Despite these first evidence, several authors reported variable responses of GPR18 following the administration of NAGly [368,369].

For the first time Penumarti and colleagues demonstrated that GPR18 is expressed in the rostral ventrolateral medulla of rats and exerts tonic restraining influence on blood pressure [370]. In particular, authors observed that the systemic administration of abnormal cannabidiol, a synthetic agonist of GPR18, induced a dose-dependent reduction of blood pressure and increased heart rate. In addition, GPR18 activation increased neuronal adiponectin and •NO, and finally reduced neuronal ROS/RNS levels. These findings suggested for the first time a sympathoinhibitory role of GPR18 (Figure 5) [370].

More recently, another study confirmed that chronic GPR18 activation with its agonist abnormal cannabidiol produced hypotension, suppressed the cardiac sympathetic dominance, and improved left ventricular function in conscious rats [371]. In the same study, ex vivo analysis of plasma, heart, and vascular tissues of treated rats revealed an increase in cardiac and plasmatic adiponectin levels, an increase in aortic eNOS expression, augmented levels of vascular and serum •NO, high levels of myocardial and plasmatic guanosine 3′,5′-cyclic monophosphate (cGMP), an increase of myocardial Akt and ERK1/2 phosphorylation, and, more importantly, reduced myocardial ROS/RNS formation [371]. These results suggest a protective role of GPR18 in cardiovascular diseases, in particular highlights the possibility to consider GPR18 as a viable molecular target for developing new antihypertensive drugs which are able to improve also the cardiac function.

Human GPR55 receptor was identified for the first time in 1999, through in silico studies, and was subsequently cloned [372]. GPR55 receptor is widely expressed, and therefore its activity was correlated with multiple physiological processes. In particular, GPR55 is expressed in the frontal cortex, striatum, hippocampus, hypothalamus, cerebellum, and brainstem [372,373]. Moreover, GPR55 was also found in peripheral organs and cells such as dorsal root ganglion [148], spleen, adrenal glands, jejunum, ileum [373], pancreas [374], bones [375] and microglia [376]. The GPR55 pharmacology and its downstream signaling are not yet certain. Nevertheless, some authors reported that ECs such as AEA, 2-AG, and virodhamine can activate both etherologous and native GPR55-expressing cells [148,273,377], while other groups reported that ECs are weak ligands [378,379], may act as partial agonists [175], or are not able to activate GPR55 receptors [380,381]. Another open debate regards the ability of PEA to activate [373] or not the GPR55 receptors [148,382]. Despite the controversial results about the ability of ECs to activate GPR55, it is well accepted that the endogenous lipid L-α-lysophosphatidylinositol (LPI) and its analogue 2-arachidonoyl-sn-glycero-3-phosphoinositol are endogenous ligand of GPR55 [379–382]. However, it is necessary to specify that LPI is not selective only for GPR55 [383]. Moreover, GPR55 may also heterodimerize with other receptors, such as CB2 [384], thus further confounding the results obtained so far.

About the mechanisms of downstream signaling, GPR55 activation was associated with an increase of intracellular Ca^{2+} levels, with the activation of RhoA and ERK1/2 pathway, and with the activation of several transcription factors, such as the nuclear factor of activated T-cells and the cAMP response element binding protein (CREB) [380,382].

The human orphan receptor GPR119 was identified for the first time in 2003 by sequence alignment tools analysis [385]. GPR119 is expressed mainly in pancreas and gut, in particular in β-cells and pancreatic polypeptide-producing PP cells, where its activity modulates the glucose-dependent insulin secretion [386,387], as well as in enteroendocrine L-cells, where it regulates the secretion of glucagon-like peptide 1 [388,389]. GPR119 is also expressed in liver [390] and skeletal muscle [Cornall et al., 2013]. In normal-weight and healthy patients it was observed that gut GPR119 expression rapidly increased following acute fat exposure [391], thus suggesting a potential involvement of GPR119 in type 2 dyabetes, metabolic disorder, and obesity.

The main endogenous ligands of GPR119 are, in order of potency, OLDA, OEA, PEA, and AEA [392,393]. Other endogenous GPR119 agonists are 2-oleoylglycerol [394] and oleoyl-lysophosphati dylcholine [386]. Clearly, also in this case, further studies are required to better characterize the pharmacological profile of GPR119.

Increasing evidence suggests that ECs may regulate ROS/RNS levels and thus reactive aldehydes formation by targeting GPR55. In this regard, Balenga and colleagues showed that GPR55 activity modulates RhoA-dependent neutrophil migration, and it may prevent oxidative damage [395]. In particular, this study, performed on neutrophils, demonstrated that 2-AG-induced ROS/RNS production, which was mediated by a CB2-dependent mechanism, appeared to be significantly decreased following the co-treatment with the GPR55 agonist LPI [395]. This negative interaction between GPR55 and CB2 was observed during neutrophil respiratory burst. Therefore, after an initial synergism in inducing chemotaxis, GPR55 and CB2 disengaged and, by a functional repression, GPR55 decreased CB2-induced oxidative damage by blocking CB2 downstream signaling [395]. Conversely, a recent study performed on human natural killer cells and monocytes unveiled a proinflammatory role of GPR55 activation (Figure 5) [396], which could be potentially correlated with an increase of ROS/RNS production and thus with oxidative stress.

8. The Role of Antioxidant System as Scavenger of ROS/RNS and Reactive Aldehydes

The "endogenous antioxidant system" relies on several enzymes, peptides, cofactors, and other molecules that are essential for the maintenance of a physiological redox homeostasis. Overall, endogenous antioxidants may be divided into two main groups, formed by enzymatic and non-enzymatic antioxidants [6,397]. The enzymatic group include CAT [398], SOD [399,400], GPx, GR, GT [401], thioredoxin (Trx) and thioredoxin reductase (TrxR) [402] while the non-enzymatic group include several antioxidant molecules such as GSH, GSSG, [403], vitamin A (retinol) [404], vitamin C (L-ascorbic acid) [405], vitamin E (tocopherols) [406], coenzyme Q10 (CoQ10) [407], carotenoids [408], flavonoids, polyphenols [409–411], minerals such as Se^{2+} [412], Cu^{2+}, and Zn^{2+} [413], as well as metabolites such as uric acid, bilirubin [414] and melatonin [415], which also possess antioxidant properties.

Briefly, CATs are Cu^{2+}/Zn^{2+}-dependent enzymes present in peroxisomes that catalyze the conversion of H_2O_2 in water and oxygen [398]. Among SOD enzymes, cytolosic SOD are Cu^{2+}/Zn^{2+}-dependent enzymes, while mitochondrial SODs are Mn^{2+}-dependent enzymes that metabolize $O_2 \bullet^-$ into H_2O_2 and oxygen. Therefore, SOD represents the first line of defense against reactive aldehydes formation [400]. GPx, GR and GT are Se^{2+}-dependent enzymes that, together with GSH and GSSG, constitute the glutathione system, which contributes to eliminate H_2O_2 and other reactive molecules [403]. Similarly, Trx, TrxR, and NADPH constitute the thioredoxin system, which is critical for redox regulation of protein function and signaling via thiol redox control [402].

Vitamin A is produced in the liver, derives from β-carotene and acts as a lipid peroxidation blocker by preventing the chaining process in the propagation phase [404]. Similarly, also vitamin E acts as a lipid peroxidation blocker by donating a hydrogen atom to peroxyl radicals, thus forming tocopheroxyl radicals which are unable to continue the propagation phase of lipid peroxidation [416]. Vitamin C is effective in scavenging several ROS/RNS as well as in the detoxification of peroxyl and hydroxyl radicals [405]. CoQ10 is involved in the neutralization of the damages induced by peroxyl radicals and also in the regeneration of vitamin E [407]. Uric acid is known to prevent protein nitrosylation, as well as lipid and protein peroxidation, and therefore it is considered as a protectant agent of the CNS [417]. Melatonin is a natural scavenger derived from tryptophan, which is involved in the neutralization of several ROS/RNS and thus reduces the generation of reactive aldehydes [415]. Finally, flavonoids and polyphenols are ubiquitous plant-derived molecules, which act as chelators and scavengers of ROS/RNS as well as of hydroxyl and peroxyl radicals [418,419].

9. Conclusions

Oxidative stress represents an underlying disturbance that is involved in many pathophysiological conditions. Increasing evidence suggests that tissues with a high oxygen consumption, such as brain and heart among others, are particularly sensitive to lipid peroxidation products and free radical accumulation, which are responsible for oxidative stress–induced damages with consequent cell death [2,4–6,218].

Thus, acting on the cellular processes that suppress the generation of these reactive small molecules or altering the expression and/or activity of enzymes involved in their formation may be crucial for the treatment of a growing number of diseases linked with redox homeostasis deregulation.

In this scenario, there is rising acknowledgment about a cross-talk between the ECS and various redox-dependent processes. Indeed, it has been observed that the redox impairment induces the enhancement of AEA and 2-AG levels, as a consequence of phospholipid hydrolysis [420,421], and the upregulation of CB1 and CB2 expression [422,423], as well as the downregulation of FAAH [422].

A large number of reports point to the involvement of ECs and their lipid analogues in regulating ROS/RNS and reactive aldehydes generation through targeting CB1 and CB2 [8–10] and thereby exerting protective effects in cardiovascular as well as renal, hepatic, neuropsychiatric, and neurodegenerative diseases.

Moreover, it has been observed that, depending on the type of cell and/or injury, cannabinoid receptors show opposite effects in oxidative stress modulation, since CB1 activation results in a redox imbalance enhancement, while CB2 stimulation is responsible for lowering oxidative stress [9,223] and may convey beneficial free radical scavenging effects.

Overall, the mechanisms by which CB2 receptors, following ECs-mediated activation, are involved in the reduction of oxidative injury seem to be primarily mediated by the reduction of NOX2 and NOX4, and the simultaneous induction of the antioxidant defense through the increase of the SOD scavenging enzymes [180,181].

Emerging evidence indicates that the neuroprotective, cardioprotective and renoprotective effects of ECs and NAEs are additionally mediated by CB1/CB2-independent mechanisms and involve the contribution of alternative intracellular targets such as PPAR-α, TRPV1, GPR55, and GPR18 [169,348,349,370,371,395].

In particular, an interplay between PPAR-α and oxidative stress has been suggested from the observation that an imbalance in the redox state may modulate several signaling pathways, including PPAR-α signaling, via transcriptional regulation and post-translational modification.

Among the PPAR-α ligands, PEA appeared to exert beneficial effects by simultaneously enhancing the antioxidant defense through the increase of SOD expression and inhibiting NOXs activity with a consequent reduction of the lipid peroxidation products such as MDA [138].

Although the huge amount of knowledge has been gained about the effects of the ECs on oxidative stress and lipid peroxidation in several pathological conditions, many ECS compounds fail during clinical trials due to inefficacy or unforeseeable safety concerns. For the treatment of the cardiovascular diseases, for instance, no cannabinoid-based drugs have been approved so far, except for those acting as PPARs agonists [348]. Among the limitations that play a role in restricting the translation of ECs studies into clinical trials, the different animal paradigms as well as the route of administration used (central vs peripheral) and the differences between species seem to be primarily involved.

Moreover, most of the studies have focused on the role of CB1, CB2, TRPV1, PPARs and less is known about other candidates such as GPR18, GPR55 and GPR119.

Despite promising goals have been achieved over the last years on ECS research, there is an urgent necessity to expand the knowledge on the ECs complex signalling in order to better identify an explanation of the serious side-effects observed in clinical studies. Lessons from clinical experience should encourage the scientific community to better clarify how to modulate the ECS thus leading to major breakthroughs in the treatment of many diseases.

Overall, the findings discussed in this review may further elucidate the complex interaction existing between ECS, oxidative stress, and lipid peroxidation, resulting in a better understanding of the multiple beneficial effects of this signaling system in several pathological conditions related to a redox status impairment.

Author Contributions: The project idea was developed by S.G., T.C. and A.M.G.; C.A.G., S.C., T.C., and A.R. wrote the first draft of the manuscript. C.A.G., S.C., A.R., J.B.K., M.d.C., T.C., S.G., D.D., R.V., and A.M.G. conducted the literature review and revised the manuscript. S.C., C.A.G., and D.D. created the figures.

Funding: This project was supported by the Italian Ministry for Education, University and Research (PRIN20153NBRS3_003 to S.G.).

Conflicts of Interest: The authors declare no conflict of interest.

Abbreviations

•NO = nitric oxide; •OH = hydroxyl radical; $^1O_2$ = oxygen singlet; 2-AG = 2-arachidonoyl-glycerol; 2-AGE = 2-arachidonoylglyceryl ether or noladin ether; 4-HHE = 4-hydroxy-hexanal; 4-HNE = 4-hydroxy-2-nonenal; 4-ONE = 4-oxo-nonenal; 6-OHDA = 6-hydroxydopamine; AC = adenylyl cyclase; ACOX1 = acyl-CoA oxidase 1; ACR = acrolein; AEA = *N*-arachidonoyl-ethanolamine or anandamide; $ApoE^{-/-}$ = Apolipoprotein E-deficient mice; ARA-S = arachidonoyl-L-serine; AT1 = angiotensin II type 1 receptor; CAMKII = Ca^{2+}/calmodulin-dependent protein kinase II; cAMP = adenosine $3',5'$-cyclic monophosphate; CAT = catalase; CB1 = cannabinoid receptor type 1; CB2 = cannabinoid receptor type 2; CD36 = cluster of differentiation 36; CGD = chronic granulomatous disorder; cGMP = guanosine $3',5'$-cyclic monophosphate; CNS = central nervous system; CoQ10 = coenzyme Q10; COX-2 = cyclooxygenase type 2; COXs = cyclooxygenases; CREB = cAMP response element binding protein; CTA = crotonaldehyde; DAGL = diacylglycerol lipase; dG = deoxyguanosine; ECS = endocannabinoid system; ECs = endocannabinoids; eNOS = endothelial nitric oxide synthase; ERK1/2 = extracellular signal-related kinases; FAAH = fatty acid amide hydrolase; GPCRs = G-protein-coupled receptors; GPR18 = G protein-coupled receptor 18; GPR55 = G protein-coupled receptor 55; GPR119 = G protein-coupled receptor 119; GPx = glutathione peroxidase; GR = glutathione reductase; GSH = glutathione; GSSG = oxidized glutathione; GT = glutathione transferase; GTPase = guanosine triphosphatase; H_2O_2 = hydrogen peroxide; HClO = hypochlorous acid; HEK = human embryonic kidney; I/R = ischemia/reperfusion; JNK = c-Jun *N*-terminal kinase; LOXs = lipooxygenases; LPI = L-α-lysophosphatidylinositol; MAG = monoacylglycerols; MAGL = monoacylglycerol lipase; MAPKs = mitogen-activated protein kinases; MAPK/ERK1/2 = mitogen-activated protein kinases/extracellular signal-regulated kinases; MDA = malondialdehyde; NADA = *N*-arachidonoyldopamine; NAEs = *N*-acylethanolamines; NAGLy = *N*-arachidonoylglycine; NAPE = *N*-acyl-phosphatidylethanolamine; NAPE-PLD = NAPE-phospholipase D; NF-κB = nuclear factor-κB; NO_2 = nitric dioxide; NO_2^- = nitrite; NO_3^- = nitrate; NOS = nitric oxide synthase; NOS1 = NOS type 1; NOX2 = NOX type 2; NOX3 = NOX type 3; NOXs = NADPH oxidase enzymes; Nrf2 = nuclear factor-erythroid 2-related factor 2; $O_2•^-$ = superoxide anion; O_3 = ozone; ODA = Cis-9,10-octadecanoamide or oleamide; OEA = oleoylethanolamide; OLDA = *N*-oleoyldopamine; $ONOO^-$ = peroxynitrite; oxLDL = oxidized low-density lipoproteins; PEA = palmitoylethanolamide; PGE2 = prostaglandin E2; PIP2 = phosphatidylinositol 4,5-bisphosphate; PIP3 = phosphatidylinositol 3,4,5-trisphosphate; PKA = protein kinase A; PKC = protein kinase C; PMP70 = peroxisomal membrane protein-70; PPARs = peroxisome proliferator-activated receptors; PPRES = peroxisome proliferator response elements; PUFAs = polyunsatured fatty acids; Rap = Ras-related protein; Rap1GAP = Rap1 GTPase activating protein; RhoGEFs = RhoGTPase nucleotide exchange factors; ROS/RNS = reactive oxygen and nitrogen species; RXR = retinoid X receptor; SEA = stearoylethanolamide; SFKs = Src family kinases; SGK1 = serum glucocorticoid-induced protein kinase-1; SOD = superoxide dismutase; TRP = transient receptor potential; TRPV = transient receptor potential vanilloid; TRPV1 = transient receptor potential vanilloid 1; Trx = thioredoxin; TrxR = thioredoxin reductase.

References

1. Pomara, C.; Cassano, T.; D'Errico, S.; Bello, S.; Romano, A.D.; Riezzo, I.; Serviddio, G. Data available on the extent of cocaine use and dependence: Biochemistry, pharmacologic effects and global burden of disease of cocaine abusers. *Curr. Med. Chem.* **2012**, *19*, 5647–5657. [CrossRef] [PubMed]

2. Matthews, A.T.; Ross, M.K. Oxyradical Stress, Endocannabinoids, and Atherosclerosis. *Toxics* **2015**, *3*, 481–498. [CrossRef] [PubMed]

3. Sureshbabu, A.; Ryter, S.W.; Choi, M.E. Oxidative stress and autophagy: Crucial modulators of kidney injury. *Redox Biol.* **2015**, *4*, 208–214. [CrossRef] [PubMed]

4. Cassano, T.; Serviddio, G.; Gaetani, S.; Romano, A.; Dipasquale, P.; Cianci, S.; Bellanti, F.; Laconca, L.; Romano, A.D.; Padalino, I.; et al. Glutamatergic alterations and mitochondrial impairment in a murine model of Alzheimer disease. *Neurobiol. Aging* **2012**, *33*, 1121-e1. [CrossRef] [PubMed]

5. Cassano, T.; Pace, L.; Bedse, G.; Lavecchia, A.M.; De Marco, F.; Gaetani, S.; Serviddio, G. Glutamate and Mitochondria: Two Prominent Players in the Oxidative Stress-Induced Neurodegeneration. *Curr. Alzheimer Res.* **2016**, *13*, 185–197. [CrossRef] [PubMed]

6. Serviddio, G.; Romano, A.D.; Cassano, T.; Bellanti, F.; Altomare, E.; Vendemiale, G. Principles and therapeutic relevance for targeting mitochondria in aging and neurodegenerative diseases. *Curr. Pharm. Des.* **2011**, *17*, 2036–2055. [CrossRef] [PubMed]

7. Guéraud, F.; Atalay, M.; Bresgen, N.; Cipak, A.; Eckl, P.M.; Huc, L.; Jouanin, I.; Siems, W.; Uchida, K. Chemistry and biochemistry of lipid peroxidation products. *Free Radic. Res.* **2010**, *44*, 1098–1124. [CrossRef] [PubMed]

8. Mukhopadhyay, P.; Rajesh, M.; Bátkai, S.; Patel, V.; Kashiwaya, Y.; Liaudet, L.; Evgenov, O.V.; Mackie, K.; Haskó, G.; Pacher, P. CB1 cannabinoid receptors promote oxidative stress and cell death in murine models of doxorubicin-induced cardiomyopathy and in human cardiomyocytes. *Cardiovasc. Res.* **2010**, *85*, 773–784. [CrossRef] [PubMed]

9. Han, K.H.; Lim, S.; Ryu, J.; Lee, C.W.; Kim, Y.; Kang, J.H.; Kang, S.S.; Ahn, Y.K.; Park, C.S.; Kim, J.J. CB1 and CB2 cannabinoid receptors differentially regulate the production of reactive oxygen species by macrophages. *Cardiovasc. Res.* **2009**, *84*, 378–386. [CrossRef] [PubMed]

10. Hao, X.; Chen, J.; Luo, Z.; He, H.; Yu, H.; Ma, L.; Ma, S.; Zhu, T.; Liu, D.; Zhu, Z. TRPV1 activation prevents high-salt diet-induced nocturnal hypertension in mice. *Pflügers Arch. Eur. J. Physiol.* **2011**, *461*, 345–353. [CrossRef] [PubMed]

11. Sies, H. Oxidative stress: A concept in redox biology and medicine. *Redox Biol.* **2015**, *4*, 180–183. [CrossRef] [PubMed]

12. Di Meo, S.; Reed, T.T.; Venditti, P.; Victor, V.M. Role of ROS and RNS Sources in Physiological and Pathological Conditions. *Oxid. Med. Cell. Longev.* **2016**, *2016*, 1245049. [CrossRef] [PubMed]

13. Weidinger, A.; Kozlov, A.V. Biological Activities of Reactive Oxygen and Nitrogen Species: Oxidative Stress versus Signal Transduction. *Biomolecules* **2015**, *5*, 472–484. [CrossRef] [PubMed]

14. Liu, Y.; Fiskum, G.; Schubert, D. Generation of reactive oxygen species by the mitochondrial electron transport chain. *J. Neurochem.* **2002**, *80*, 780–787. [CrossRef] [PubMed]

15. Aguirre, J.; Lambeth, J.D. Nox enzymes from fungus to fly to fish and what they tell us about Nox function in mammals. *Free Radic. Biol. Med.* **2010**, *49*, 1342–1353. [CrossRef] [PubMed]

16. Panday, A.; Sahoo, M.K.; Osorio, D.; Batra, S. NADPH oxidases: An overview from structure to innate immunity-associated pathologies. *Cell. Mol. Immunol.* **2015**, *12*, 5–23. [CrossRef] [PubMed]

17. Sevier, C.S.; Kaiser, C.A. Ero1 and redox homeostasis in the endoplasmic reticulum. *Biochim. Biophys. Acta* **2008**, *1783*, 549–556. [CrossRef] [PubMed]

18. Wang, W.; Wang, S.; Yan, L.; Madara, P.; Del Pilar Cintron, A.; Wesley, R.A.; Danner, R.L. Superoxide production and reactive oxygen species signaling by endothelial nitric-oxide synthase. *J. Biol. Chem.* **2000**, *275*, 16899–16903. [CrossRef] [PubMed]

19. Hrycay, E.G.; Bandiera, S.M. Involvement of Cytochrome P450 in Reactive Oxygen Species Formation and Cancer. *Adv. Pharmacol.* **2015**, *74*, 35–84. [CrossRef] [PubMed]

20. Nathan, C.; Cunningham-Bussel, A. Beyond oxidative stress: An immunologist's guide to reactive oxygen species. *Nat. Rev. Immunol.* **2013**, *13*, 349–361. [CrossRef] [PubMed]

21. Vergeade, A.; Mulder, P.; Vendeville, C.; Ventura-Clapier, R.; Thuillez, C.; Monteil, C. Xanthine oxidase contributes to mitochondrial ROS generation in an experimental model of cocaine-induced diastolic dysfunction. *J. Cardiovasc. Pharmacol.* **2012**, *60*, 538–543. [CrossRef] [PubMed]

22. McGrath, A.P.; Hilmer, K.M.; Collyer, C.A.; Shepard, E.M.; Elmore, B.O.; Brown, D.E.; Dooley, D.M.; Guss, J.M. Structure and inhibition of human diamine oxidase. *Biochemistry* **2009**, *48*, 9810–9822. [CrossRef] [PubMed]

23. Marnett, L.J. Prostaglandin synthase-mediated metabolism of carcinogens and a potential role for peroxyl radicals as reactive intermediates. *Environ. Health Perspect.* **1990**, *88*, 5–12. [CrossRef] [PubMed]

24. Schröder, P.; Krutmann, J. Environmental Oxidative Stress—Environmental Sources of ROS. In *Reactions, Processes*; Grune, T., Ed.; Springer: Berlin/Heidelberg, Germany, 2005; Volume 2, ISBN 3540235876.

25. Dupré-Crochet, S.; Erard, M.; Nüβe, O. ROS production in phagocytes: Why, when, and where? *J. Leukoc. Biol.* **2013**, *94*, 657–670. [CrossRef] [PubMed]

26. Görlach, A.; Bertram, K.; Hudecova, S.; Krizanova, O. Calcium and ROS: A mutual interplay. *Redox Biol.* **2015**, *6*, 260–271. [CrossRef] [PubMed]

27. Winkelstein, J.A.; Marino, M.C.; Johnston, R.B., Jr.; Boyle, J.; Curnutte, J.; Gallin, J.I.; Malech, H.L.; Holland, S.M.; Ochs, H.; Quie, P.; et al. Chronic granulomatous disease. Report on a national registry of 368 patients. *Medicine* **2000**, *79*, 155–169. [CrossRef]

28. Quie, P.G.; White, J.G.; Holmes, B.; Good, R.A. In vitro bactericidal capacity of human polymorphonuclear leukocytes: Diminished activity in chronic granulomatous disease of childhood. *J. Clin. Investig.* **1967**, *46*, 668–679. [CrossRef] [PubMed]

29. Holmes, B.; Quie, P.G.; Windhorst, D.B.; Good, R.A. Fatal granulomatous disease of childhood. An inborn abnormality of phagocytic function. *Lancet* **1966**, *1*, 1225–1228. [CrossRef]

30. Bylund, J.; Goldblatt, D.; Speert, D.P. Chronic granulomatous disease: From genetic defect to clinical presentation. *Adv. Exp. Med. Biol.* **2005**, *568*, 67–87. [CrossRef] [PubMed]

31. Quinn, M.T.; Ammons, M.C.; Deleo, F.R. The expanding role of NADPH oxidases in health and disease: No longer just agents of death and destruction. *Clin. Sci.* **2006**, *111*, 1–20. [CrossRef] [PubMed]

32. Cifuentes, M.E.; Pagano, P.J. Targeting reactive oxygen species in hypertension. *Curr. Opin. Nephrol. Hypertens.* **2006**, *15*, 179–186. [CrossRef] [PubMed]

33. Moncada, S.; Higgs, E.A. The discovery of nitric oxide and its role in vascular biology. *Br. J. Pharmacol.* **2006**, *147* (Suppl. 1), S193–S201. [CrossRef] [PubMed]

34. Drummond, G.R.; Sobey, C.G. Endothelial NADPH oxidases: Which NOX to target in vascular disease? *Trends Endocrinol. Metab.* **2014**, *25*, 452–463. [CrossRef] [PubMed]

35. Wilcox, C.S. Redox regulation of the afferent arteriole and tubuloglomerular feedback. *Acta Physiol. Scand.* **2003**, *179*, 217–223. [CrossRef] [PubMed]

36. Wilcox, C.S. Oxidative stress and nitric oxide deficiency in the kidney: A critical link to hypertension? *Am. J. Physiol. Regul. Integr. Comp. Physiol.* **2005**, *289*, R913–R935. [CrossRef] [PubMed]

37. Zou, A.P.; Cowley, A.W., Jr. Reactive oxygen species and molecular regulation of renal oxygenation. *Acta Physiol. Scand.* **2003**, *179*, 233–241. [CrossRef] [PubMed]

38. Juncos, R.; Hong, N.J.; Garvin, J.L. Differential effects of superoxide on luminal and basolateral Na^+/H^+ exchange in the thick ascending limb. *Am. J. Physiol. Regul. Integr. Comp. Physiol.* **2006**, *290*, R79–R83. [CrossRef] [PubMed]

39. Hoidal, J.R.; Brar, S.S.; Sturrock, A.B.; Sanders, K.A.; Dinger, B.; Fidone, S.; Kennedy, T.P. The role of endogenous NADPH oxidases in airway and pulmonary vascular smooth muscle function. *Antioxid. Redox Signal.* **2003**, *5*, 751–758. [CrossRef] [PubMed]

40. Brar, S.S.; Kennedy, T.P.; Sturrock, A.B.; Huecksteadt, T.P.; Quinn, M.T.; Murphy, T.M.; Chitano, P.; Hoidal, J.R. NADPH oxidase promotes NF-kappaB activation and proliferation in human airway smooth muscle. *Am. J. Physiol. Lung Cell. Mol. Physiol.* **2002**, *282*, L782L795. [CrossRef] [PubMed]

41. Piao, Y.J.; Seo, Y.H.; Hong, F.; Kim, J.H.; Kim, Y.J.; Kang, M.H.; Kim, B.S.; Jo, S.A.; Jo, I.; Jue, D.M.; et al. Nox 2 stimulates muscle differentiation via NF-kappaB/iNOS pathway. *Free Radic. Biol. Med.* **2005**, *38*, 989–1001. [CrossRef] [PubMed]

42. Kojim, S.; Ikeda, M.; Shibukawa, A.; Kamikawa, Y. Modification of 5-hydroxytryptophan-evoked 5-hydroxytryptamine formation of guinea pig colonic mucosa by reactive oxygen species. *Jpn. J. Pharmacol.* **2002**, *88*, 114–118. [CrossRef] [PubMed]

43. Wang, G.; Anrather, J.; Huang, J.; Speth, R.C.; Pickel, V.M.; Iadecola, C. NADPH oxidase contributes to angiotensin II signaling in the nucleus tractus solitarius. *J. Neurosci.* **2004**, *24*, 5516–5524. [CrossRef] [PubMed]

44. Erdös, B.; Broxson, C.S.; King, M.A.; Scarpace, P.J.; Tümer, N. Acute pressor effect of central angiotensin II is mediated by NAD(P)H-oxidase-dependent production of superoxide in the hypothalamic cardiovascular regulatory nuclei. *J. Hypertens.* **2006**, *24*, 109–116. [CrossRef] [PubMed]

45. Mander, P.K.; Jekabsone, A.; Brown, G.C. Microglia proliferation is regulated by hydrogen peroxide from NADPH oxidase. *J. Immunol.* **2006**, *176*, 1046–1052. [CrossRef] [PubMed]

46. Fritz, K.S.; Petersen, D.R. An overview of the chemistry and biology of reactive aldehydes. *Free Radic. Biol. Med.* **2013**, *59*, 85–91. [CrossRef] [PubMed]

47. Yin, H.; Xu, L.; Porter, N.A. Free radical lipid peroxidation: Mechanisms and analysis. *Chem. Rev.* **2011**, *111*, 5944–5972. [CrossRef] [PubMed]

48. Niki, E.; Yoshida, Y.; Saito, Y.; Noguchi, N. Lipid peroxidation: Mechanisms, inhibition, and biological effects. *Biochem. Biophys. Res. Commun.* **2005**, *338*, 668–676. [CrossRef] [PubMed]

49. Porter, N.A.; Caldwell, S.E.; Mills, K.A. Mechanisms of free radical oxidation of unsaturated lipids. *Lipids* **1995**, *30*, 277–290. [CrossRef] [PubMed]

50. Forman, H.J.; Fukuto, J.M.; Miller, T.; Zhang, H.; Rinna, A.; Levy, S. The chemistry of cell signaling by reactive oxygen and nitrogen species and 4-hydroxynonenal. *Arch. Biochem. Biophys.* **2008**, *477*, 183–195. [CrossRef] [PubMed]

51. Poli, G.; Schaur, R.J.; Siems, W.G.; Leonarduzzi, G. 4-hydroxynonenal: A membrane lipid oxidation product of medicinal interest. *Med. Res. Rev.* **2008**, *28*, 569–631. [CrossRef] [PubMed]

52. Noguchi, N. Role of oxidative stress in adaptive responses in special reference to atherogenesis. *J. Clin. Biochem. Nutr.* **2008**, *43*, 131–138. [CrossRef] [PubMed]

53. Zmijewski, J.W.; Landar, A.; Watanabe, N.; Dickinson, D.A.; Noguchi, N.; Darley-Usmar, V.M. Cell signalling by oxidized lipids and the role of reactive oxygen species in the endothelium. *Biochem. Soc. Trans.* **2005**, *33*, 1385–1389. [CrossRef] [PubMed]

54. Romano, A.; Serviddio, G.; Calcagnini, S.; Villani, R.; Giudetti, A.M.; Cassano, T.; Gaetani, S. Linking lipid peroxidation and neuropsychiatric disorders: Focus on 4-hydroxy-2-nonenal. *Free Radic. Biol. Med.* **2017**, *111*, 281–293. [CrossRef] [PubMed]

55. Negre-Salvayre, A.; Coatrieux, C.; Ingueneau, C.; Salvayre, R. Advanced lipid peroxidation end products in oxidative damage to proteins. Potential role in diseases and therapeutic prospects for the inhibitors. *Br. J. Pharmacol.* **2008**, *153*, 6–20. [CrossRef] [PubMed]

56. Winczura, A.; Zdżalik, D.; Tudek, B. Damage of DNA and proteins by major lipid peroxidation products in genome stability. *Free Radic. Res.* **2012**, *46*, 442–459. [CrossRef] [PubMed]

57. Winter, C.K.; Segall, H.J.; Haddon, W.F. Formation of cyclic adducts of deoxyguanosine with the aldehydes trans-4-hydroxy-2-hexenal and trans-4-hydroxy-2-nonenal in vitro. *Cancer Res.* **1986**, *46*, 5682–5686. [PubMed]

58. Chung, F.L.; Young, R.; Hecht, S.S. Formation of cyclic $1,N^2$-propanodeoxyguanosine adducts in DNA upon reaction with acrolein or crotonaldehyde. *Cancer Res.* **1984**, *44*, 990–995. [PubMed]

59. Cohn, J.A.; Tsai, L.; Friguet, B.; Szweda, L.I. Chemical characterization of a protein-4-hydroxy-2-nonenal cross-link: Immunochemical detection in mitochondria exposed to oxidative stress. *Arch. Biochem. Biophys.* **1996**, *328*, 158–164. [CrossRef] [PubMed]

60. Seto, H.; Okuda, T.; Takesue, T.; Ikemura, T. Reaction of malonaldehyde with nucleic acid. I. Formation of fluorescent pyrimido[1,2-a]purin-10-one nucleosides. *Bull. Chem. Soc. Jpn.* **1983**, *56*, 1799–1802. [CrossRef]

61. Esterbauer, H. Cytotoxicity and genotoxicity of lipid-oxidation products. *Am. J. Clin. Nutr.* **1993**, *57*, 779S–785S. [CrossRef] [PubMed]

62. Schaur, R.J. Basic aspects of the biochemical reactivity of 4-hydroxynonenal. *Mol. Asp. Med.* **2003**, *24*, 149–159. [CrossRef]

63. Zarkovic, N.; Cipak, A.; Jaganjac, M.; Borovic, S.; Zarkovic, K. Pathophysiological relevance of aldehydic protein modifications. *J. Proteom.* **2013**, *92*, 239–247. [CrossRef] [PubMed]

64. Rindgen, D.; Nakajima, M.; Wehrli, S.; Xu, K.; Blair, I.A. Covalent modifications to 2′-deoxyguanosine by 4-oxo-2-nonenal, a novel product of lipid peroxidation. *Chem. Res. Toxicol.* **1999**, *12*, 1195–1204. [CrossRef] [PubMed]

65. Lee, S.H.; Rindgen, D.; Bible, R.H., Jr.; Hajdu, E.; Blair, I.A. Characterization of 2′-deoxyadenosine adducts derived from 4-oxo-2-nonenal, a novel product of lipid peroxidation. *Chem. Res. Toxicol.* **2000**, *13*, 565–574. [CrossRef] [PubMed]

66. Pollack, M.; Oe, T.; Lee, S.H.; Silva Elipe, M.V.; Arison, B.H.; Blair, I.A. Characterization of 2′-deoxycytidine adducts derived from 4-oxo-2-nonenal, a novel lipid peroxidation product. *Chem. Res. Toxicol.* **2003**, *16*, 893–900. [CrossRef] [PubMed]

67. Williams, M.V.; Lee, S.H.; Pollack, M.; Blair, I.A. Endogenous lipid hydroperoxide-mediated DNA-adduct formation in min mice. *J. Biol. Chem.* **2006**, *281*, 10127–10133. [CrossRef] [PubMed]

68. Del Rio, D.; Stewart, A.J.; Pellegrini, N. A review of recent studies on malondialdehyde as toxic molecule and biological marker of oxidative stress. *Nutr. Metab. Cardiovasc. Dis.* **2005**, *15*, 316–328. [CrossRef] [PubMed]

69. Zhao, J.; Chen, J.; Zhu, H.; Xiong, Y.L. Mass spectrometric evidence of malonaldehyde and 4-hydroxynonenal adductions to radical-scavenging soy peptides. *J. Agric. Food Chem.* **2012**, *60*, 9727–9736. [CrossRef] [PubMed]

70. Mukai, F.H.; Goldstein, B.D. Mutagenicity of malonaldehyde, a decomposition product of peroxidized polyunsaturated fatty acids. *Science* **1976**, *191*, 868–869. [CrossRef] [PubMed]

71. Basu, A.K.; Marnett, L.J. Unequivocal demonstration that malondialdehyde is a mutagen. *Carcinogenesis* **1983**, *4*, 331–333. [CrossRef] [PubMed]

72. Marnett, L.J.; Hurd, H.K.; Hollstein, M.C.; Levin, D.E.; Esterbauer, H.; Ames, B.N. Naturally occurring carbonyl compounds are mutagens in Salmonella tester strain TA104. *Mutat. Res.* **1985**, *148*, 25–34. [CrossRef]

73. Vasanthi, P.; Nalini, G.; Rajasekhar, G. Status of oxidative stress in rheumatoid arthritis. *Int. J. Rheum. Dis.* **2009**, *12*, 29–33. [CrossRef] [PubMed]

74. Mishra, R.; Singh, A.; Chandra, V.; Negi, M.P.; Tripathy, B.C.; Prakash, J.; Gupta, V. A comparative analysis of serological parameters and oxidative stress in osteoarthritis and rheumatoid arthritis. *Rheumatol. Int.* **2012**, *32*, 2377–2382. [CrossRef] [PubMed]

75. Mateen, S.; Moin, S.; Khan, A.Q.; Zafar, A.; Fatima, N. Increased Reactive Oxygen Species Formation and Oxidative Stress in Rheumatoid Arthritis. *PLoS ONE* **2016**, *11*, e0152925. [CrossRef] [PubMed]

76. Shah, D.; Wanchu, A.; Bhatnagar, A. Interaction between oxidative stress and chemokines: Possible pathogenic role in systemic lupus erythematosus and rheumatoid arthritis. *Immunobiology* **2011**, *216*, 1010–1017. [CrossRef] [PubMed]

77. Liao, C.C.; Chang, Y.S.; Cheng, C.W.; Chi, W.M.; Tsai, K.L.; Chen, W.J.; Kung, T.S.; Tai, C.C.; Lin, Y.F.; Lin, H.T.; et al. Isotypes of autoantibodies against differentially expressed novel malondialdehyde-modified peptide adducts in serum of Taiwanese women with rheumatoid arthritis. *J. Proteom.* **2018**, *170*, 141–150. [CrossRef] [PubMed]

78. Cvetkovic, J.T.; Wållberg-Jonsson, S.; Ahmed, E.; Rantapää-Dahlqvist, S.; Lefvert, A.K. Increased levels of autoantibodies against copper-oxidized low density lipoprotein, malondialdehyde-modified low density lipoprotein and cardiolipin in patients with rheumatoid arthritis. *Rheumatology* **2002**, *41*, 988–995. [CrossRef] [PubMed]

79. Wållberg-Jonsson, S.; Cvetkovic, J.T.; Sundqvist, K.G.; Lefvert, A.K.; Rantapää-Dahlqvist, S. Activation of the immune system and inflammatory activity in relation to markers of atherothrombotic disease and atherosclerosis in rheumatoid arthritis. *J. Rheumatol.* **2002**, *29*, 875–882. [PubMed]

80. Chung, F.L.; Chen, H.J.; Nath, R.G. Lipid peroxidation as a potential endogenous source for the formation of exocyclic DNA adducts. *Carcinogenesis* **1996**, *17*, 2105–2111. [CrossRef] [PubMed]

81. Chung, F.L.; Nath, R.G.; Nagao, M.; Nishikawa, A.; Zhou, G.D.; Randerath, K. Endogenous formation and significance of $1,N^2$-propanodeoxyguanosine adducts. *Mutat. Res.* **1999**, *424*, 71–81. [CrossRef]

82. Wang, M.Y.; Chung, F.L.; Hecht, S.S. Identification of crotonaldehyde as a hepatic microsomal metabolite formed by alpha-hydroxylation of the carcinogen N-nitrosopyrrolidine. *Chem. Res. Toxicol.* **1988**, *1*, 28–31. [CrossRef] [PubMed]

83. Wang, M.; McIntee, E.J.; Cheng, G.; Shi, Y.; Villalta, P.W.; Hecht, S.S. Identification of paraldol-deoxyguanosine adducts in DNA reacted with crotonaldehyde. *Chem. Res. Toxicol.* **2000**, *13*, 1065–1074. [CrossRef] [PubMed]

84. International Agency for Research on Cancer. IARC Monographs on the Evaluation of Carcinogenic Risks to Humans. In *Dry Cleaning, Some Chlorinated Solvents and Other Industrial Chemicals*; International Agency for Research on Cancer: Lyon, France, 1995; Volume 63, pp. 373–391.

85. Chung, F.L.; Tanaka, T.; Hecht, S.S. Induction of liver tumors in F344 rats by crotonaldehyde. *Cancer Res.* **1986**, *46*, 1285–1289. [PubMed]

86. Ichihashi, K.; Osawa, T.; Toyokuni, S.; Uchida, K. Endogenous formation of protein adducts with carcinogenic aldehydes: Implications for oxidative stress. *J. Biol. Chem.* **2001**, *276*, 23903–23913. [CrossRef] [PubMed]

87. Furuhata, A.; Nakamura, M.; Osawa, T.; Uchida, K. Thiolation of protein-bound carcinogenic aldehyde. An electrophilic acrolein-lysine adduct that covalently binds to thiols. *J. Biol. Chem.* **2002**, *277*, 27919–27926. [CrossRef] [PubMed]

88. Uchida, K.; Kanematsu, M.; Morimitsu, Y.; Osawa, T.; Noguchi, N.; Niki, E. Acrolein is a product of lipid peroxidation reaction. Formation of free acrolein and its conjugate with lysine residues in oxidized low density lipoproteins. *J. Biol. Chem.* **1998**, *273*, 16058–16066. [CrossRef] [PubMed]

89. Esterbauer, H.; Schaur, R.J.; Zollner, H. Chemistry and biochemistry of 4-hydroxynonenal, malonaldehyde and related aldehydes. *Free Radic. Biol. Med.* **1991**, *11*, 81–128. [CrossRef]

90. Cohen, G.; Riahi, Y.; Sunda, V.; Deplano, S.; Chatgilialoglu, C.; Ferreri, C.; Kaiser, N.; Sasson, S. Signaling properties of 4-hydroxyalkenals formed by lipid peroxidation in diabetes. *Free Radic. Biol. Med.* **2013**, *65*, 978–987. [CrossRef] [PubMed]

91. Higdon, A.; Diers, A.R.; Oh, J.Y.; Landar, A.; Darley-Usmar, V.M. Cell signalling by reactive lipid species: New concepts and molecular mechanisms. *Biochem. J.* **2012**, *442*, 453–464. [CrossRef] [PubMed]

92. Riahi, Y.; Cohen, G.; Shamni, O.; Sasson, S. Signaling and cytotoxic functions of 4-hydroxyalkenals. *Am. J. Physiol. Endocrinol. Metab.* **2010**, *299*, E879–E886. [CrossRef] [PubMed]

93. Schaur, R.J.; Siems, W.; Bresgen, N.; Eckl, P.M. 4-Hydroxy-nonenal-A Bioactive Lipid Peroxidation Product. *Biomolecules* **2015**, *5*, 2247–2337. [CrossRef] [PubMed]

94. Calabrese, E.J.; Bachmann, K.A.; Bailer, A.J.; Bolger, P.M.; Borak, J.; Cai, L.; Cedergreen, N.; Cherian, M.G.; Chiueh, C.C.; Clarkson, T.W.; et al. Biological stress response terminology: Integrating the concepts of adaptive response and preconditioning stress within a hormetic dose-response framework. *Toxicol. Appl. Pharmacol.* **2007**, *222*, 122–128. [CrossRef] [PubMed]

95. Pizzimenti, S.; Barrera, G.; Dianzani, M.U.; Brüsselbach, S. Inhibition of D1, D2, and A-cyclin expression in HL-60 cells by the lipid peroxydation product 4-hydroxynonenal. *Free Radic. Biol. Med.* **1999**, *26*, 1578–1586. [CrossRef]

96. Huang, Y.; Li, W.; Kong, A.N. Anti-oxidative stress regulator NF-E2-related factor 2 mediates the adaptive induction of antioxidant and detoxifying enzymes by lipid peroxidation metabolite 4-hydroxynonenal. *Cell Biosci.* **2012**, *2*, 40. [CrossRef] [PubMed]

97. Zhang, Y.; Sano, M.; Shinmura, K.; Tamaki, K.; Katsumata, Y.; Matsuhashi, T.; Morizane, S.; Ito, H.; Hishiki, T.; Endo, J.; et al. 4-hydroxy-2-nonenal protects against cardiac ischemia-reperfusion injury via the Nrf2-dependent pathway. *J. Mol. Cell. Cardiol.* **2010**, *49*, 576–586. [CrossRef] [PubMed]

98. Siow, R.C.; Ishii, T.; Mann, G.E. Modulation of antioxidant gene expression by 4-hydroxynonenal: Atheroprotective role of the Nrf2/ARE transcription pathway. *Redox Rep.* **2007**, *12*, 11–15. [CrossRef] [PubMed]

99. Tanito, M.; Agbaga, M.P.; Anderson, R.E. Upregulation of thioredoxin system via Nrf2-antioxidant responsive element pathway in adaptive-retinal neuroprotection in vivo and in vitro. *Free Radic. Biol. Med.* **2007**, *42*, 1838–1850. [CrossRef] [PubMed]

100. Ishii, T.; Itoh, K.; Ruiz, E.; Leake, D.S.; Unoki, H.; Yamamoto, M.; Mann, G.E. Role of Nrf2 in the regulation of CD36 and stress protein expression in murine macrophages: Activation by oxidatively modified LDL and 4-hydroxynonenal. *Circ. Res.* **2004**, *94*, 609–616. [CrossRef] [PubMed]

101. De Petrocellis, L.; Cascio, M.G.; Di Marzo, V. The endocannabinoid system: A general view and latest additions. *Br. J. Pharmacol.* **2004**, *141*, 765–774. [CrossRef] [PubMed]

102. Mechoulam, R.; Fride, E.; Di Marzo, V. Endocannabinoids. *Eur. J. Pharmacol.* **1998**, *359*, 1–18. [CrossRef]

103. Devane, W.A.; Hanus, L.; Breuer, A.; Pertwee, R.G.; Stevenson, L.A.; Griffin, G.; Gibson, D.; Mandelbaum, A.; Etinger, A.; Mechoulam, R. Isolation and structure of a brain constituent that binds to the cannabinoid receptor. *Science* **1992**, *258*, 1946–1949. [CrossRef] [PubMed]

104. Sierra, S.; Luquin, N.; Navarro-Otano, J. The endocannabinoid system in cardiovascular function: Novel insights and clinical implications. *Clin. Auton Res.* **2018**, *28*, 35–52. [CrossRef] [PubMed]

105. Fonseca, B.M.; Costa, M.A.; Almada, M.; Correia-da-Silva, G.; Teixeira, N.A. Endogenous cannabinoids revisited: A biochemistry perspective. *Prostaglandins Other Lipid Mediat.* **2013**, *102–103*, 13–30. [CrossRef] [PubMed]

106. Sugiura, T.; Kondo, S.; Sukagawa, A.; Nakane, S.; Shinoda, A.; Itoh, K.; Yamashita, A.; Waku, K. 2-Arachidonoylglycerol: A possible endogenous cannabinoid receptor ligand in brain. *Biochem. Biophys. Res. Commun.* **1995**, *215*, 89–97. [CrossRef] [PubMed]

107. Stella, N.; Schweitzer, P.; Piomelli, D. A second endogenous cannabinoid that modulates long-term potentiation. *Nature* **1997**, *388*, 773–778. [CrossRef] [PubMed]

108. Fonseca, B.M.; Correia-da-Silva, G.; Taylor, A.H.; Lam, P.M.; Marczylo, T.H.; Bell, S.C.; Konje, J.C.; Teixeira, N.A. The endocannabinoid 2-arachidonoylglycerol (2-AG) and metabolizing enzymes during rat fetoplacental development: A role in uterine remodelling. *Int. J. Biochem. Cell Biol.* **2010**, *42*, 1884–1892. [CrossRef] [PubMed]

109. Fonseca, B.M.; Correia-da-Silva, G.; Taylor, A.H.; Lam, P.M.; Marczylo, T.H.; Konje, J.C.; Bell, S.C.; Teixeira, N.A. N-acylethanolamine levels and expression of their metabolizing enzymes during pregnancy. *Endocrinology* **2010**, *151*, 3965–3974. [CrossRef] [PubMed]

110. Di Marzo, V. The endocannabinoid system: Its general strategy of action, tools for its pharmacological manipulation and potential therapeutic exploitation. *Pharmacol. Res.* **2009**, *60*, 77–84. [CrossRef] [PubMed]

111. Gonsiorek, W.; Lunn, C.; Fan, X.; Narula, S.; Lundell, D.; Hipkin, R.W. Endocannabinoid 2-arachidonyl glycerol is a full agonist through human type 2 cannabinoid receptor: Antagonism by anandamide. *Mol. Pharmacol.* **2000**, *57*, 1045–1050. [PubMed]

112. Pertwee, R.G. The pharmacology of cannabinoid receptors and their ligands: An overview. *Int. J. Obes.* **2006**, *30* (Suppl. 1), S13–S18. [CrossRef] [PubMed]

113. Matsuda, L.A.; Lolait, S.J.; Brownstein, M.J.; Young, A.C.; Bonner, T.I. Structure of a cannabinoid receptor and functional expression of the cloned cDNA. *Nature* **1990**, *346*, 561–564. [CrossRef] [PubMed]

114. Freund, T.F.; Katona, I.; Piomelli, D. Role of endogenous cannabinoids in synaptic signaling. *Physiol. Rev.* **2003**, *83*, 1017–1066. [CrossRef] [PubMed]

115. Pacher, P.; Steffens, S. The emerging role of the endocannabinoid system in cardiovascular disease. *Semin. Immunopathol.* **2009**, *31*, 63–77. [CrossRef] [PubMed]

116. Tam, J.; Hinden, L.; Drori, A.; Udi, S.; Azar, S.; Baraghithy, S. The therapeutic potential of targeting the peripheral endocannabinoid/CB$_1$ receptor system. *Eur. J. Intern. Med.* **2018**, *49*, 23–29. [CrossRef] [PubMed]

117. Hu, S.S.; Mackie, K. Distribution of the Endocannabinoid System in the Central Nervous System. *Handb. Exp. Pharmacol.* **2015**, *231*, 59–93. [CrossRef] [PubMed]

118. Bénard, G.; Massa, F.; Puente, N.; Lourenço, J.; Bellocchio, L.; Soria-Gómez, E.; Matias, I.; Delamarre, A.; Metna-Laurent, M.; Cannich, A.; et al. Mitochondrial CB$_1$ receptors regulate neuronal energy metabolism. *Nat. Neurosci.* **2012**, *15*, 558–564. [CrossRef] [PubMed]

119. Pacher, P.; Kunos, G. Modulating the endocannabinoid system in human health and disease—Successes and failures. *FEBS J.* **2013**, *280*, 1918–1943. [CrossRef] [PubMed]

120. Pertwee, R.G. Cannabinoid receptor ligands: Clinical and neuropharmacological considerations, relevant to future drug discovery and development. *Expert Opin. Investig. Drugs* **2000**, *9*, 1553–1571. [CrossRef] [PubMed]

121. Jordan, J.D.; Carey, K.D.; Stork, P.J.; Iyengar, R. Modulation of rap activity by direct interaction of Galpha(o) with Rap1 GTPase-activating protein. *J. Biol. Chem.* **1999**, *274*, 21507–21510. [CrossRef] [PubMed]

122. Glass, M.; Felder, C.C. Concurrent stimulation of cannabinoid CB1 and dopamine D2 receptors augments cAMP accumulation in striatal neurons: Evidence for a G$_s$ linkage to the CB1 receptor. *J. Neurosci.* **1997**, *17*, 5327–5333. [CrossRef] [PubMed]

123. Maneuf, Y.P.; Brotchie, J.M. Paradoxical action of the cannabinoid WIN 55, 212–212 in stimulated and basal cyclic AMP accumulation in rat globus pallidus slices. *Br. J. Pharmacol.* **1997**, *120*, 1397–1398. [CrossRef] [PubMed]

124. Howlett, A.C. Cannabinoid receptor signaling. *Handb. Exp. Pharmacol.* **2005**, *168*, 53–79. [CrossRef]

125. Turu, G.; Hunyady, L. Signal transduction of the CB1 cannabinoid receptor. *J. Mol. Endocrinol.* **2010**, *44*, 75–85. [CrossRef] [PubMed]

126. Pacher, P.; Batkai, S.; Kunos, G. The endocannabinoid system as an emerging target of pharmacotherapy. *Pharmacol. Rev.* **2006**, *58*, 389–462. [CrossRef] [PubMed]

127. Zygmunt, P.M.; Petersson, J.; Andersson, D.A.; Chuang, H.; Sørgård, M.; Di Marzo, V.; Julius, D.; Högestätt, E.D. Vanilloid receptors on sensory nerves mediate the vasodilator action of anandamide. *Nature* **1999**, *400*, 452–457. [CrossRef] [PubMed]

128. Starowicz, K.; Nigam, S.; Di Marzo, V. Biochemistry and pharmacology of endovanilloids. *Pharmacol. Ther.* **2007**, *114*, 13–33. [CrossRef] [PubMed]

129. Di Marzo, V.; De Petrocellis, L. Endocannabinoids as regulators of transient receptor potential (TRP) channels: A further opportunity to develop new endocannabinoid-based therapeutic drugs. *Curr. Med. Chem.* **2010**, *17*, 1430–1449. [CrossRef] [PubMed]

130. O'Sullivan, S.E. Cannabinoids go nuclear: Evidence for activation of peroxisome proliferator-activated receptors. *Br. J. Pharmacol.* **2007**, *152*, 576–582. [CrossRef] [PubMed]

131. Lu, H.C.; Mackie, K. An introduction to the endogenous cannabinoid system. *Biol. Psychiatry* **2016**, *79*, 516–525. [CrossRef] [PubMed]

132. Haugh, O.; Penman, J.; Irving, A.J.; Campbell, V.A. The emerging role of the cannabinoid receptor family in peripheral and neuro-immune interactions. *Curr. Drug Targets* **2016**, *17*, 1834–1840. [CrossRef] [PubMed]

133. Yang, D.; Luo, Z.; Ma, S.; Wong, W.T.; Ma, L.; Zhong, J.; He, H.; Zhao, Z.; Cao, T.; Yan, Z.; et al. Activation of TRPV1 by dietary capsaicin improves endothelium-dependent vasorelaxation and prevents hypertension. *Cell Metab.* **2010**, *12*, 130–141. [CrossRef] [PubMed]

134. Poblete, I.M.; Orliac, M.L.; Briones, R.; Adler-Graschinsky, E.; Huidobro-Toro, J.P. Anandamide elicits an acute release of nitric oxide through endothelial TRPV1 receptor activation in the rat arterial mesenteric bed. *J. Physiol.* **2005**, *568*, 539–551. [CrossRef] [PubMed]

135. Randhawa, P.K.; Jaggi, A.S. TRPV1 channels in cardiovascular system: A double edged sword? *Int. J. Cardiol.* **2017**, *228*, 103–113. [CrossRef] [PubMed]

136. Di Marzo, V.; Bisogno, T.; De Petrocellis, L. Anandamide: Some like it hot. *Trends Pharmacol. Sci.* **2001**, *22*, 346–349. [CrossRef]

137. Starowicz, K.; Makuch, W.; Osikowicz, M.; Piscitelli, F.; Petrosino, S.; Di Marzo, V.; Przewlocka, B. Spinal anandamide produces analgesia in neuropathic rats: Possible CB_1- and TRPV1-mediated mechanisms. *Neuropharmacology* **2012**, *62*, 1746–1755. [CrossRef] [PubMed]

138. Mattace Raso, G.; Simeoli, R.; Russo, R.; Santoro, A.; Pirozzi, C.; d'Emmanuele di Villa Bianca, R.; Mitidieri, E.; Paciello, O.; Pagano, T.B.; Orefice, N.S.; et al. N-Palmitoylethanolamide protects the kidney from hypertensive injury in spontaneously hypertensive rats via inhibition of oxidative stress. *Pharmacol. Res.* **2013**, *76*, 67–76. [CrossRef] [PubMed]

139. Zoete, V.; Grosdidier, A.; Michielin, O. Peroxisome proliferator-activated receptor structures: Ligand specificity, molecular switch and interactions with regulators. *Biochim. Biophys. Acta* **2007**, *1771*, 915–925. [CrossRef] [PubMed]

140. Shin, S.J.; Lim, J.H.; Chung, S.; Youn, D.Y.; Chung, H.W.; Kim, H.W.; Lee, J.-H.; Chang, Y.S.; Park, C.W.; et al. Peroxisome proliferator-activated receptor-alpha activator fenofibrate prevents high-fat diet-induced renal lipotoxicity in spontaneously hypertensive rats. *Hypertens. Res.* **2009**, *32*, 835–845. [CrossRef] [PubMed]

141. Gelosa, P.; Banfi, C.; Gianella, A.; Brioschi, M.; Pignieri, A.; Nobili, E.; Castiglioni, L.; Cimino, M.; Tremoli, E.; Sironi, L. Peroxisome proliferator-activated receptor α agonism prevents renal damage and the oxidative stress and inflammatory processes affecting the brains of stroke-prone rats. *J. Pharmacol. Exp. Ther.* **2010**, *335*, 324–331. [CrossRef] [PubMed]

142. Rosenson, R.S.; Wolff, D.A.; Huskin, A.L.; Helenowski, I.B.; Rademaker, A.W. Fenofibrate therapy ameliorates fasting and postprandial lipoproteinemia, oxidative stress, and the inflammatory response in subjects with hypertriglyceridemia and the metabolic syndrome. *Diabetes Care* **2007**, *30*, 1945–1951. [CrossRef] [PubMed]

143. Sun, Y.; Alexander, S.P.H.; Kendall, D.A.; Bennett, A.J. Cannabinoids and PPARalpha signalling. *Biochem. Soc. Trans.* **2006**, *34*, 1095–1097. [CrossRef] [PubMed]
144. Bouaboula, M.; Hilairet, S.; Marchand, J.; Fajas, L.; Le Fur, G.; Casellas, P. Anandamide induced PPARgamma transcriptional activation and 3T3-L1 preadipocyte differentiation. *Eur. J. Pharmacol.* **2005**, *517*, 174–181. [CrossRef] [PubMed]
145. Du, H.; Chen, X.; Zhang, J.; Chen, C. Inhibition of COX-2 expression by endocannabinoid 2-arachidonoylglycerol is mediated via PPAR-γ. *Br. J. Pharmacol.* **2011**, *163*, 1533–1549. [CrossRef] [PubMed]
146. Baker, D.; Pryce, G.; Davies, W.L.; Hiley, C.R. In silico patent searching reveals a new cannabinoid receptor. *Trends Pharmacol. Sci.* **2006**, *27*, 1–4. [CrossRef] [PubMed]
147. Shah, U.; Kowalski, T.J. GPR119 agonists for the potential treatment of type 2 diabetes and related metabolic disorders. *Vitam. Horm.* **2010**, *84*, 415–448. [CrossRef] [PubMed]
148. Lauckner, J.E.; Jensen, J.B.; Chen, H.Y.; Lu, H.C.; Hille, B.; Mackie, K. GPR55 is a cannabinoid receptor that increases intracellular calcium and inhibits M current. *Proc. Natl. Acad. Sci. USA* **2008**, *105*, 2699–2704. [CrossRef] [PubMed]
149. Ohno-Shosaku, T.; Maejima, T.; Kano, M. Endogenous cannabinoids mediate retrograde signals from depolarized postsynaptic neurons to presynaptic terminals. *Neuron* **2001**, *29*, 729–738. [CrossRef]
150. Hashimotodani, Y.; Ohno-Shosaku, T.; Kano, M. Endocannabinoids and synaptic function in the CNS. *Neuroscientist* **2007**, *13*, 127–137. [CrossRef] [PubMed]
151. Oddi, S.; Fezza, F.; Pasquariello, N.; De Simone, C.; Rapino, C.; Dainese, E.; Finazzi-Agrò, A.; Maccarrone, M. Evidence for the intracellular accumulation of anandamide in adiposomes. *Cell. Mol. Life Sci.* **2008**, *65*, 840–850. [CrossRef] [PubMed]
152. Di Marzo, V.; Fontana, A.; Cadas, H.; Schinelli, S.; Cimino, G.; Schwartz, J.C.; Piomelli, D. Formation and inactivation of endogenous cannabinoid anandamide in central neurons. *Nature* **1994**, *372*, 686–691. [CrossRef] [PubMed]
153. Wang, J.; Ueda, N. Biology of endocannabinoid synthesis system. *Prostaglandins Other Lipid Mediat.* **2009**, *89*, 112–119. [CrossRef] [PubMed]
154. Cravatt, B.F.; Giang, D.K.; Mayfield, S.P.; Boger, D.L.; Lerner, R.A.; Gilula, N.B. Molecular characterization of an enzyme that degrades neuromodulatory fatty-acid amides. *Nature* **1996**, *384*, 83–87. [CrossRef] [PubMed]
155. Seillier, A.; Advani, T.; Cassano, T.; Hensler, J.G.; Giuffrida, A. Inhibition of fatty-acid amide hydrolase and CB1 receptor antagonism differentially affect behavioural responses in normal and PCP-treated rats. *Int. J. Neuropsychopharmacol.* **2010**, *13*, 373–386. [CrossRef] [PubMed]
156. Bedse, G.; Colangeli, R.; Lavecchia, A.M.; Romano, A.; Altieri, F.; Cifani, C.; Cassano, T.; Gaetani, S. Role of the basolateral amygdala in mediating the effects of the fatty acid amide hydrolase inhibitor URB597 on HPA axis response to stress. *Eur. Neuropsychopharmacol.* **2014**, *24*, 1511–1523. [CrossRef] [PubMed]
157. Di Marzo, V. Targeting the endocannabinoid system: To enhance or reduce? *Nat. Rev. Drug Discov.* **2008**, *7*, 438–455. [CrossRef] [PubMed]
158. Ueda, N.; Tsuboi, K.; Uyama, T. N-acylethanolamine metabolism with special reference to N-acylethanolamine-hydrolyzing acid amidase (NAAA). *Prog. Lipid Res.* **2010**, *49*, 299–315. [CrossRef] [PubMed]
159. Ueda, N.; Tsuboi, K.; Uyama, T.; Ohnishi, T. Biosynthesis and degradation of the endocannabinoid 2-arachidonoylglycerol. *Biofactors* **2011**, *37*, 1–7. [CrossRef] [PubMed]
160. Rouzer, C.A.; Marnett, L.J. Endocannabinoid oxygenation by cyclooxygenases, lipoxygenases, andcytochromes P450: Cross-talk between the eicosanoid and endocannabinoid signaling pathways. *Chem. Rev.* **2011**, *111*, 5899–5921. [CrossRef] [PubMed]
161. Bradshaw, H.B.; Walker, J.M. The expanding field of cannabimimetic and related lipid mediators. *Br. J. Pharmacol.* **2009**, *144*, 459–465. [CrossRef] [PubMed]
162. Berdyshev, E.V.; Schmid, P.C.; Krebsbach, R.J.; Hillard, C.J.; Huang, C.; Chen, N.; Dong, Z.; Schmid, H.H. Cannabinoid-receptor-independent cell signalling by N-acylethanolamines. *Biochem. J.* **2001**, *360*, 67–75. [CrossRef] [PubMed]
163. Azari, E.K.; Ramachandran, D.; Weibel, S.; Arnold, M.; Romano, A.; Gaetani, S.; Langhans, W.; Mansouri, A. Vagal afferents are not necessary for the satiety effect of the gut lipid messenger oleoylethanolamide. *Am. J. Physiol. Regul. Integr. Comp. Physiol.* **2014**, *307*, R167–R178. [CrossRef] [PubMed]

164. Provensi, G.; Coccurello, R.; Umehara, H.; Munari, L.; Giacovazzo, G.; Galeotti, N.; Nosi, D.; Gaetani, S.; Romano, A.; Moles, A.; et al. Satiety factor oleoylethanolamide recruits the brain histaminergic system to inhibit food intake. *Proc. Natl. Acad. Sci. USA* **2014**, *111*, 11527–11532. [CrossRef] [PubMed]

165. Romano, A.; Cassano, T.; Tempesta, B.; Cianci, S.; Dipasquale, P.; Coccurello, R.; Cuomo, V.; Gaetani, S. The satiety signal oleoylethanolamide stimulates oxytocin neurosecretion from rat hypothalamic neurons. *Peptides* **2013**, *49*, 21–26. [CrossRef] [PubMed]

166. Romano, A.; Karimian Azari, E.; Tempesta, B.; Mansouri, A.; Micioni Di Bonaventura, M.V.; Ramachandran, D.; Lutz, T.A.; Bedse, G.; Langhans, W.; Gaetani, S. High dietary fat intake influences the activation of specific hindbrain and hypothalamic nuclei by the satiety factor oleoylethanolamide. *Physiol. Behav.* **2014**, *136*, 55–62. [CrossRef] [PubMed]

167. Romano, A.; Coccurello, R.; Giacovazzo, G.; Bedse, G.; Moles, A.; Gaetani, S. Oleoylethanolamide: A novel potential pharmacological alternative to cannabinoid antagonists for the control of appetite. *Biomed. Res. Int.* **2014**, *2014*, 203425. [CrossRef] [PubMed]

168. Gaetani, S.; Kaye, W.H.; Cuomo, V.; Piomelli, D. Role of endocannabinoids and their analogues in obesity and eating disorders. *Eat. Weight Disord.* **2008**, *13*, e42–e48. [CrossRef] [PubMed]

169. Fidaleo, M.; Fanelli, F.; Ceru, M.P.; Moreno, S. Neuroprotective properties of peroxisome proliferator-activated receptor alpha (PPARα) and its lipid ligands. *Curr. Med. Chem.* **2014**, *21*, 2803–2821. [CrossRef] [PubMed]

170. Bedse, G.; Romano, A.; Lavecchia, A.M.; Cassano, T.; Gaetani, S. The role of endocannabinoid signaling in the molecular mechanisms of neurodegeneration in Alzheimer's disease. *J. Alzheimer Dis.* **2015**, *43*, 1115–1136. [CrossRef] [PubMed]

171. Scuderi, C.; Bronzuoli, M.R.; Facchinetti, R.; Pace, L.; Ferraro, L.; Broad, K.D.; Serviddio, G.; Bellanti, F.; Palombelli, G.; Carpinelli, G.; et al. Ultramicronized palmitoylethanolamide rescues learning and memory impairments in a triple transgenic mouse model of Alzheimer's disease by exerting anti-inflammatory and neuroprotective effects. *Transl. Psychiatry* **2018**, *8*, 32. [CrossRef] [PubMed]

172. Bronzuoli, M.R.; Facchinetti, R.; Steardo, L., Jr.; Romano, A.; Stecca, C.; Passarella, S.; Steardo, L.; Cassano, T.; Scuderi, C. Palmitoylethanolamide Dampens Reactive Astrogliosis and Improves Neuronal Trophic Support in a Triple Transgenic Model of Alzheimer's Disease: In Vitro and In Vivo Evidence. *Oxid. Med. Cell. Longev.* **2018**, *2018*, 4720532. [CrossRef] [PubMed]

173. Hanus, L.; Abu-Lafi, S.; Fride, E.; Breuer, A.; Vogel, Z.; Shalev, D.E.; Kustanovich, I.; Mechoulam, R. 2-arachidonyl glyceryl ether, an endogenous agonist of the cannabinoid CB1 receptor. *Proc. Natl. Acad. Sci. USA* **2001**, *98*, 3662–3665. [CrossRef] [PubMed]

174. Porter, A.C.; Sauer, J.M.; Knierman, M.D.; Becker, G.W.; Berna, M.J.; Bao, J.; Nomikos, G.G.; Carter, P.; Bymaster, F.P.; Leese, A.B.; et al. Characterization of a novel endocannabinoid, virodhamine, with antagonist activity at the CB1 receptor. *J. Pharmacol. Exp. Ther.* **2002**, *301*, 1020–1024. [CrossRef] [PubMed]

175. Sharir, H.; Console-Bram, L.; Mundy, C.; Popoff, S.N.; Kapur, A.; Abood, M.E. The Endocannabinoids Anandamide and Virodhamine Modulate the Activity of the Candidate Cannabinoid Receptor GPR55. *J. Neuroimmune Pharmacol.* **2012**, *7*, 856–865. [CrossRef] [PubMed]

176. Kohno, M.; Hasegawa, H.; Inoue, A.; Muraoka, M.; Miyazaki, T.; Oka, K.; Yasukawa, M. Identification of N-arachidonylglycine as the endogenous ligand for orphan G-protein-coupled receptor GPR18. *Biochem. Biophys. Res. Commun.* **2006**, *347*, 827–832. [CrossRef] [PubMed]

177. Lipina, C.; Hundal, H.S. Modulation of cellular redox homeostasis by the endocannabinoid system. *Open Biol.* **2016**, *6*, 150276. [CrossRef] [PubMed]

178. Horváth, B.; Mukhopadhyay, P.; Kechrid, M.; Patel, V.; Tanchian, G.; Wink, D.A.; Gertsch, J.; Pacher, P. β-Caryophyllene ameliorates cisplatin-induced nephrotoxicity in a cannabinoid 2 receptor-dependent manner. *Free Radic. Biol. Med.* **2012**, *52*, 1325–1333. [CrossRef] [PubMed]

179. Mukhopadhyay, P.; Pan, H.; Rajesh, M.; Batkai, S.; Patel, V.; Harvey-White, J.; Mukhopadhyay, B.; Hasko, G.; Gao, B.; Mackie, K.; et al. CB1 cannabinoid receptors promote oxidative/nitrosative stress, inflammation and cell death in a murine nephropathy model. *Br. J. Pharmacol.* **2010**, *160*, 657–668. [CrossRef] [PubMed]

180. Mukhopadhyay, P.; Rajesh, M.; Pan, H.; Patel, V.; Mukhopadhyay, B.; Bátkai, S.; Gao, B.; Haskó, G.; Pacher, P. Cannabinoid-2 receptor limits inflammation, oxidative/nitrosative stress, and cell death in nephropathy. *Free Radic. Biol. Med.* **2010**, *48*, 457–467. [CrossRef] [PubMed]

181. Mukhopadhyay, P.; Baggelaar, M.; Erdelyi, K.; Cao, Z.; Cinar, R.; Fezza, F.; Ignatowska-Janlowska, B.; Wilkerson, J.; van Gils, N.; Hansen, T.; et al. The novel, orally available and peripherally restricted selective cannabinoid CB2 receptor agonist LEI-101 prevents cisplatin-induced nephrotoxicity. *Br. J. Pharmacol.* **2016**, *173*, 446–458. [CrossRef] [PubMed]

182. Bedse, G.; Di Domenico, F.; Serviddio, G.; Cassano, T. Aberrant insulin signaling in Alzheimer's disease: Current knowledge. *Front. Neurosci.* **2015**, *9*, 204. [CrossRef] [PubMed]

183. Barone, E.; Di Domenico, F.; Cassano, T.; Arena, A.; Tramutola, A.; Lavecchia, M.A.; Coccia, R.; Butterfield, D.A.; Perluigi, M. Impairment of biliverdin reductase-A promotes brain insulin resistance in Alzheimer disease: A new paradigm. *Free Radic. Biol. Med.* **2016**, *91*, 127–142. [CrossRef] [PubMed]

184. Pardeshi, R.; Bolshette, N.; Gadhave, K.; Ahire, A.; Ahmed, S.; Cassano, T.; Gupta, V.B.; Lahkar, M. Insulin signaling: An opportunistic target to minify the risk of Alzheimer's disease. *Psychoneuroendocrinology* **2017**, *83*, 159–171. [CrossRef] [PubMed]

185. Gruden, G.; Barutta, F.; Kunos, G.; Pacher, P. Role of the endocannabinoid system in diabetes and diabetic complications. *Br. J. Pharmacol.* **2016**, *173*, 1116–1127. [CrossRef] [PubMed]

186. Pacher, P.; Hasko, G. Endocannabinoids and cannabinoid receptors in ischaemia-reperfusion injury and preconditioning. *Br. J. Pharmacol.* **2008**, *153*, 252–262. [CrossRef] [PubMed]

187. Silvestri, C.; Ligresti, A.; Di Marzo, V. Peripheral effects of the endocannabinoid system in energy homeostasis: Adipose tissue, liver and skeletal muscle. *Rev. Endocr. Metab. Disord.* **2011**, *12*, 153–162. [CrossRef] [PubMed]

188. Muriel, P. Role of free radicals in liver diseases. *Hepatol. Int.* **2009**, *3*, 526–536. [CrossRef] [PubMed]

189. Mukhopadhyay, P.; Horváth, B.; Rajesh, M.; Matsumoto, S.; Saito, K.; Bátkai, S.; Patel, V.; Tanchian, G.; Gao, R.Y.; Cravatt, B.F.; et al. Fatty acid amide hydrolase is a key regulator of endocannabinoid-induced myocardial tissue injury. *Free Radic. Biol. Med.* **2011**, *50*, 179–195. [CrossRef] [PubMed]

190. Rajesh, M.; Mukhopadhyay, P.; Haskó, G.; Liaudet, L.; Mackie, K.; Pacher, P. Cannabinoid-1 receptor activation induces reactive oxygen species-dependent and -independent mitogen-activated protein kinase activation and cell death in human coronary artery endothelial cells. *Br. J. Pharmacol.* **2010**, *160*, 688–700. [CrossRef] [PubMed]

191. Bátkai, S.; Osei-Hyiaman, D.; Pan, H.; El-Assal, O.; Rajesh, M.; Mukhopadhyay, P.; Hong, F.; Harvey-White, J.; Jafri, A.; Haskó, G.; et al. Cannabinoid-2 receptor mediates protection against hepatic ischemia/reperfusion injury. *FASEB J.* **2007**, *21*, 1788–1800. [CrossRef] [PubMed]

192. Rajesh, M.; Pan, H.; Mukhopadhyay, P.; Batkai, S.; Osei-Hyiaman, D.; Hasko, G.; Liaudet, L.; Gao, B.; Pacher, P. Cannabinoid-2 receptor agonist HU-308 protects against hepatic ischemia/reperfusion injury by attenuating oxidative stress, inflammatory response, and apoptosis. *J. Leukoc. Biol.* **2007**, *82*, 1382–1389. [CrossRef] [PubMed]

193. Montecucco, F.; Lenglet, S.; Braunersreuther, V.; Burger, F.; Pelli, G.; Bertolotto, M.; Mach, F.; Steffens, S. CB2 cannabinoid receptor activation is cardioprotective in a mouse model of ischemia/reperfusion. *J. Mol. Cell. Cardiol.* **2009**, *46*, 612–620. [CrossRef] [PubMed]

194. Zhang, M.; Adler, M.W.; Abood, M.E.; Ganea, D.; Jallo, J.; Tuma, R.F. CB2 receptor activation attenuates microcirculatory dysfunction during cerebral ischemic/reperfusion injury. *Microvasc. Res.* **2009**, *78*, 86–94. [CrossRef] [PubMed]

195. Murikinati, S.; Juttler, E.; Keinert, T.; Ridder, D.A.; Muhammad, S.; Waibler, Z.; Ledent, C.; Zimmer, A.; Kalinke, U.; Schwaninger, M. Activation of cannabinoid 2 receptors protects against cerebral ischemia by inhibiting neutrophil recruitment. *FASEB J.* **2009**. [CrossRef] [PubMed]

196. Steffens, S.; Veillard, N.R.; Arnaud, C.; Pelli, G.; Burger, F.; Staub, C.; Karsak, M.; Zimmer, A.; Frossard, J.L.; Mach, F. Low dose oral cannabinoid therapy reduces progression of atherosclerosis in mice. *Nature* **2005**, *434*, 782–786. [CrossRef] [PubMed]

197. Rajesh, M.; Mukhopadhyay, P.; Batkai, S.; Hasko, G.; Liaudet, L.; Huffman, J.W.; Csiszar, A.; Ungvari, Z.; Mackie, K.; Chatterjee, S.; et al. CB2-receptor stimulation attenuates TNF-alpha-induced human endothelial cell activation, transendothelial migration of monocytes, and monocyte–endothelial adhesion. *Am. J. Physiol. Heart Circ. Physiol.* **2007**, *293*, H2210–H2218. [CrossRef] [PubMed]

198. Pacher, P.; Mechoulam, R. Is lipid signaling through cannabinoid 2 receptors part of a protective system? *Prog. Lipid Res.* **2011**, *50*, 193–211. [CrossRef] [PubMed]

199. Matthews, A.T.; Lee, J.H.; Borazjani, A.; Mangum, L.C.; Hou, X.; Ross, M.K. Oxyradical stress increases the biosynthesis of 2-arachidonoylglycerol: Involvement of NADPH oxidase. *Am. J. Physiol. Cell Physiol.* **2016**, *311*, C960–C974. [CrossRef] [PubMed]

200. Hoyer, F.F.; Steinmetz, M.; Zimmer, S.; Becker, A.; Lütjohann, D.; Buchalla, R.; Zimmer, A.; Nickenig, G. Atheroprotection via cannabinoid receptor-2 is mediated by circulating and vascular cells in vivo. *J. Mol. Cell. Cardiol.* **2011**, *51*, 1007–1014. [CrossRef] [PubMed]

201. Tiyerili, V.; Zimmer, S.; Jung, S.; Wassmann, K.; Naehle, C.P.; Lütjohann, D.; Zimmer, A.; Nickenig, G.; Wassmann, S. CB$_1$ receptor inhibition leads to decreased vascular AT1 receptor expression, inhibition of oxidative stress and improved endothelial function. *Basic Res. Cardiol.* **2010**, *105*, 465–477. [CrossRef] [PubMed]

202. Zolese, G.; Bacchetti, T.; Masciangelo, S.; Ragni, L.; Ambrosi, S.; Ambrosini, A.; Marini, M.; Ferretti, G. Effect of acylethanolamides on lipid peroxidation and paraoxonase activity. *Biofactors* **2008**, *33*, 201–209. [CrossRef] [PubMed]

203. Gulaya, N.M.; Kuzmenko, A.I.; Margitich, V.M.; Govseeva, N.M.; Melnichuk, S.D.; Goridko, T.M.; Zhukov, A.D. Long-chain *N*-acylethanolamines inhibit lipid peroxidation in rat liver mitochondria under acute hypoxic hypoxia. *Chem. Phys. Lipids* **1998**, *97*, 49–54. [CrossRef]

204. Poli, G.; Albano, E.; Dianzani, M.U. The role of lipid peroxidation in liver damage. *Chem. Phys. Lipids* **1987**, *45*, 117–142. [CrossRef]

205. Emerit, J.; Chaudiere, J. *Free Radicals and Lipid Peroxidation in Cell Biology*; Handbook of Free Radicals and Antioxidants in Biomedicine; CRC Press: Boca Raton, FL, USA, 1989; pp. 177–185.

206. Gulaya, N.M.; Melnik, A.A.; Balkov, D.I.; Volkov, G.L.; Vysotskiy, M.V.; Vaskovsky, V.E. The effect of long-chain *N*-acylethanolamines on some membrane-associated functions of neuroblastoma C1300 N18 cells. *Biochim. Biophys. Acta* **1993**, *1152*, 280–288. [CrossRef]

207. Parinandi, N.L.; Schmid, H.H. Effects of long-chain *N*-acylethanolamines on lipid peroxidation in cardiac mitochondria. *FEBS Lett.* **1988**, *237*, 49–52. [CrossRef]

208. Zolese, G.; Bacchetti, T.; Ambrosini, A.; Wozniak, M.; Bertoli, E.; Ferretti, G. Increased plasma concentrations of palmitoylethanolamide, an endogenous fatty acid amide, affect oxidative damage of human low-density lipoproteins: An in vitro study. *Atherosclerosis* **2005**, *182*, 47–55. [CrossRef] [PubMed]

209. Biernacki, M.; Łuczaj, W.; Gęgotek, A.; Toczek, M.; Bielawska, K.; Skrzydlewska, E. Crosstalk between liver antioxidant and the endocannabinoid systems after chronic administration of the FAAH inhibitor, URB597, to hypertensive rats. *Toxicol. Appl. Pharmacol.* **2016**, *301*, 31–41. [CrossRef] [PubMed]

210. Di Marzo, V.; Maccarrone, M. FAAH and anandamide: Is 2-AG really the odd one out? *Trends Pharmacol. Sci.* **2008**, *29*, 229–233. [CrossRef] [PubMed]

211. Basu, P.P.; Aloysius, M.M.; Shah, N.J.; Brown, R.S., Jr. Review article: The endocannabinoid system in liver disease, a potential therapeutic target. *Aliment. Pharmacol. Ther.* **2014**, *39*, 790–801. [CrossRef] [PubMed]

212. DeLeve, L.D.; Wang, X.; Kanel, G.C.; Atkinson, R.D.; McCuskey, R.S. Prevention of hepatic fibrosis in a murine model of metabolic syndrome with nonalcoholic steatohepatitis. *Am. J. Pathol.* **2008**, *173*, 993–1001. [CrossRef] [PubMed]

213. Mallat, A.; Teixeira-Clerc, F.; Lotersztajn, S. Cannabinoid signaling and liver therapeutics. *J. Hepatol.* **2013**, *59*, 891–896. [CrossRef] [PubMed]

214. Simonian, N.A.; Coyle, J.T. Oxidative stress in neurodegenerative diseases. *Annu. Rev. Pharmacol. Toxicol.* **1996**, *36*, 83–106. [CrossRef] [PubMed]

215. Behl, C. Alzheimer's disease and oxidative stress: Implications for novel therapeutic approaches. *Prog. Neurobiol.* **1999**, *57*, 301–323. [CrossRef]

216. Bedse, G.; Romano, A.; Cianci, S.; Lavecchia, A.M.; Lorenzo, P.; Elphick, M.R.; Laferla, F.M.; Vendemiale, G.; Grillo, C.; Altieri, F.; et al. Altered expression of the CB1 cannabinoid receptor in the triple transgenic mouse model of Alzheimer's disease. *J. Alzheimer Dis.* **2014**, *40*, 701–712. [CrossRef] [PubMed]

217. Gatta, E.; Lefebvre, T.; Gaetani, S.; dos Santos, M.; Marrocco, J.; Mir, A.M.; Cassano, T.; Maccari, S.; Nicoletti, F.; Mairesse, J. Evidence for an imbalance between tau O-GlcNAcylation and phosphorylation in the hippocampus of a mouse model of Alzheimer's disease. *Pharmacol. Res.* **2016**, *105*, 186–197. [CrossRef] [PubMed]

218. Milton, N.G. Role of hydrogen peroxide in the aetiology of Alzheimer's disease: Implications for treatment. *Drugs Aging* **2004**, *21*, 81–100. [CrossRef] [PubMed]

219. Ano, Y.; Sakudo, A.; Kimata, T.; Uraki, R.; Sugiura, K.; Onodera, T. Oxidative damage to neurons caused by the induction of microglial NADPH oxidase in encephalomyocarditis virus infection. *Neurosci. Lett.* **2010**, *469*, 39–43. [CrossRef] [PubMed]

220. Ye, Q.; Huang, B.; Zhang, X.; Zhu, Y.; Chen, X. Astaxanthin protects against MPP+-induced oxidative stress in PC12 cells via the HO-1/NOX2 axis. *BMC Neurosci.* **2012**, *13*, 156. [CrossRef] [PubMed]

221. Jia, J.; Ma, L.; Wu, M.; Zhang, L.; Zhang, X.; Zhai, Q.; Jiang, T.; Wang, Q.; Xiong, L. Anandamide protects HT22 cells exposed to hydrogen peroxide by inhibiting CB1 receptor-mediated type 2 NADPH oxidase. *Oxid. Med. Cell. Longev.* **2014**, *2014*, 893516. [CrossRef] [PubMed]

222. Chung, Y.C.; Bok, E.; Huh, S.H.; Park, J.Y.; Yoon, S.H.; Kim, S.R.; Kim, Y.S.; Maeng, S.; Park, S.H.; Jin, B.K. Cannabinoid receptor type 1 protects nigrostriatal dopaminergic neurons against MPTP neurotoxicity by inhibiting microglial activation. *J. Immunol.* **2011**, *187*, 6508–6517. [CrossRef] [PubMed]

223. Cassano, T.; Calcagnini, S.; Pace, L.; De Marco, F.; Romano, A.; Gaetani, S. Cannabinoid Receptor 2 Signaling in Neurodegenerative Disorders: From Pathogenesis to a Promising Therapeutic Target. *Front. Neurosci.* **2017**, *11*, 30. [CrossRef] [PubMed]

224. Jayant, S.; Sharma, B.M.; Bansal, R.; Sharma, B. Pharmacological benefits of selective modulation of cannabinoid receptor type 2 (CB2) in experimental Alzheimer's disease. *Pharmacol. Biochem. Behav.* **2016**, *140*, 39–50. [CrossRef] [PubMed]

225. Koppel, J.; Vingtdeux, V.; Marambaud, P.; d'Abramo, C.; Jimenez, H.; Stauber, M.; Friedman, R.; Davies, P. CB2 receptor deficiency increases amyloid pathology and alters tau processing in a transgenic mouse model of Alzheimer's disease. *Mol. Med.* **2014**, *20*, 29–36. [CrossRef] [PubMed]

226. Chen, R.; Zhang, J.; Wu, Y.; Wang, D.; Feng, G.; Tang, Y.P.; Teng, Z.; Chen, C. Monoacylglycerol lipase is a therapeutic target for Alzheimer's disease. *Cell Rep.* **2012**, *2*, 1329–1339. [CrossRef] [PubMed]

227. Paloczi, J.; Varga, Z.V.; Hasko, G.; Pacher, P. Neuroprotection in Oxidative Stress-Related Neurodegenerative Diseases: Role of Endocannabinoid System Modulation. *Antioxid. Redox Signal.* **2018**, *29*, 75–108. [CrossRef] [PubMed]

228. Pihlaja, R.; Takkinen, J.; Eskola, O.; Vasara, J.; López-Picón, F.R.; Haaparanta-Solin, M.; Rinne, J.O. Monoacylglycerol lipase inhibitor JZL184 reduces neuroinflammatory response in APdE9 mice and in adult mouse glial cells. *J. Neuroinflamm.* **2015**, *12*, 81. [CrossRef] [PubMed]

229. Kessiova, M.; Alexandrova, A.; Georgieva, A.; Kirkova, M.; Todorov, S. In vitro effects of CB1 receptor ligands on lipid peroxidation and antioxidant defense systems in the rat brain. *Pharmacol. Rep.* **2006**, *58*, 870–875. [PubMed]

230. Sun, H.J.; Lu, Y.; Wang, H.W.; Zhang, H.; Wang, S.R.; Xu, W.Y.; Fu, H.L.; Yao, X.Y.; Yang, F.; Yuan, H.B. Activation of Endocannabinoid Receptor 2 as a Mechanism of Propofol Pretreatment-Induced Cardioprotection against Ischemia-Reperfusion Injury in Rats. *Oxid. Med. Cell. Longev.* **2017**, *2017*, 2186383. [CrossRef] [PubMed]

231. Hayn, M.H.; Ballesteros, I.; de Miguel, F.; Coyle, C.H.; Tyagi, S.; Yoshimura, N.; Chancellor, M.B.; Tyagi, P. Functional and immunohistochemical characterization of CB1 and CB2 receptors in rat bladder. *Urology* **2008**, *72*, 1174–1178. [CrossRef] [PubMed]

232. Wang, Z.Y.; Wang, P.; Bjorling, D.E. Activation of cannabinoid receptor 2 inhibits experimental cystitis. *Am. J. Physiol. Regul. Integr. Comp. Physiol.* **2013**, *304*, R846–R853. [CrossRef] [PubMed]

233. Hermanson, D.J.; Marnett, L.J. Cannabinoids endocannabinoids, and cancer. *Cancer Metastasis Rev.* **2011**, *30*, 599–612. [CrossRef] [PubMed]

234. Ravi, J.; Sneh, A.; Shilo, K.; Nasser, M.W.; Ganju, R.K. FAAH inhibition enhances anandamide mediated anti-tumorigenic effects in non-small cell lung cancer by downregulating the EGF/EGFR pathway. *Oncotarget* **2014**, *5*, 2475–2486. [CrossRef] [PubMed]

235. Clapham, D.E.; Runnels, L.W.; Strübing, C. The TRP ion channel family. *Nat. Rev. Neurosci.* **2001**, *2*, 387–396. [CrossRef] [PubMed]

236. Harteneck, C.; Plant, T.D.; Schultz, G. From worm to man: Three subfamilies of TRP channels. *Trends Neurosci.* **2000**, *23*, 159–166. [CrossRef]

237. Wu, L.J.; Sweet, T.B.; Clapham, D.E. International Union of Basic and Clinical Pharmacology. LXXVI. Current progress in the mammalian TRP ion channel family. *Pharmacol. Rev.* **2010**, *62*, 381–404. [CrossRef] [PubMed]

238. Venkatachalam, K.; Montell, C. TRP channels. *Annu. Rev. Biochem.* **2007**, *76*, 387–417. [CrossRef] [PubMed]

239. Moran, M.M.; Xu, H.; Clapham, D.E. TRP ion channels in the nervous system. *Curr. Opin. Neurobiol.* **2004**, *14*, 362–369. [CrossRef] [PubMed]

240. Desai, B.N.; Clapham, D.E. TRP channels and mice deficient in TRP channels. *Pflugers Arch.* **2005**, *451*, 11–18. [CrossRef] [PubMed]

241. Vriens, J.; Appendino, G.; Nilius, B. Pharmacology of vanilloid transient receptor potential cation channels. *Mol. Pharmacol.* **2009**, *75*, 1262–1279. [CrossRef] [PubMed]

242. Cavanaugh, D.J.; Chesler, A.T.; Jackson, A.C.; Sigal, Y.M.; Yamanaka, H.; Grant, R.; O'Donnell, D.; Nicoll, R.A.; Shah, N.M.; Julius, D.; et al. *Trpv1* reporter mice reveal highly restricted brain distribution and functional expression in arteriolar smooth muscle cells. *J. Neurosci.* **2011**, *31*, 5067–5077. [CrossRef] [PubMed]

243. Caterina, M.J.; Schumacher, M.A.; Tominaga, M.; Rosen, T.A.; Levine, J.D.; Julius, D. The capsaicin receptor: A heat-activated ion channel in the pain pathway. *Nature* **1997**, *389*, 816–824. [CrossRef] [PubMed]

244. Tominaga, M.; Caterina, M.J.; Malmberg, A.B.; Rosen, T.A.; Gilbert, H.; Skinner, K.; Raumann, B.E.; Basbaum, A.I.; Julius, D. The cloned capsaicin receptor integrates multiple pain-producing stimuli. *Neuron* **1998**, *21*, 531–543. [CrossRef]

245. Jordt, S.E.; Tominaga, M.; Julius, D. Acid potentiation of the capsaicin receptor determined by a key extracellular site. *Proc. Natl. Acad. Sci. USA* **2000**, *97*, 8134–8139. [CrossRef] [PubMed]

246. Iida, T.; Moriyama, T.; Kobata, K.; Morita, A.; Murayama, N.; Hashizume, S.; Fushiki, T.; Yazawa, S.; Watanabe, T.; Tominaga, M. TRPV1 activation and induction of nociceptive response by a non-pungent capsaicin-like compound, capsiate. *Neuropharmacology* **2003**, *44*, 958–967. [CrossRef]

247. Blednov, Y.A.; Harris, R.A. Deletion of vanilloid receptor (TRPV1) in mice alters behavioral effects of ethanol. *Neuropharmacology* **2009**, *56*, 814–820. [CrossRef] [PubMed]

248. Ellingson, J.M.; Silbaugh, B.C.; Brasser, S.M. Reduced oral ethanol avoidance in mice lacking transient receptor potential channel vanilloid receptor 1. *Behav. Genet.* **2009**, *39*, 62–72. [CrossRef] [PubMed]

249. Ahern, G.P. Activation of TRPV1 by the satiety factor oleoylethanolamide. *J. Biol. Chem.* **2003**, *278*, 30429–30434. [CrossRef] [PubMed]

250. Huang, S.M.; Bisogno, T.; Trevisani, M.; Al-Hayani, A.; De Petrocellis, L.; Fezza, F.; Tognetto, M.; Petros, T.J.; Krey, J.F.; Chu, C.J.; et al. An endogenous capsaicin-like substance with high potency at recombinant and native vanilloid VR1 receptors. *Proc. Natl. Acad. Sci. USA* **2002**, *99*, 8400–8405. [CrossRef] [PubMed]

251. Chu, C.J.; Huang, S.M.; De Petrocellis, L.; Bisogno, T.; Ewing, S.A.; Miller, J.D.; Zipkin, R.E.; Daddario, N.; Appendino, G.; Di Marzo, V.; et al. *N*-oleoyldopamine, a novel endogenous capsaicin-like lipid that produces hyperalgesia. *J. Biol. Chem.* **2003**, *278*, 13633–13639. [CrossRef] [PubMed]

252. Hwang, S.W.; Cho, H.; Kwak, J.; Lee, S.Y.; Kang, C.J.; Jung, J.; Cho, S.; Min, K.H.; Suh, Y.G.; Kim, D.; et al. Direct activation of capsaicin receptors by products of lipoxygenases: Endogenous capsaicin-like substances. *Proc. Natl. Acad. Sci. USA* **2000**, *97*, 6155–6160. [CrossRef] [PubMed]

253. Numazaki, M.; Tominaga, T.; Takeuchi, K.; Murayama, N.; Toyooka, H.; Tominaga, M. Structural determinant of TRPV1 desensitization interacts with calmodulin. *Proc. Natl. Acad. Sci. USA* **2003**, *100*, 8002–8006. [CrossRef] [PubMed]

254. Rosenbaum, T.; Gordon-Shaag, A.; Munari, M.; Gordon, S.E. Ca^{2+}/calmodulin modulates TRPV1 activation by capsaicin. *J. Gen. Physiol.* **2004**, *123*, 53–62. [CrossRef] [PubMed]

255. Lishko, P.V.; Procko, E.; Jin, X.; Phelps, C.B.; Gaudet, R. The ankyrin repeats of TRPV1 bind multiple ligands and modulate channel sensitivity. *Neuron* **2007**, *54*, 905–918. [CrossRef] [PubMed]

256. Kwon, Y.; Hofmann, T.; Montell, C. Integration of phosphoinositide- and calmodulin-mediated regulation of TRPC6. *Mol. Cell* **2007**, *25*, 491–503. [CrossRef] [PubMed]

257. Premkumar, L.S.; Ahern, G.P. Induction of vanilloid receptor channel activity by protein kinase C. *Nature* **2000**, *408*, 985–990. [CrossRef] [PubMed]

258. De Petrocellis, L.; Harrison, S.; Bisogno, T.; Tognetto, M.; Brandi, I.; Smith, G.D.; Creminon, C.; Davis, J.B.; Geppetti, P.; Di Marzo, V. The vanilloid receptor (VR1)-mediated effects of anandamide are potently enhanced by the cAMP-dependent protein kinase. *J. Neurochem.* **2001**, *77*, 1660–1663. [CrossRef] [PubMed]

259. Docherty, R.J.; Yeats, J.C.; Bevan, S.; Boddeke, H.W. Inhibition of calcineurin inhibits the desensitization of capsaicin-evoked currents in cultured dorsal root ganglion neurones from adult rats. *Pflugers Arch.* **1996**, *431*, 828–837. [CrossRef] [PubMed]

260. Birder, L.A.; Nakamura, Y.; Kiss, S.; Nealen, M.L.; Barrick, S.; Kanai, A.J.; Wang, E.; Ruiz, G.; De Groat, W.C.; Apodaca, G.; et al. Altered urinary bladder function in mice lacking the vanilloid receptor TRPV1. *Nat. Neurosci.* **2002**, *5*, 856–860. [CrossRef] [PubMed]

261. Rong, W.; Hillsley, K.; Davis, J.B.; Hicks, G.; Winchester, W.J.; Grundy, D. Jejunal afferent nerve sensitivity in wild-type and TRPV1 knockout mice. *J. Physiol.* **2004**, *560*, 867–881. [CrossRef] [PubMed]

262. Treesukosol, Y.; Lyall, V.; Heck, G.L.; DeSimone, J.A.; Spector, A.C. A psychophysical and electrophysiological analysis of salt taste in *Trpv1* null mice. *Am. J. Physiol. Regul. Integr. Comp. Physiol.* **2007**, *292*, R1799–R1809. [CrossRef] [PubMed]

263. Cagiano, R.; Cassano, T.; Coluccia, A.; Gaetani, S.; Giustino, A.; Steardo, L.; Tattoli, M.; Trabace, L.; Cuomo, V. Genetic factors involved in the effects of developmental low-level alcohol induced behavioral alterations in rats. *Neuropsychopharmacology* **2002**, *26*, 191–203. [CrossRef]

264. Geppetti, P.; Materazzi, S.; Nicoletti, P. The transient receptor potential vanilloid 1: Role in airway inflammation and disease. *Eur. J. Pharmacol.* **2006**, *533*, 207–214. [CrossRef] [PubMed]

265. Gupta, S.; Sharma, B.; Singh, P.; Sharma, B.M. Modulation of transient receptor potential vanilloid subtype 1 (TRPV1) and norepinephrine transporters (NET) protect against oxidative stress, cellular injury, and vascular dementia. *Curr. Neurovasc. Res.* **2014**, *11*, 94–106. [CrossRef] [PubMed]

266. Gupta, S.; Sharma, B. Pharmacological benefits of agomelatine and vanillin in experimental model of Huntington's disease. *Pharmacol. Biochem. Behav.* **2014**, *122*, 122–135. [CrossRef] [PubMed]

267. Cao, Z.; Balasubramanian, A.; Marrelli, S.P. Pharmacologically induced hypothermia via TRPV1 channel agonism provides neuroprotection following ischemic stroke when initiated 90 min after reperfusion. *Am. J. Physiol. Regul. Integr. Comp. Physiol.* **2014**, *306*, R149–R156. [CrossRef] [PubMed]

268. Neeper, M.P.; Liu, Y.; Hutchinson, T.L.; Wang, Y.; Flores, C.M.; Qin, N. Activation properties of heterologously expressed mammalian TRPV2: Evidence for species dependence. *J. Biol. Chem.* **2007**, *282*, 15894–15902. [CrossRef] [PubMed]

269. Caterina, M.J.; Rosen, T.A.; Tominaga, M.; Brake, A.J.; Julius, D. A capsaicin-receptor homologue with a high threshold for noxious heat. *Nature* **1999**, *398*, 436–441. [CrossRef] [PubMed]

270. Szöllősi, A.G.; Oláh, A.; Tóth, I.B.; Papp, F.; Czifra, G.; Panyi, G.; Bíró, T. Transient receptor potential vanilloid-2 mediates the effects of transient heat shock on endocytosis of human monocyte-derived dendritic cells. *FEBS Lett.* **2013**, *587*, 1440–1445. [CrossRef] [PubMed]

271. Saito, M.; Hanson, P.I.; Schlesinger, P. Luminal chloride-dependent activation of endosome calcium channels: Patch clamp study of enlarged endosomes. *J. Biol. Chem.* **2007**, *282*, 27327–27333. [CrossRef] [PubMed]

272. Abe, K.; Puertollano, R. Role of TRP channels in the regulation of the endosomal pathway. *Physiology* **2011**, *26*, 14–22. [CrossRef] [PubMed]

273. Lévêque, M.; Penna, A.; Le Trionnaire, S.; Belleguic, C.; Desrues, B.; Brinchault, G.; Jouneau, S.; Lagadic-Gossmann, D.; Martin-Chouly, C. Phagocytosis depends on TRPV2-mediated calcium influx and requires TRPV2 in lipids rafts: Alteration in macrophages from patients with cystic fibrosis. *Sci. Rep.* **2018**, *8*, 4310. [CrossRef] [PubMed]

274. Link, T.M.; Park, U.; Vonakis, B.M.; Raben, D.M.; Soloski, M.J.; Caterina, M.J. TRPV2 has a pivotal role in macrophage particle binding and phagocytosis. *Nat. Immunol.* **2010**, *11*, 232–239. [CrossRef] [PubMed]

275. Peier, A.M.; Reeve, A.J.; Andersson, D.A.; Moqrich, A.; Earley, T.J.; Hergarden, A.C.; Story, G.M.; Colley, S.; Hogenesch, J.B.; McIntyre, P.; et al. A heat-sensitive TRP channel expressed in keratinocytes. *Science* **2002**, *296*, 2046–2049. [CrossRef] [PubMed]

276. Smith, G.D.; Gunthorpe, M.J.; Kelsell, R.E.; Hayes, P.D.; Reilly, P.; Facer, P.; Wright, J.E.; Jerman, J.C.; Walhin, J.P.; Ooi, L.; Egerton, J.; et al. TRPV3 is a temperature-sensitive vanilloid receptor-like protein. *Nature* **2002**, *418*, 186–190. [CrossRef] [PubMed]

277. Moqrich, A.; Hwang, S.W.; Earley, T.J.; Petrus, M.J.; Murray, A.N.; Spencer, K.S.; Andahazy, M.; Story, G.M.; Patapoutian, A. Impaired thermosensation in mice lacking TRPV3, a heat and camphor sensor in the skin. *Science* **2005**, *307*, 1468–1472. [CrossRef] [PubMed]

278. Xu, H.; Delling, M.; Jun, J.C.; Clapham, D.E. Oregano, thyme and clove-derived flavors and skin sensitizers activate specific TRP channels. *Nat. Neurosci.* **2006**, *9*, 628–635. [CrossRef] [PubMed]

279. Phelps, C.B.; Wang, R.R.; Choo, S.S.; Gaudet, R. Differential regulation of TRPV1, TRPV3, and TRPV4 sensitivity through a conserved binding site on the ankyrin repeat domain. *J. Biol. Chem.* **2010**, *285*, 731–740. [CrossRef] [PubMed]

280. Mandadi, S.; Sokabe, T.; Shibasaki, K.; Katanosaka, K.; Mizuno, A.; Moqrich, A.; Patapoutian, A.; Fukumi-Tominaga, T.; Mizumura, K.; Tominaga, M. TRPV3 in keratinocytes transmits temperature information to sensory neurons via ATP. *Pflugers Arch.* **2009**, *458*, 1093–1102. [CrossRef] [PubMed]

281. Huang, S.M.; Lee, H.; Chung, M.K.; Park, U.; Yu, Y.Y.; Bradshaw, H.B.; Coulombe, P.A.; Walker, J.M.; Caterina, M.J. Overexpressed transient receptor potential vanilloid 3 ion channels in skin keratinocytes modulate pain sensitivity via prostaglandin E2. *J. Neurosci.* **2008**, *28*, 13727–13737. [CrossRef] [PubMed]

282. Chung, M.K.; Lee, H.; Mizuno, A.; Suzuki, M.; Caterina, M.J. TRPV3 and TRPV4 mediate warmth-evoked currents in primary mouse keratinocytes. *J. Biol. Chem.* **2004**, *279*, 21569–21575. [CrossRef] [PubMed]

283. Liedtke, W.; Choe, Y.; Martí-Renom, M.A.; Bell, A.M.; Denis, C.S.; Sali, A.; Hudspeth, A.J.; Friedman, J.M.; Heller, S. Vanilloid receptor-related osmotically activated channel (VR-OAC), a candidate vertebrate osmoreceptor. *Cell* **2000**, *103*, 525–535. [CrossRef]

284. Güler, A.D.; Lee, H.; Iida, T.; Shimizu, I.; Tominaga, M.; Caterina, M. Heat-evoked activation of the ion channel, TRPV4. *J. Neurosci.* **2002**, *22*, 6408–6414. [CrossRef] [PubMed]

285. Watanabe, H.; Vriens, J.; Prenen, J.; Droogmans, G.; Voets, T.; Nilius, B. Anandamide and arachidonic acid use epoxyeicosatrienoic acids to activate TRPV4 channels. *Nature* **2003**, *424*, 434–438. [CrossRef] [PubMed]

286. Vriens, J.; Owsianik, G.; Fisslthaler, B.; Suzuki, M.; Janssens, A.; Voets, T.; Morisseau, C.; Hammock, B.D.; Fleming, I.; Busse, R.; et al. Modulation of the Ca^2 permeable cation channel TRPV4 by cytochrome P450 epoxygenases in vascular endothelium. *Circ. Res.* **2005**, *97*, 908–915. [CrossRef] [PubMed]

287. Strotmann, R.; Semtner, M.; Kepura, F.; Plant, T.D.; Schöneberg, T. Interdomain interactions control Ca^{2+}-dependent potentiation in the cation channel TRPV4. *PLoS ONE* **2010**, *5*, e10580. [CrossRef] [PubMed]

288. White, J.P.; Cibelli, M.; Urban, L.; Nilius, B.; McGeown, J.G.; Nagy, I. TRPV4: Molecular Conductor of a Diverse Orchestra. *Physiol. Rev.* **2016**, *96*, 911–973. [CrossRef] [PubMed]

289. Takahashi, N.; Hamada-Nakahara, S.; Itoh, Y.; Takemura, K.; Shimada, A.; Ueda, Y.; Kitamata, M.; Matsuoka, R.; Hanawa-Suetsugu, K.; Senju, Y.; et al. TRPV4 channel activity is modulated by direct interaction of the ankyrin domain to PI(4,5)P$_2$. *Nat. Commun.* **2014**, *5*, 4994. [CrossRef] [PubMed]

290. Gao, X.; Wu, L.; O'Neil, R.G. Temperature-modulated diversity of TRPV4 channel gating: Activation by physical stresses and phorbol ester derivatives through protein kinase C-dependent and -independent pathways. *J. Biol. Chem.* **2003**, *278*, 27129–27137. [CrossRef] [PubMed]

291. Fan, H.C.; Zhang, X.; McNaughton, P.A. Activation of the TRPV4 ion channel is enhanced by phosphorylation. *J. Biol. Chem.* **2009**, *284*, 27884–27891. [CrossRef] [PubMed]

292. Wegierski, T.; Lewandrowski, U.; Müller, B.; Sickmann, A.; Walz, G. Tyrosine phosphorylation modulates the activity of TRPV4 in response to defined stimuli. *J. Biol. Chem.* **2009**, *284*, 2923–2933. [CrossRef] [PubMed]

293. Lee, E.J.; Shin, S.H.; Chun, J.; Hyun, S.; Kim, Y.; Kang, S.S. The modulation of TRPV4 channel activity through its Ser 824 residue phosphorylation by SGK1. *Anim. Cells Syst.* **2010**, *14*, 99–114. [CrossRef]

294. Delany, N.S.; Hurle, M.; Facer, P.; Alnadaf, T.; Plumpton, C.; Kinghorn, I.; See, C.G.; Costigan, M.; Anand, P.; Woolf, C.J.; et al. Identification and characterization of a novel human vanilloid receptor-like protein, VRL-2. *Physiol. Genom.* **2001**, *4*, 165–174. [CrossRef] [PubMed]

295. Schumacher, M.A.; Jong, B.E.; Frey, S.L.; Sudanagunta, S.P.; Capra, N.F.; Levine, J.D. The stretch-inactivated channel, a vanilloid receptor variant, is expressed in small-diameter sensory neurons in the rat. *Neurosci. Lett.* **2000**, *287*, 215–218. [CrossRef]

296. Suzuki, M.; Mizuno, A.; Kodaira, K.; Imai, M. Impaired pressure sensation in mice lacking TRPV4. *J. Biol. Chem.* **2003**, *278*, 22664–22668. [CrossRef] [PubMed]

297. Birder, L.; Kullmann, F.A.; Lee, H.; Barrick, S.; de Groat, W.; Kanai, A.; Caterina, M. Activation of urothelial transient receptor potential vanilloid 4 by 4alpha-phorbol 12,13-didecanoate contributes to altered bladder reflexes in the rat. *J. Pharmacol. Exp. Ther.* **2007**, *323*, 227–235. [CrossRef] [PubMed]

298. Gevaert, T.; Vriens, J.; Segal, A.; Everaerts, W.; Roskams, T.; Talavera, K.; Owsianik, G.; Liedtke, W.; Daelemans, D.; Dewachter, I.; et al. Deletion of the transient receptor potential cation channel TRPV4 impairs murine bladder voiding. *J. Clin. Investig.* **2007**, *117*, 3453–3462. [CrossRef] [PubMed]

299. Alvarez, D.F.; King, J.A.; Weber, D.; Addison, E.; Liedtke, W.; Townsley, M.I. Transient receptor potential vanilloid 4-mediated disruption of the alveolar septal barrier: A novel mechanism of acute lung injury. *Circ. Res.* **2006**, *99*, 988–995. [CrossRef] [PubMed]

300. Hamanaka, K.; Jian, M.Y.; Weber, D.S.; Alvarez, D.F.; Townsley, M.I.; Al-Mehdi, A.B.; King, J.A.; Liedtke, W.; Parker, J.C. TRPV4 initiates the acute calcium-dependent permeability increase during ventilator-induced lung injury in isolated mouse lungs. *Am. J. Physiol. Lung Cell. Mol. Physiol.* **2007**, *293*, L923–L932. [CrossRef] [PubMed]

301. Alessandri-Haber, N.; Dina, O.A.; Joseph, E.K.; Reichling, D.; Levine, J.D. A transient receptor potential vanilloid 4-dependent mechanism of hyperalgesia is engaged by concerted action of inflammatory mediators. *J. Neurosci.* **2006**, *26*, 3864–3874. [CrossRef] [PubMed]

302. Vennekens, R.; Hoenderop, J.G.; Prenen, J.; Stuiver, M.; Willems, P.H.; Droogmans, G.; Nilius, B.; Bindels, R.J. Permeation and gating properties of the novel epithelial Ca^{2+} channel. *J. Biol. Chem.* **2000**, *275*, 3963–3969. [CrossRef] [PubMed]

303. Yue, L.; Peng, J.B.; Hediger, M.A.; Clapham, D.E. CaT1 manifests the pore properties of the calcium-release-activated calcium channel. *Nature* **2001**, *410*, 705–709. [CrossRef] [PubMed]

304. Clapham, D.E. TRP channels as cellular sensors. *Nature* **2003**, *426*, 517–524. [CrossRef] [PubMed]

305. Hoenderop, J.G.; Vennekens, R.; Müller, D.; Prenen, J.; Droogmans, G.; Bindels, R.J.; Nilius, B. Function and expression of the epithelial Ca^{2+} channel family: Comparison of mammalian ECaC1 and 2. *J. Physiol.* **2001**, *537*, 747–761. [CrossRef] [PubMed]

306. Lambers, T.T.; Weidema, A.F.; Nilius, B.; Hoenderop, J.G.; Bindels, R.J. Regulation of the mouse epithelial Ca^{2+} channel TRPV6 by the Ca^{2+}-sensor calmodulin. *J. Biol. Chem.* **2004**, *279*, 28855–28861. [CrossRef] [PubMed]

307. Niemeyer, B.A.; Bergs, C.; Wissenbach, U.; Flockerzi, V.; Trost, C. Competitive regulation of CaT-like-mediated Ca^{2+} entry by protein kinase C and calmodulin. *Proc. Natl. Acad. Sci. USA* **2001**, *98*, 3600–3605. [CrossRef] [PubMed]

308. Nilius, B.; Vennekens, R.; Prenen, J.; Hoenderop, J.G.; Bindels, R.J.; Droogmans, G. Whole-cell and single channel monovalent cation currents through the novel rabbit epithelial Ca^{2+} channel ECaC. *J. Physiol.* **2000**, *527 Pt 2*, 239–248. [CrossRef] [PubMed]

309. Voets, T.; Janssens, A.; Prenen, J.; Droogmans, G.; Nilius, B. Mg^{2+}-dependent gating and strong inward rectification of the cation channel TRPV6. *J. Gen. Physiol.* **2003**, *121*, 245–260. [CrossRef] [PubMed]

310. Lee, J.; Cha, S.K.; Sun, T.J.; Huang, C.L. PIP2 activates TRPV5 and releases its inhibition by intracellular Mg^{2+}. *J. Gen. Physiol.* **2005**, *126*, 439–451. [CrossRef] [PubMed]

311. Rohács, T.; Lopes, C.M.; Michailidis, I.; Logothetis, D.E. PI(4,5)P2 regulates the activation and desensitization of TRPM8 channels through the TRP domain. *Nat. Neurosci.* **2005**, *8*, 626–634. [CrossRef] [PubMed]

312. Thyagarajan, B.; Lukacs, V.; Rohacs, T. Hydrolysis of phosphatidylinositol 4,5-bisphosphate mediates calcium-induced inactivation of TRPV6 channels. *J. Biol. Chem.* **2008**, *283*, 14980–14987. [CrossRef] [PubMed]

313. Thyagarajan, B.; Benn, B.S.; Christakos, S.; Rohacs, T. Phospholipase C-mediated regulation of transient receptor potential vanilloid 6 channels: Implications in active intestinal Ca^{2+} transport. *Mol. Pharmacol.* **2009**, *75*, 608–616. [CrossRef] [PubMed]

314. Hoenderop, J.G.; van der Kemp, A.W.; Hartog, A.; van de Graaf, S.F.; van Os, C.H.; Willems, P.H.; Bindels, R.J. Molecular identification of the apical Ca^{2+} channel in 1, 25-dihydroxyvitamin D3-responsive epithelia. *J. Biol. Chem.* **1999**, *274*, 8375–8378. [CrossRef] [PubMed]

315. Hoenderop, J.G.; van Leeuwen, J.P.; van der Eerden, B.C.; Kersten, F.F.; van der Kemp, A.W.; Mérillat, A.M.; Waarsing, J.H.; Rossier, B.C.; Vallon, V.; Hummler, E.; et al. Renal Ca^{2+} wasting, hyperabsorption, and reduced bone thickness in mice lacking TRPV5. *J. Clin. Investig.* **2003**, *112*, 1906–1914. [CrossRef] [PubMed]

316. Van der Eerden, B.C.; Hoenderop, J.G.; de Vries, T.J.; Schoenmaker, T.; Buurman, C.J.; Uitterlinden, A.G.; Pols, H.A.; Bindels, R.J.; van Leeuwen, J.P. The epithelial Ca^{2+} channel TRPV5 is essential for proper osteoclastic bone resorption. *Proc. Natl. Acad. Sci. USA* **2005**, *102*, 17507–17512. [CrossRef] [PubMed]

317. Peng, J.B.; Chen, X.Z.; Berger, U.V.; Weremowicz, S.; Morton, C.C.; Vassilev, P.M.; Brown, E.M.; Hediger, M.A. Human calcium transport protein CaT1. *Biochem. Biophys. Res. Commun.* **2000**, *278*, 326–332. [CrossRef] [PubMed]

318. Hirnet, D.; Olausson, J.; Fecher-Trost, C.; Bödding, M.; Nastainczyk, W.; Wissenbach, U.; Flockerzi, V.; Freichel, M. The TRPV6 gene, cDNA and protein. *Cell Calcium* **2003**, *33*, 509–518. [CrossRef]

319. Peng, J.B.; Chen, X.Z.; Berger, U.V.; Vassilev, P.M.; Tsukaguchi, H.; Brown, E.M.; Hediger, M.A. Molecular cloning and characterization of a channel-like transporter mediating intestinal calcium absorption. *J. Biol. Chem.* **1999**, *274*, 22739–22746. [CrossRef] [PubMed]

320. Zhuang, L.; Peng, J.B.; Tou, L.; Takanaga, H.; Adam, R.M.; Hediger, M.A.; Freeman, M.R. Calcium-selective ion channel, CaT1, is apically localized in gastrointestinal tract epithelia and is aberrantly expressed in human malignancies. *Lab. Investig.* **2002**, *82*, 1755–1764. [CrossRef] [PubMed]

321. Nijenhuis, T.; Hoenderop, J.G.; van der Kemp, A.W.; Bindels, R.J. Localization and regulation of the epithelial Ca^{2+} channel TRPV6 in the kidney. *J. Am. Soc. Nephrol.* **2003**, *14*, 2731–2740. [CrossRef] [PubMed]

322. Suzuki, Y.; Kovacs, C.S.; Takanaga, H.; Peng, J.B.; Landowski, C.P.; Hediger, M.A. Calcium channel TRPV6 is involved in murine maternal-fetal calcium transport. *J. Bone Min. Res.* **2008**, *23*, 1249–1256. [CrossRef] [PubMed]

323. Fernandes, E.S.; Vong, C.T.; Quek, S.; Cheong, J.; Awal, S.; Gentry, C.; Aubdool, A.A.; Liang, L.; Bodkin, J.V.; Bevan, S.; et al. Superoxide generation and leukocyte accumulation: Key elements in the mediation of leukotriene B$_4$-induced itch by transient receptor potential ankyrin 1 and transient receptor potential vanilloid 1. *FASEB J.* **2013**, *27*, 1664–1673. [CrossRef] [PubMed]

324. Sisignano, M.; Park, C.K.; Angioni, C.; Zhang, D.D.; von Hehn, C.; Cobos, E.J.; Ghasemlou, N.; Xu, Z.Z.; Kumaran, V.; Lu, R.; et al. 5,6-EET is released upon neuronal activity and induces mechanical pain hypersensitivity via TRPA1 on central afferent terminals. *J. Neurosci.* **2012**, *32*, 6364–6372. [CrossRef] [PubMed]

325. Ross, R.A.; Craib, S.J.; Stevenson, L.A.; Pertwee, R.G.; Henderson, A.; Toole, J.; Ellington, H.C. Pharmacological characterization of the anandamide cyclooxygenase metabolite: Prostaglandin E2 ethanolamide. *J. Pharmacol. Exp. Ther.* **2002**, *301*, 900–907. [CrossRef] [PubMed]

326. Yu, M.; Ives, D.; Ramesha, C.S. Synthesis of prostaglandin E2 ethanolamide from anandamide by cyclooxygenase-2. *J. Biol. Chem.* **1997**, *272*, 21181–21186. [CrossRef] [PubMed]

327. Ueda, N.; Yamamoto, K.; Yamamoto, S.; Tokunaga, T.; Shirakawa, E.; Shinkai, H.; Ogawa, M.; Sato, T.; Kudo, I.; Inoue, K.; et al. Lipoxygenase-catalyzed oxygenation of arachidonylethanolamide, a cannabinoid receptor agonist. *Biochim. Biophys. Acta* **1995**, *1254*, 127–134. [CrossRef]

328. Ueda, N.; Kurahashi, Y.; Yamamoto, K.; Yamamoto, S.; Tokunaga, T. Enzymes for anandamide biosynthesis and metabolism. *J. Lipid Mediat. Cell Signal.* **1996**, *14*, 57–61. [CrossRef]

329. Craib, S.J.; Ellington, H.C.; Pertwee, R.G.; Ross, R.A. A possible role of lipoxygenases in the activation of vanilloid receptors by anandamide in the guinea-pig bronchus. *Br. J. Pharmacol.* **2001**, *134*, 30–37. [CrossRef] [PubMed]

330. Smart, D.; Gunthorpe, M.J.; Jerman, J.C.; Nasir, S.; Gray, J.; Muir, A.I.; Chambers, J.K.; Randall, A.D.; Davis, J.B. The endogenous lipid anandamide is a full agonist at the human vanilloid receptor (hVR1). *Br. J. Pharmacol.* **2000**, *129*, 227–230. [CrossRef] [PubMed]

331. Jacobsson, S.O.; Wallin, T.; Fowler, C.J. Inhibition of rat C6 glioma cell proliferation by endogenous and synthetic cannabinoids. Relative involvement of cannabinoid and vanilloid receptors. *J. Pharmacol. Exp. Ther.* **2001**, *299*, 951–959. [PubMed]

332. Sarker, K.P.; Obara, S.; Nakata, M.; Kitajima, I.; Maruyama, I. Anandamide induces apoptosis of PC-12 cells: Involvement of superoxide and caspase-3. *FEBS Lett.* **2000**, *472*, 39–44. [CrossRef]

333. Maccarrone, M.; Lorenzon, T.; Bari, M.; Melino, G.; Finazzi-Agro, A. Anandamide induces apoptosis in human cells via vanilloid receptors. Evidence for a protective role of cannabinoid receptors. *J. Biol. Chem.* **2000**, *275*, 31938–31945. [CrossRef] [PubMed]

334. Yang, Z.H.; Wang, X.H.; Wang, H.P.; Hu, L.Q.; Zheng, X.M.; Li, S.W. Capsaicin mediates cell death in bladder cancer T24 cells through reactive oxygen species production and mitochondrial depolarization. *Urology* **2010**, *75*, 735–741. [CrossRef] [PubMed]

335. Ma, F.; Zhang, L.; Westlund, K.N. Reactive oxygen species mediate TNFR1 increase after TRPV1 activation in mouse DRG neurons. *Mol. Pain* **2009**, *5*, 31. [CrossRef] [PubMed]

336. Talbot, S.; Dias, J.P.; Lahjouji, K.; Bogo, M.R.; Campos, M.M.; Gaudreau, P.; Couture, R. Activation of TRPV1 by capsaicin induces functional kinin B(1) receptor in rat spinal cord microglia. *J. Neuroinflamm.* **2012**, *9*, 16. [CrossRef] [PubMed]

337. Adam-Vizi, V.; Starkov, A.A. Calcium and mitochondrial reactive oxygen species generation: How to read the facts. *J. Alzheimer Dis.* **2010**, *20* (Suppl. 2), S413–S426. [CrossRef] [PubMed]

338. Nishio, S.; Teshima, Y.; Takahashi, N.; Thuc, L.C.; Saito, S.; Fukui, A.; Kume, O.; Fukunaga, N.; Hara, M.; Nakagawa, M.; et al. Activation of CaMKII as a key regulator of reactive oxygen species production in diabetic rat heart. *J. Mol. Cell. Cardiol.* **2012**, *52*, 1103–1111. [CrossRef] [PubMed]

339. Pandey, D.; Gratton, J.P.; Rafikov, R.; Black, S.M.; Fulton, D.J. Calcium/calmodulin-dependent kinase II mediates the phosphorylation and activation of NADPH oxidase 5. *Mol. Pharmacol.* **2011**, *80*, 407–415. [CrossRef] [PubMed]

340. Kishimoto, E.; Naito, Y.; Handa, O.; Okada, H.; Mizushima, K.; Hirai, Y.; Nakabe, N.; Uchiyama, K.; Ishikawa, T.; Takagi, T.; et al. Oxidative stress-induced posttranslational modification of TRPV1 expressed in esophageal epithelial cells. *Am. J. Physiol. Gastrointest. Liver Physiol.* **2011**, *301*, G230–G238. [CrossRef] [PubMed]

341. Yoshida, T.; Inoue, R.; Morii, T.; Takahashi, N.; Yamamoto, S.; Hara, Y.; Tominaga, M.; Shimizu, S.; Sato, Y.; Mori, Y. Nitric oxide activates TRP channels by cysteine S-nitrosylation. *Nat. Chem. Biol.* **2006**, *2*, 596–607. [CrossRef] [PubMed]

342. Takahashi, N.; Mori, Y. TRP Channels as Sensors and Signal Integrators of Redox Status Changes. *Front. Pharmacol.* **2011**, *2*, 58. [CrossRef] [PubMed]

343. Susankova, K.; Tousova, K.; Vyklicky, L.; Teisinger, J.; Vlachova, V. Reducing and oxidizing agents sensitize heat-activated vanilloid receptor (TRPV1) current. *Mol. Pharmacol.* **2006**, *70*, 383–394. [CrossRef] [PubMed]

344. Taylor-Clark, T.E.; McAlexander, M.A.; Nassenstein, C.; Sheardown, S.A.; Wilson, S.; Thornton, J.; Carr, M.J.; Undem, B.J. Relative contributions of TRPA1 and TRPV1 channels in the activation of vagal bronchopulmonary C-fibres by the endogenous autacoid 4-oxononenal. *J. Physiol.* **2008**, *586*, 347–359. [CrossRef] [PubMed]

345. Pei, Z.; Zhuang, Z.; Sang, H.; Wu, Z.; Meng, R.; He, E.Y.; Scott, G.I.; Maris, J.R.; Li, R.; Ren, J. α,β-Unsaturated aldehyde crotonaldehyde triggers cardiomyocyte contractile dysfunction: Role of TRPV1 and mitochondrial function. *Pharmacol. Res.* **2014**, *82*, 40–50. [CrossRef] [PubMed]

346. Inoue, I.; Goto, S.; Matsunaga, T.; Nakajima, T.; Awata, T.; Hokari, S.; Komoda, T.; Katayama, S. The ligands/activators for peroxisome proliferator-activated receptor alpha (PPARalpha) and PPARgamma increase Cu^{2+},Zn^{2+}-superoxide dismutase and decrease p22phox message expressions in primary endothelial cells. *Metabolism* **2001**, *50*, 3–11. [CrossRef] [PubMed]

347. Girnun, G.D.; Domann, F.E.; Moore, S.A.; Robbins, M.E. Identification of a functional peroxisome proliferator-activated receptor response element in the rat catalase promoter. *Mol. Endocrinol.* **2002**, *16*, 2793–2801. [CrossRef] [PubMed]

348. Diep, Q.N.; Amiri, F.; Touyz, R.M.; Cohn, J.S.; Endemann, D.; Neves, M.F.; Schiffrin, E.L. PPARalpha activator effects on Ang II-induced vascular oxidative stress and inflammation. *Hypertension* **2002**, *40*, 866–871. [CrossRef] [PubMed]

349. Landmesser, U.; Cai, H.; Dikalov, S.; McCann, L.; Hwang, J.; Jo, H.; Holland, S.M.; Harrison, D.G. Role of p47 in vascular oxidative stress and hypertension caused by angiotensin II. *Hypertension* **2002**, *40*, 511–515. [CrossRef] [PubMed]

350. Martinez, A.A.; Morgese, M.G.; Pisanu, A.; Macheda, T.; Paquette, M.A.; Seillier, A.; Cassano, T.; Carta, A.R.; Giuffrida, A. Activation of PPAR gamma receptors reduces levodopa-induced dyskinesias in 6-OHDA-lesioned rats. *Neurobiol. Dis.* **2015**, *74*, 295–304. [CrossRef] [PubMed]

351. Impellizzeri, D.; Esposito, E.; Attley, J.; Cuzzocrea, S. Targeting inflammation: New therapeutic approaches in chronic kidney disease (CKD). *Pharmacol. Res.* **2014**, *81*, 91–102. [CrossRef] [PubMed]

352. Di Paola, R.; Impellizzeri, D.; Mondello, P.; Velardi, E.; Aloisi, C.; Cappellani, A.; Esposito, E.; Cuzzocrea, S. Palmitoylethanolamide reduces early renal dysfunction and injury caused by experimental ischemia and reperfusion in mice. *Shock* **2012**, *38*, 356–366. [CrossRef] [PubMed]

353. Moreno, S.; Mugnaini, E.; Ceru, M.P. Immunocytochemical localization of catalase in the central nervous system of the rat. *J. HistoChem. Cytochem.* **1995**, *43*, 1253–1267. [CrossRef] [PubMed]

354. Moreno, S.; Nardacci, R.; Ceru, M.P. Regional and ultrastructural immunolocalization of copper-zinc superoxide dismutase in rat central nervous system. *J. Histochem. Cytochem.* **1997**, *45*, 1611–1622. [CrossRef] [PubMed]

355. Farioli-Vecchioli, S.; Moreno, S.; Ceru, M.P. lmmunocytochemical localization of acyl-CoA oxidase in the rat central nervous system. *J. Neurocytol.* **2001**, *30*, 21–33. [CrossRef] [PubMed]

356. Poynter, M.E.; Daynes, R.A. Peroxisome proliferator-activated receptor alpha activation modulates cellular redox status, represses nuclear factor-kappaB signaling, and reduces inflammatory cytokine production in aging. *J. Biol. Chem.* **1998**, *273*, 32833–32841. [CrossRef] [PubMed]

357. Cimini, A.; Benedetti, E.; Cristiano, L.; Sebastiani, P.; D'Amico, M.A.; D'Angelo, B.; Di Loreto, S. Expression of peroxisome proliferator-activated receptors (PPARs) and retinoic acid receptors (RXRs) in rat cortical neurons. *Neuroscience* **2005**, *130*, 325–337. [CrossRef] [PubMed]

358. Sayre, L.M.; Perry, G.; Smith, M.A. Oxidative Stress and Neurotoxicity. *Chem. Res. Toxicol.* **2008**, *21*, 172–188. [CrossRef] [PubMed]

359. Sultana, R.; Butterfield, D.A. Role of oxidative stress in the progression of Alzheimer's disease. *J. Alzheimer Dis.* **2010**, *19*, 341–353. [CrossRef] [PubMed]

360. Fanelli, F.; Sepe, S.; D'Amelio, M.; Bernardi, C.; Cristiano, L.; Cimini, A.; Cecconi, F.; Ceru', M.P.; Moreno, S. Age-dependent roles of peroxisomes in the hippocampus of a transgenic mouse model of Alzheimer's disease. *Mol. Neurodegener.* **2013**, *8*, 8. [CrossRef] [PubMed]

361. Duncan, R.S.; Chapman, K.D.; Koulen, P. The neuroprotective properties of palmitoylethanolamine against oxidative stress in a neuronal cell line. *Mol. Neurodegener.* **2009**, *4*, 50. [CrossRef] [PubMed]

362. Raso, G.M.; Esposito, E.; Vitiello, S.; Iacono, A.; Santoro, A.; D'Agostino, G.; Sasso, O.; Russo, R.; Piazza, P.V.; Calignano, A.; et al. Palmitoylethanolamide stimulation induces allopregnanolone synthesis in C6 Cells and primary astrocytes: Involvement of peroxisome-proliferator activated receptor-a. *J. Neuroendocrinol.* **2011**, *23*, 591–600. [CrossRef] [PubMed]

363. Avagliano, C.; Russo, R.; De Caro, C.; Cristiano, C.; La Rana, G.; Piegari, G.; Paciello, O.; Citraro, R.; Russo, E.; De Sarro, G.; et al. Palmitoylethanolamide protects mice against 6-OHDA-induced neurotoxicity and endoplasmic reticulum stress: In vivo and in vitro evidence. *Pharmacol. Res.* **2016**, *113*, 276–289. [CrossRef] [PubMed]

364. Manea, A.; Manea, S.A.; Todirita, A.; Albulescu, I.C.; Raicu, M.; Sasson, S.; Simionescu, M. High-glucose-increased expression and activation of NADPH oxidase in human vascular smooth muscle cells is mediated by 4-hydroxynonenal-activated PPARα and PPARβ/δ. *Cell Tissue Res.* **2015**, *361*, 593–604. [CrossRef] [PubMed]

365. Irving, A.; Abdulrazzaq, G.; Chan, S.L.F.; Penman, J.; Harvey, J.; Alexander, S.P.H. Cannabinoid Receptor-Related Orphan G Protein-Coupled Receptors. *Adv. Pharmacol.* **2017**, *80*, 223–247. [CrossRef] [PubMed]

366. Gantz, I.; Muraoka, A.; Yang, Y.K.; Samuelson, L.C.; Zimmerman, E.M.; Cook, H.; Yamada, T. Cloning and chromosomal localization of a gene (GPR18) encoding a novel seven transmembrane receptor highly expressed in spleen and testis. *Genomics* **1997**, *42*, 462–466. [CrossRef] [PubMed]

367. Vassilatis, D.K.; Hohmann, J.G.; Zeng, H.; Li, F.; Ranchalis, J.E.; Mortrud, M.T.; Brown, A.; Rodriguez, S.S.; Weller, J.R.; Wright, A.C.; et al. The G protein-coupled receptor repertoires of human and mouse. *Proc. Natl. Acad. Sci. USA* **2003**, *100*, 4903–4908. [CrossRef] [PubMed]

368. Lu, V.B.; Puhl, H.L., 3rd; Ikeda, S.R. N-Arachidonyl glycine does not activate G protein-coupled receptor 18 signaling via canonical pathways. *Mol. Pharmacol.* **2013**, *83*, 267–282. [CrossRef] [PubMed]

369. Finlay, D.B.; Joseph, W.R.; Grimsey, N.L.; Glass, M. GPR18 undergoes a high degree of constitutive trafficking but is unresponsive to N-Arachidonoyl Glycine. *PeerJ* **2016**, *4*, e1835. [CrossRef] [PubMed]

370. Penumarti, A.; Abdel-Rahman, A.A. The novel endocannabinoid receptor GPR18 is expressed in the rostral ventrolateral medulla and exerts tonic restraining influence on blood pressure. *J. Pharmacol. Exp. Ther.* **2014**, *349*, 29–38. [CrossRef] [PubMed]

371. Matouk, A.I.; Taye, A.; El-Moselhy, M.A.; Heeba, G.H.; Abdel-Rahman, A.A. The Effect of Chronic Activation of the Novel Endocannabinoid Receptor GPR18 on Myocardial Function and Blood Pressure in Conscious Rats. *J. Cardiovasc. Pharmacol.* **2017**, *69*, 23–33. [CrossRef] [PubMed]

372. Sawzdargo, M.; Nguyen, T.; Lee, D.K.; Lynch, K.R.; Cheng, R.; Heng, H.H.; George, S.R.; O'Dowd, B.F. Identification and cloning of three novel human G protein-coupled receptor genes GPR52, PsiGPR53 and GPR55: GPR55 is extensively expressed in human brain. *Brain Res. Mol. Brain Res.* **1999**, *64*, 193–198. [CrossRef]

373. Ryberg, E.; Larsson, N.; Sjögren, S.; Hjorth, S.; Hermansson, N.O.; Leonova, J.; Elebring, T.; Nilsson, K.; Drmota, T.; Greasley, P.J. The orphan receptor GPR55 is a novel cannabinoid receptor. *Br. J. Pharmacol.* **2007**, *152*, 1092–1101. [CrossRef] [PubMed]

374. Romero-Zerbo, S.Y.; Rafacho, A.; Díaz-Arteaga, A.; Suárez, J.; Quesada, I.; Imbernon, M.; Ross, R.A.; Dieguez, C.; Rodríguez de Fonseca, F.; Nogueiras, R.; et al. A role for the putative cannabinoid receptor GPR55 in the islets of Langerhans. *J. Endocrinol.* **2011**, *211*, 177–185. [CrossRef] [PubMed]

375. Whyte, L.S.; Ryberg, E.; Sims, N.A.; Ridge, S.A.; Mackie, K.; Greasley, P.J.; Ross, R.A.; Rogers, M.J. The putative cannabinoid receptor GPR55 affects osteoclast function in vitro and bone mass in vivo. *Proc. Natl. Acad. Sci. USA* **2009**, *106*, 16511–16516. [CrossRef] [PubMed]

376. Pietr, M.; Kozela, E.; Levy, R.; Rimmerman, N.; Lin, Y.H.; Stella, N.; Vogel, Z.; Juknat, A. Differential changes in GPR55 during microglial cell activation. *FEBS Lett.* **2009**, *583*, 2071–2076. [CrossRef] [PubMed]

377. Waldeck-Weiermair, M.; Zoratti, C.; Osibow, K.; Balenga, N.; Goessnitzer, E.; Waldhoer, M.; Malli, R.; Graier, W.F. Integrin clustering enables anandamide-induced Ca^{2+} signaling in endothelial cells via GPR55 by protection against CB1-receptor-triggered repression. *J. Cell Sci.* **2008**, *121*, 1704–1717. [CrossRef] [PubMed]

378. Kapur, A.; Zhao, P.; Sharir, H.; Bai, Y.; Caron, M.G.; Barak, L.S.; Abood, M.E. Atypical responsiveness of the orphan receptor GPR55 to cannabinoid ligands. *J. Biol. Chem.* **2009**, *284*, 29817–29827. [CrossRef] [PubMed]

379. Yin, H.; Chu, A.; Li, W.; Wang, B.; Shelton, F.; Otero, F.; Nguyen, D.G.; Caldwell, J.S.; Chen, Y.A. Lipid G protein-coupled receptor ligand identification using beta-arrestin PathHunter assay. *J. Biol. Chem.* **2009**, *284*, 12328–12338. [CrossRef] [PubMed]

380. Henstridge, C.M.; Balenga, N.A.; Ford, L.A.; Ross, R.A.; Waldhoer, M.; Irving, A.J. The GPR55 ligand L-alpha-lysophosphatidylinositol promotes RhoA-dependent Ca^{2+} signaling and NFAT activation. *FASEB J.* **2009**, *23*, 183–193. [CrossRef] [PubMed]

381. Oka, S.; Toshida, T.; Maruyama, K.; Nakajima, K.; Yamashita, A.; Sugiura, T. 2-Arachidonoyl-sn-glycero-3-phosphoinositol: A possible natural ligand for GPR55. *J. Biochem.* **2009**, *145*, 13–20. [CrossRef] [PubMed]

382. Oka, S.; Nakajima, K.; Yamashita, A.; Kishimoto, S.; Sugiura, T. Identification of GPR55 as a lysophosphatidylinositol receptor. *Biochem. Biophys. Res. Commun.* **2007**, *362*, 928–934. [CrossRef] [PubMed]

383. Bondarenko, A.; Waldeck-Weiermair, M.; Naghdi, S.; Poteser, M.; Malli, R.; Graier, W.F. GPR55-dependent and -independent ion signalling in response to lysophosphatidylinositol in endothelial cells. *Br. J. Pharmacol.* **2010**, *161*, 308–320. [CrossRef] [PubMed]

384. Moreno, E.; Andradas, C.; Medrano, M.; Caffarel, M.M.; Pérez-Gómez, E.; Blasco-Benito, S.; Gómez-Cañas, M.; Pazos, M.R.; Irving, A.J.; Lluís, C.; et al. Targeting CB2-GPR55 receptor heteromers modulates cancer cell signaling. *J. Biol. Chem.* **2014**, *289*, 21960–21972. [CrossRef] [PubMed]

385. Fredriksson, R.; Höglund, P.J.; Gloriam, D.E.; Lagerström, M.C.; Schiöth, H.B. Seven evolutionarily conserved human rhodopsin G protein-coupled receptors lacking close relatives. *FEBS Lett.* **2003**, *554*, 381–388. [CrossRef]

386. Soga, T.; Ohishi, T.; Matsui, T.; Saito, T.; Matsumoto, M.; Takasaki, J.; Matsumoto, S.; Kamohara, M.; Hiyama, H.; Yoshida, S.; et al. Lysophosphatidylcholine enhances glucose-dependent insulin secretion via an orphan G-protein-coupled receptor. *Biochem. Biophys. Res. Commun.* **2005**, *326*, 744–751. [CrossRef] [PubMed]

387. Sakamoto, Y.; Inoue, H.; Kawakami, S.; Miyawaki, K.; Miyamoto, T.; Mizuta, K.; Itakura, M. Expression and distribution of Gpr119 in the pancreatic islets of mice and rats: Predominant localization in pancreatic polypeptide-secreting PP-cells. *Biochem. Biophys. Res. Commun.* **2006**, *351*, 474–480. [CrossRef] [PubMed]

388. Chu, Z.L.; Carroll, C.; Alfonso, J.; Gutierrez, V.; He, H.; Lucman, A.; Pedraza, M.; Mondala, H.; Gao, H.; Bagnol, D.; et al. A role for intestinal endocrine cell-expressed g protein-coupled receptor 119 in glycemic control by enhancing glucagon-like Peptide-1 and glucose-dependent insulinotropic Peptide release. *Endocrinology* **2008**, *149*, 2038–2047. [CrossRef] [PubMed]

389. Lauffer, L.M.; Iakoubov, R.; Brubaker, P.L. GPR119 is essential for oleoylethanolamide-induced glucagon-like peptide-1 secretion from the intestinal enteroendocrine L-cell. *Diabetes* **2009**, *58*, 1058–1066. [CrossRef] [PubMed]

390. Yang, J.W.; Kim, H.S.; Im, J.H.; Kim, J.W.; Jun, D.W.; Lim, S.C.; Lee, K.; Choi, J.M.; Kim, S.K.; Kang, K.W. GPR119: A promising target for nonalcoholic fatty liver disease. *FASEB J.* **2016**, *30*, 324–335. [CrossRef] [PubMed]

391. Cvijanovic, N.; Isaacs, N.J.; Rayner, C.K.; Feinle-Bisset, C.; Young, R.L.; Little, T.J. Duodenal fatty acid sensor and transporter expression following acute fat exposure in healthy lean humans. *Clin. Nutr.* **2017**, *36*, 564–569. [CrossRef] [PubMed]

392. Overton, H.A.; Babbs, A.J.; Doel, S.M.; Fyfe, M.C.; Gardner, L.S.; Griffin, G.; Jackson, H.C.; Procter, M.J.; Rasamison, C.M.; Tang-Christensen, M.; et al. Deorphanization of a G protein-coupled receptor for oleoylethanolamide and its use in the discovery of small-molecule hypophagic agents. *Cell Metab.* **2006**, *3*, 167–175. [CrossRef] [PubMed]

393. Chu, Z.L.; Carroll, C.; Chen, R.; Alfonso, J.; Gutierrez, V.; He, H.; Lucman, A.; Xing, C.; Sebring, K.; Zhou, J.; et al. *N*-oleoyldopamine enhances glucose homeostasis through the activation of GPR119. *Mol. Endocrinol.* **2010**, *24*, 161–170. [CrossRef] [PubMed]

394. Hansen, K.B.; Rosenkilde, M.M.; Knop, F.K.; Wellner, N.; Diep, T.A.; Rehfeld, J.F.; Andersen, U.B.; Holst, J.J.; Hansen, H.S. 2-Oleoyl glycerol is a GPR119 agonist and signals GLP-1 release in humans. *J. Clin. Endocrinol. Metab.* **2011**, *96*, E1409–E1417. [CrossRef] [PubMed]

395. Balenga, N.A.; Aflaki, E.; Kargl, J.; Platzer, W.; Schröder, R.; Blättermann, S.; Kostenis, E.; Brown, A.J.; Heinemann, A.; Waldhoer, M. GPR55 regulates cannabinoid 2 receptor-mediated responses in human neutrophils. *Cell Res.* **2011**, *21*, 1452–1469. [CrossRef] [PubMed]

396. Chiurchiù, V.; Lanuti, M.; De Bardi, M.; Battistini, L.; Maccarrone, M. The differential characterization of GPR55 receptor in human peripheral blood reveals a distinctive expression in monocytes and NK cells and a proinflammatory role in these innate cells. *Int. Immunol.* **2015**, *27*, 153–160. [CrossRef] [PubMed]

397. He, L.; He, T.; Farrar, S.; Ji, L.; Liu, T.; Ma, X. Antioxidants Maintain Cellular Redox Homeostasis by Elimination of Reactive Oxygen Species. *Cell. Physiol. Biochem.* **2017**, *44*, 532–553. [CrossRef] [PubMed]

398. Chelikani, P.; Fita, I.; Loewen, P.C. Diversity of structures and properties among catalases. *Cell. Mol. Life Sci.* **2004**, *61*, 192–208. [CrossRef] [PubMed]

399. Bannister, J.V.; Bannister, W.H.; Rotilio, G. Aspects of the structure, function, and applications of superoxide dismutase. *CRC Crit. Rev. Biochem.* **1987**, *22*, 111–180. [CrossRef] [PubMed]

400. Zelko, I.N.; Mariani, T.J.; Folz, R.J. Superoxide dismutase multigene family: A comparison of the CuZn-SOD (SOD1), Mn-SOD (SOD2), and EC-SOD (SOD3) gene structures, evolution, and expression. *Free Radic. Biol. Med.* **2002**, *33*, 337–349. [CrossRef]

401. Brigelius-Flohé, R. Tissue-specific functions of individual glutathione peroxidases. *Free Radic. Biol. Med.* **1999**, *27*, 951–965. [CrossRef]

402. Arnér, E.S.; Holmgren, A. Physiological functions of thioredoxin and thioredoxin reductase. *Eur. J. Biochem.* **2000**, *267*, 6102–6109. [CrossRef] [PubMed]

403. Meister, A.; Anderson, M.E. Glutathione. *Annu. Rev. Biochem.* **1983**, *52*, 711–760. [CrossRef] [PubMed]

404. Palace, V.P.; Khaper, N.; Qin, Q.; Singal, P.K. Antioxidant potentials of vitamin A and carotenoids and their relevance to heart disease. *Free Radic. Biol. Med.* **1999**, *26*, 746–761. [CrossRef]

405. Rice, M.E. Ascorbate regulation and its neuroprotective role in the brain. *Trends Neurosci.* **2000**, *23*, 209–216. [CrossRef]

406. McCay, P.B. Vitamin E: Interactions with free radicals and ascorbate. *Annu. Rev. Nutr.* **1985**, *5*, 323–340. [CrossRef] [PubMed]

407. Turunen, M.; Olsson, J.; Dallner, G. Metabolism and function of coenzyme Q. *Biochim. Biophys. Acta* **2004**, *1660*, 171–199. [CrossRef] [PubMed]

408. Paiva, S.A.; Russell, R.M. Beta-carotene and other carotenoids as antioxidants. *J. Am. Coll. Nutr.* **1999**, *18*, 426–433. [CrossRef] [PubMed]

409. Giudetti, A.M.; Salzet, M.; Cassano, T. Oxidative Stress in Aging Brain: Nutritional and Pharmacological Interventions for Neurodegenerative Disorders. *Oxid. Med. Cell. Longev.* **2018**, *2018*, 3416028. [CrossRef] [PubMed]

410. Heim, K.E.; Tagliaferro, A.R.; Bobilya, D.J. Flavonoid antioxidants: Chemistry, metabolism and structure-activity relationships. *J. Nutr. Biochem.* **2002**, *13*, 572–584. [CrossRef]

411. Sandoval-Acuña, C.; Ferreira, J.; Speisky, H. Polyphenols and mitochondria: An update on their increasingly emerging ROS-scavenging independent actions. *Arch. Biochem. Biophys.* **2014**, *559*, 75–90. [CrossRef] [PubMed]

412. Tabassum, A.; Bristow, R.G.; Venkateswaran, V. Ingestion of selenium and other antioxidants during prostate cancer radiotherapy: A good thing? *Cancer Treat. Rev.* **2010**, *36*, 230–234. [CrossRef] [PubMed]

413. Prasad, A.S.; Bao, B.; Beck, F.W.; Kucuk, O.; Sarkar, F.H. Antioxidant effect of zinc in humans. *Free Radic. Biol. Med.* **2004**, *37*, 1182–1190. [CrossRef] [PubMed]

414. Rizzo, A.M.; Berselli, P.; Zava, S.; Montorfano, G.; Negroni, M.; Corsetto, P.; Berra, B. Endogenous antioxidants and radical scavengers. *Adv. Exp. Med. Biol.* **2010**, *698*, 52–67. [CrossRef] [PubMed]

415. Reiter, R.J.; Tan, D.X.; Poeggeler, B.; Menendez-Pelaez, A.; Chen, L.D.; Saarela, S. Melatonin as a free radical scavenger: Implications for aging and age-related diseases. *Ann. N. Y. Acad. Sci.* **1994**, *719*, 1–12. [CrossRef] [PubMed]

416. Burton, G.W.; Traber, M.G. Vitamin E: Antioxidant activity, biokinetics, and bioavailability. *Annu. Rev. Nutr.* **1990**, *10*, 357–382. [CrossRef] [PubMed]

417. Bowman, G.L.; Shannon, J.; Frei, B.; Kaye, J.A.; Quinn, J.F. Uric acid as a CNS antioxidant. *J. Alzheimer Dis.* **2010**, *19*, 1331–1336. [CrossRef] [PubMed]

418. Cotelle, N. Role of flavonoids in oxidative stress. *Curr. Top. Med. Chem.* **2001**, *1*, 569–590. [CrossRef] [PubMed]

419. Hussain, T.; Tan, B.; Yin, Y.; Blachier, F.; Tossou, M.C.; Rahu, N. Oxidative Stress and Inflammation: What Polyphenols Can Do for Us? *Oxid. Med. Cell. Longev.* **2016**, *2016*, 7432797. [CrossRef] [PubMed]

420. Martinez, J.; Moreno, J.J. Role of Ca^{2+}-Independent Phospholipase A_2 on Arachidonic Acid Release Induced by Reactive Oxygen Species. *Arch. Biochem. Biophys.* **2001**, *392*, 257–262. [CrossRef] [PubMed]

421. Bátkai, S.; Rajesh, M.; Mukhopadhyay, P.; Haskó, G.; Liaudet, L.; Cravatt, B.F.; Csiszár, A.; Ungvári, Z.; Pacher, P. Decreased age-related cardiac dysfunction, myocardial nitrative stress, inflammatory gene expression, and apoptosis in mice lacking fatty acid amide hydrolase. *Am. J. Physiol. Heart Circ. Physiol.* **2007**, *293*, H909–H918. [CrossRef] [PubMed]

422. Wei, Y.; Wang, X.; Wang, L. Presence and regulation of cannabinoid receptors in human retinal pigment epithelial cells. *Mol. Vis.* **2009**, *15*, 1243–1251. [PubMed]

423. Wang, M.; Abais, J.M.; Meng, N.; Zhang, Y.; Ritter, J.K.; Li, P.L.; Tang, W.X. Upregulation of cannabinoid receptor-1 and fibrotic activation of mouse hepatic stellate cells during Schistosoma J. infection: Role of NADPH oxidase. *Free Radic. Biol. Med.* **2014**, *71*, 109–120. [CrossRef] [PubMed]

antioxidants

MDPI

Review

Evening Primrose (*Oenothera biennis*) Biological Activity Dependent on Chemical Composition

Magdalena Timoszuk, Katarzyna Bielawska and Elżbieta Skrzydlewska *

Department of Inorganic and Analytical Chemistry, Medical University of Bialystok, 15-089 Bialystok, Poland; magdalena.timoszuk@umb.edu.pl (M.T.); katarzyna.bielawska@umb.edu.pl (K.B.)
* Correspondence: elzbieta.skrzydlewska@umb.edu.pl; Tel./Fax: +48-857-485-882

Received: 11 July 2018; Accepted: 8 August 2018; Published: 14 August 2018

Abstract: Evening primrose (*Oenothera* L.) is a plant belonging to the family Onagraceae, in which the most numerous species is *Oenothera biennis*. Some plants belonging to the genus *Oenothera* L. are characterized by biological activity. Therefore, studies were conducted to determine the dependence of biological activity on the chemical composition of various parts of the evening primrose, mainly leaves, stems, and seeds. Common components of all parts of the *Oenothera biennis* plants are fatty acids, phenolic acids, and flavonoids. In contrast, primrose seeds also contain proteins, carbohydrates, minerals, and vitamins. Therefore, it is believed that the most interesting sources of biologically active compounds are the seeds and, above all, evening primrose seed oil. This oil contains mainly aliphatic alcohols, fatty acids, sterols, and polyphenols. Evening primrose oil (EPO) is extremely high in linoleic acid (LA) (70–74%) and γ-linolenic acid (GLA) (8–10%), which may contribute to the proper functioning of human tissues because they are precursors of anti-inflammatory eicosanoids. EPO supplementation results in an increase in plasma levels of γ-linolenic acid and its metabolite dihomo-γ-linolenic acid (DGLA). This compound is oxidized by lipoxygenase (15-LOX) to 15-hydroxyeicosatrienoic acid (15-HETrE) or, under the influence of cyclooxygenase (COX), DGLA is metabolized to series 1 prostaglandins. These compounds have anti-inflammatory and anti-proliferative properties. Furthermore, 15-HETrE blocks the conversion of arachidonic acid (AA) to leukotriene A_4 (LTA_4) by direct inhibition of 5-LOX. In addition, γ-linolenic acid suppresses inflammation mediators such as interleukin 1β (IL-1β), interleukin 6 (IL-6), and cytokine - tumor necrosis factor α (TNF-α). The beneficial effects of EPO have been demonstrated in the case of atopic dermatitis, psoriasis, Sjögren's syndrome, asthma, and anti-cancer therapy.

Keywords: evening primrose oil; γ-linolenic acid; linoleic acid; omega-6 fatty acids; eicosanoids

---

1. Introduction

Evening primrose (*Oenothera* L.) is a plant belonging to the Onagraceae family. There are about 145 species in the genus *Oenothera* L., occurring in the temperate and tropical climate zones of North and South America. Some species have adapted to new areas, inhabiting the countries of the European continent, and about 70 species are now present in Europe. The most numerous species in the *Oenothera* L. family is *Oenothera biennis*, which also has the best-studied biological activity. It has been indicated that *Oenothera biennis* is beneficial in the treatment of many diseases. Therefore, research is ongoing to determine the chemical composition of these plants and how it relates to the biological activity of evening primrose. This research mainly concerns extracts from various parts of evening primrose (e.g., the leaves, stems, and seeds) [1].

2. Chemical Composition of Evening Primrose (*Oenothera biennis*)

Methanolic extracts prepared from the aerial parts of *Oenothera biennis* contain mainly phenolic acids and flavonoids. Phenolic acids present in the analyzed extracts include the following compounds: gallic acid and its ester derivatives (e.g., methyl gallate, galloylglucose, digalloylglucose, and tris-galloylglucose), 3-*p*-feruloylquinic acid, 3-*p*-coumaroylquinic acid, 4-*p*-feruloylquinic acid, caffeic acid pentoside, ellagic acid and its ester derivatives (e.g., ellagic acid hexoside and ellagic acid pentoside), and valoneic acid dilactone. Flavonoids present in the extracts include the following compounds: myricetin 3-*O*-glucuronide, quercetin 3-*O*-galactoside, quercetin 3-*O*-glucuronide, quercetin 3-*O*-glucoside, quercetin pentoside, quercetin dihexoside, quercetin glucuronylhexoside, quercetin 3-*O*-(2'-galloyl)-glucuronide, kaempferol 3-*O*-rhamnoglucoside, kaempferol 3-*O*-glucoside, kaempferol 3-*O*-glucuronide, kaempferol 3-*O*-(2'-galloyl)-glucuronide, and kaempferol pentoside [2].

The aqueous leaf extract of *Oenothera biennis* contains phenolic compounds (e.g., ellagitannins and caffeoyl tartaric acid) and flavonoids (quercetin glucuronide and kaempferol glucuronide) [3]. Among the tannins contained in the leaves of the evening primrose are oenothein A and oenothein B. The carbohydrates present in the extracts include arabinose, galactose, glucose, mannose, galacturonic acid, and glucuronic acid.

The roots of evening primrose contain the following sterols: sitosterol, oenotheralanosterol A, and oenotheralanosterol B. The triterpenes maslinic acid and oleanolic acid are also present in the root, along with the following carbohydrates: arabinose, galactose, glucose, mannose, galacturonic acid, and glucuronic acid. The following tannins are also found: gallic acid, tetramethylellagic acid, oenostacin, and 2,7,8-trimethylellagic acid [4]. The methanolic extract of the *Oenothera biennis* root also possesses significant amounts of xanthone (9H-xanthen-9-one) and its derivatives, such as dihydroxyprenyl xanthone and cetoleilyl diglucoside, which possess diverse biological and pharmacological properties [5].

Evening primrose seeds contain about 20% oil. The amount of oil depends on various factors, such as the age of the seed, the cultivar, and the growth conditions [6].

Generally, evening primrose oil is obtained from *Oenothera biennis* seeds using the cold-pressing method. The oil is a blend of about 13 triacylglycerol fractions, where the dominant combinations consist of the following fatty acids: linoleic–linoleic–linoleic (LLL, 40%), linoleic–linoleic–γ-linolenic (LLLnγ, approximately 15%), linoleic–linoleic–palmitic (LLP, approximately 8%), and linoleic–linoleic–oleic (LLO, approximately 8%) [7]. The oil consists of triacylglycerols—about 98%, with a small amount of other lipids and about 1–2% non-saponifiable fraction [6].

Evening primrose oil is very high in linoleic (70–74%) and γ-linolenic (8–10%) acids, and also contains other fatty acids: palmitic acid, oleic acid, stearic acid, and (in smaller amounts) myristic acid, oleopalmitic acid, vaccenic acid, eicosanoic acid, and eicosenoic acid (see Table 1) [8]. The phospholipid fraction comprises only 0.05% of the oil, and the following phospholipids have been identified in it: phosphatidylcholines (31.9%), phosphatidylinositols (27.1%), phosphatidylethanolamines (17.6%), phosphatidylglycerols (16.7%), and phosphatidic acids (6.7%) [6].

Table 1. Fatty acid composition of evening primrose oil (EPO) (*Oenothera biennis* L.) [8].

Compound Name	Contents (%)
linoleic acid	73.88 ± 0.09
γ-linolenic acid	9.24 ± 0.05
oleic acid	6.93 ± 0.02
palmitic acid	6.31 ± 0.14
stearic acid	1.88 ± 0.02
vaccenic acid	0.81 ± 0.03
eicosenoic acid	0.55 ± 0.01
eicosanoic acid	0.31 ± 0.03
behenic acid	0.10 ± 0.01

Evening primrose oil includes aliphatic alcohols, which make up about 798 mg/kg of the oil, 1-tetracosanol (about 237 mg/kg oil), and 1-hexacosanol (about 290 mg/kg oil) being present in the largest amount. The main triterpenes present are β-amyrin (about 996 mg/kg oil) and squalene (about 0.40 mg/kg oil) [8]. The oil contains a small amount of tocopherols: α-tocopherol (76 mg/kg oil), γ-tocopherol (187 mg/kg oil), and δ-tocopherol (15 mg/kg oil) [6].

Evening primrose seeds also contain phenolic acids, which are present in free acid form and as ester and glycoside derivatives (see Table 2) [9]. It has been shown that the seeds contain about 15% protein and 43% carbohydrates (in the form of cellulose, along with starch and dextrin). Lignin is also found in the seeds. In addition, the seeds contain amino acids: tryptophan (1.60%), lysine (0.31%), threonine (0.35%), cysteine (1.68%), valine (0.52%), isoleucine (0.41%), leucine (0.87%), and tyrosine (1.05%). Moreover, the seeds contain minerals, mainly calcium, potassium, and magnesium, and vitamins A, B, C, and E [10].

Table 2. Phenolic acid composition (mg/kg) in *Oenothera biennis* L. seed [9].

Acid Name	Included in			
	Free	Esters	Glycosides	Total
p-hydroxyphenyl acetic	n/a	1.03 ± 0.18	0.26 ± 0.05	1.29 ± 0.19
p-hydroxybenzoic	4.12 ± 0.25	0.38 ± 0.07	0.29 ± 0.10	4.79 ± 0.26
2-hydroxy-4-methoxybenzoic	6.52 ± 0.30	n/a	0.83 ± 0.28	7.35 ± 0.41
caffeic	6.48 ± 0.29	0.80 ± 0.14	n/a	7.51 ± 0.33
hydroxycaffeic	n/a	0.77 ± 0.18	n/a	0.77 ± 0.18
m-coumaric	4.90 ± 0.45	0.83 ± 0.21	n/a	5.73 ± 0.50
p-coumaric	1.32 ± 0.10	1.96 ± 0.23	0.06 ± 0.06	3.34 ± 0.25
ferulic	4.08 ± 0.30	0.72 ± 0.09	0.22 ± 0.06	5.02 ± 0.32
gallic	1.87 ± 0.22	7.03 ± 0.82	5.91 ± 1.56	14.81 ± 1.78
protocatechuic	50.28 ± 0.77	10.96 ± 0.34	2.16 ± 2.42	63.40 ± 2.56
vanillic	5.22 ± 0.28	0.06 ± 0.02	0.83 ± 0.28	7.35 ± 0.41
veratric	n/a	0.41 ± 0.03	0.47 ± 0.15	0.88 ± 0.15
homoveratric	n/a	0.43 ± 0.06	n/a	0.43 ± 0.06
salicylic	1.15 ± 0.04	1.40 ± 0.18	n/a	2.55 ± 0.18

n/a—not available.

Evening primrose oil also contains polyphenols, such as hydroxytyrosol (1.11 mg/kg oil), vanillic acid (3.27 mg/kg oil), vanillin (17.37 mg/kg oil), *p*-coumaric acid (1.75 mg/kg oil), and ferulic acid (25.23 mg/kg oil) [8].

The unsaponifiable matter of oil is composed partially of sterols, which comprise 53.16% of this fraction (see Table 3) [8].

Table 3. Sterol content of EPO (*Oenothera biennis* L.) [8].

Compound Name	Contents (mg/kg of Oil)
β-sitosterol	7952.00 ± 342.25
kampesterol	883.32 ± 0.45
Δ5-avenasterol	429.65 ± 75.20
sitostanol	167.01 ± 39.77
clerosterol	120.44 ± 0.12
Δ5-24-estigmastadienol	94.60 ± 5.68
Δ7-estigmasterol	38.17 ± 14.33
Δ7-avenasterol	27.80 ± 16.07

The seed ash contains a group of macroelements and microelements, including calcium, magnesium, potassium, phosphorus, manganese, iron, sodium, zinc, and copper (see Table 4) [10].

Table 4. Macroelements and microelements contributing to seed ash [10].

Macroelements	Contents (mg/100g of ash)
calcium	1800
magnesium	530
potassium	460
sodium	18
phosphorus	410

Microelements	Contents (mg/100g of ash)
iron	39
zinc	7
copper	1.1
manganese	0.5

3. Biological Activity of Evening Primrose Oil (*Oenothera biennis*)

The biological effect of evening primrose oil is a result of its composition and the biological properties of its components. Since the most important components in terms of quantity are polyunsaturated fatty acids (PUFAs), mainly linoleic acid (LA) and γ-linolenic acid (GLA) which belong to the group of omega-6 acids. The biological significance, especially of these acids, will be discussed in more detail.

Linoleic acid belongs to the group of essential fatty acids. These are also called exogenous fatty acids, because the human body does not synthesize them and it is necessary to obtain them from food [11]. Evening primrose oil contains over 70% linoleic acid (LA) and about 9% γ-linolenic acid (GLA) [10]. Linoleic acid and γ-linolenic acid contribute to the proper functioning of many tissues of the human body, because they are precursors of compounds that lead to the generation of anti-inflammatory eicosanoids, such as the series 1 prostaglandins and 15-hydroxyeicosatrienoic acid (15-HETrE). On the other hand, the enzymatic conversion of linoleic acid to arachidonic acid (AA) may form pro-inflammatory compounds, such as series 2 prostaglandins and series 4 leukotrienes [11]. With reference to the above, it is suggested that evening primrose oil may influence inflammatory diseases, including skin problems.

Linoleic acid [12], among others, plays an important role in the proper functioning of the skin, especially the stratum corneum, in which it is one of the main components of the ceramides building the lipid layer. It has been shown that the presence of this acid prevents the skin from peeling and the loss of water through the epidermis, while at the same time improving skin softness and elasticity and regulating the process of epidermal keratinization. A deficiency of linoleic acid, which is contained in large quantities in ceramide 1, leads to its replacement by oleic acid. This causes a deterioration in the protective properties of the epidermis [12,13].

Under the influence of Δ-6-desaturase (D6D), linoleic acid undergoes dehydrogenation to form γ-linolenic acid. The activity of Δ-6-desaturase in human cells depends on various factors and is reduced under the influence of nicotine, alcohol, magnesium deficiency, and a poor diet rich in saturated fatty acids, and under conditions of physiological aging of the body [12,14]. The basic component of evening primrose oil is linoleic acid, and the possibility of its metabolism to γ-linolenic acid due to the action of Δ-6-desaturase is an important point. Δ-6-Desaturase (D6D) activity is highest in the liver, the brain, the heart, and lung cells [15]. D6D activity is several times higher in the fetal human liver than in the adult human liver [11]. The fatty acid desaturase 2 (FADS2) gene encoding the Δ-6-desaturase enzyme is also expressed in skin cells: within the sebaceous gland, D6D desaturates palmitic acid to sapienic acid, which is the major fatty acid of human sebum [16,17]. On the other hand, the main epidermal cells—keratinocytes—are characterized by their lack of D6D and D5D activity [18] and the dermal fibroblasts express the D6D mRNA, which is capable of desaturation of the essential fatty acid (EFA) in the skin [19].

Δ-6-Desaturase is also present in evening primrose seeds. Huang et al. (2010) report that the cDNA sequence of the D6D gene was obtained from the developing seeds. The transformation of the plasmid DNA of a *Saccharomyces cerevisiae* strain then showed that after the addition of a medium containing linoleic acid to the yeast cells, a signal was obtained from γ-linolenic acid, indicating the presence of Δ-6-desaturase in seeds of the species *Oenothera biennis* [20,21].

A deficiency of γ-linolenic acid and other metabolites of linoleic acid was demonstrated in the plasma of patients with atopic dermatitis. This is linked to a decrease in Δ-6-desaturase activity, which makes the conversion of linoleic acid to γ-linolenic acid and the formation of its metabolites impossible [12,22]. It was found that oral treatment with evening primrose oil, which contains γ-linolenic acid, may lead to a reduction in the symptoms of atopic dermatitis [22]. However, subsequent studies have not confirmed that oral supplementation with evening primrose oil improves the skin condition in patients with atopic dermatitis [23].

The hydrocarbon chain of the emerging or supplied γ-linolenic acid under the influence of an elongase is elongated to dihomo-γ-linolenic acid (DGLA). The GLA elongation is faster than the desaturation of linoleic acid [24,25]. Dihomo-γ-linolenic acid is metabolized by cyclooxygenase (COX) to series 1 prostaglandins, which are eicosanoids with anti-inflammatory activity (see Table 5) [24]. Under the influence of 15-lipoxygenase (15-LOX), DGLA is oxidized to 15-hydroxyeicosatrienoic (15-HETrE) acid, which has anti-inflammatory and anti-proliferative properties [24,26]. Therefore, increased levels of GLA and DGLA, which are metabolized to anti-inflammatory compounds, suppress the inflammatory reaction. However, a decrease in these acids' levels may lead to the development of inflammatory diseases [24,27]. In addition, GLA suppresses inflammation mediators such as IL-1β, IL-6, and TNF-α cytokines [12,28]. In contrast, 15-HETrE acid, which is a product of the oxidation of dihomo-γ-linolenic acid by 15-LOX, has the ability to inhibit the synthesis of series 4 leukotrienes, whose elevated levels cause intensified pathological cell hyperproliferation [24]. This contributes to the inhibition of the pro-inflammatory action of leukotrienes, which are involved, among others, in the development of asthma [24,29] (see Figure 1).

Regardless of the metabolism of DGLA catalyzed by COXs and 15-LOX, dihomo-γ-linolenic acid under the influence of Δ-5-desaturase (D5D) is converted to arachidonic acid (AA), which is the precursor of many lipid mediators in the body, mainly pro-inflammatory [18,24]. Under physiological conditions, the sources of AA are membrane phospholipids, from which it is released by the hydrolysis of ester bonds, mainly via the action of phospholipase A_2 (PLA$_2$). There are two main ways that lead to the formation of free arachidonic acid. One pathway leads to the hydrolytic release of AA by the cytosolic isoform of PLA$_2$. The second way leads to the release of AA by the indirect action of phospholipase C (PLC) and diacylglycerol (DAG) lipase. DAG lipase and PLC result in the formation of inositol 1,4,5-triphosphate and DAG. The latter is then hydrolyzed by DAG lipase to form free arachidonic acid and monoacylglycerol (MAG) [18,30].

In pathological conditions and, for example, in the case of excessive exposure of the skin to UV radiation, the redox balance is disturbed and oxidative stress occurs, which results in the activation of cytosolic phospholipase A2 (cPLA$_2$) and PLC in the skin cells [18]. This leads to the excessive release of AA and the increased production of eicosanoids via cyclooxygenases and lipoxygenases. COX-1 and COX-2 catalyze the transformation of arachidonic acid into the series 2 prostanoids (PGE$_2$, PGD$_2$, PGI$_2$, TXA$_2$, and TXB$_2$). Moreover, 5-lipoxygenase (5-LOX) metabolizes arachidonic acid to series 4 leukotrienes (LTB$_4$, LTC$_4$, LTD$_4$, and LTE$_4$) [18,31]. Prostaglandins, series 2 thromboxanes, and series 4 leukotrienes belong to the pro-inflammatory eicosanoids [31]. However, 15-lipoxygenase (15-LOX) catalyzes the conversion of arachidonic acid to 15-hydroxyeicosatetraenoic acid (15-HETE), whose metabolites are lipoxins. These compounds have anti-inflammatory properties [18,31]. Moreover, 15-HETE can inhibit the formation of 12-HETE, which is a metabolite of 12-LOX's catalytic action on arachidonic acid [18].

Because DGLA can be metabolically converted via three different enzyme pathways, it is important to determine which enzyme has a higher affinity for arachidonic acid and which metabolites will dominate—pro-inflammatory or anti-inflammatory. Note that γ-linolenic acid, which is one of the main acids contained in evening primrose oil, is an important precursor of DGLA, which is a precursor of anti-inflammatory eicosanoids [32]. It has been shown that GLA or DGLA supplementation causes a modest increase in the prostaglandin E1 (PGE_1) level in tissues in relation to PGE_2, but the biological properties of PGE_1 are about 20 times stronger in comparison to PGE_2 [24]. However, GLA or DGLA supplementation may cause their conversion to AA and pro-inflammatory eicosanoids. Therefore, it is suggested that the metabolism should be directed to anti-inflammatory eicosanoids. An effective solution is an application of selective Δ-5-desaturase inhibitors, which may stop DGLA's conversion to AA and its further pro-inflammatory metabolites [24]. Moreover, as a polyunsaturated fatty acid, arachidonic acid undergoes peroxidation to form electrophilic aldehydes with low molecular weight and high reactivity, which may be a cause of the modification of both nucleophilic small molecules and high-molecular weight compounds, such as proteins, lipids, and DNA, which disturbs cellular metabolism [33].

Evening primrose oil has a high content of LA and GLA, and strengthens the epidermal barrier, normalizes the excessive loss of water through the epidermis, regenerates skin, and improves smoothness, after both topical and oral applications [12].

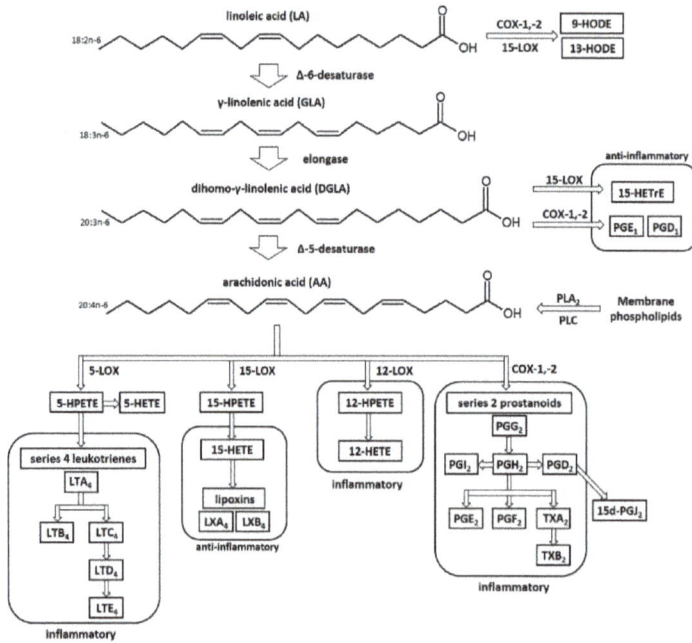

Figure 1. Linoleic acid (LA) metabolism. 9-HODE: 9-hydroxyoctadecadienoic acid; 13-HODE: 13-hydroxyoctadecadienoic acid; 15-HETrE: 15-hydroxyeicosatrienoic; PGE_1: prostaglandin E1; PGD_1: prostaglandin D1; PLA_2: phospholipase A2; PLC: phospholipase C; 5-HPETE: 5-hydroperoxyeicosatetraenoic acid; 5-HETE: 5-hydroxyeicosatetraenoic acid; LTA_4: leukotriene A4; LTB_4: leukotriene B4; LTC_4: leukotriene C4; LTD_4: leukotriene D4; LTE_4: leukotriene E4; 15-HPETE: 15-hydroperoxyeicosatetraenoic acid; 15-HETE: 15-hydroxyeicosatetraenoic acid; LXA4: lipoxin A_4; LXB_4: lipoxin B4; 12-HPETE: 12-hydroperoxyeicosatetraenoic acid; 12-HETE: 12-hydroxyeicosatetraenoic acid; PGG_2: prostaglandin G2; PGH_2: prostaglandin H2; PGI_2: prostaglandin I_2; PGD2: prostaglandin D2; 15-d-PGJ2: 15-deoxy-delta-12,14-prostaglandin J2; PGE_2: prostaglandin E2; PGF2: prostaglandin F2; TXA2: thromboxane A2; TXB2: thromboxane B2.

In addition, due to its linoleic acid content, evening primrose oil has a beneficial moisturizing effect on the mucous membrane in acne patients treated with isotretinoin [34]. This implies that skin supplementation with evening primrose oil improves the skin's water balance, which is weakened by treatment with isotretinoin. Moreover, γ-linolenic acid, contained in large amounts in the oil, is a source of the anti-inflammatory eicosanoids 15-HETrE and PGE$_1$, which have anti-proliferative properties that effectively prevent epidermal hyperproliferation [24,34]. In addition, these compounds inhibit the proliferation of smooth muscle cells and prevent the development of atherosclerotic plaque [24].

In recent years, it has been found that γ-linolenic acid is cytotoxic to glioma cells, and it can enhance gamma radiosensitivity [35,36]. It has been suggested that this effect is related to the accumulation of the toxic products of lipid peroxidation, which are cytotoxic to glioma cells. In cancer cells responsible for various types of cancer, the over-expression of the human epidermal growth factor receptor 2 (HER-2/neu) oncogene has been observed. This oncogene causes rapid and uncontrolled cell growth. However, γ-linolenic acid leads to an increase in the levels of polyomavirus enhancer activator 3 (PEA3), a transcriptional repressor of human epidermal growth factor receptor 2 (HER-2/neu) in cells, and a decrease in Her-2/neu promoter activity, thus reducing the likelihood of developing breast cancer [37]. Due to the inhibition of Her-2 expression, GLA administered together with transtuzumab, which as a monoclonal antibody binds to the Her-2 receptor, increases the process of the apoptosis of cancer cells and thus increases the effectiveness of pharmacotherapy with transtuzumab [37]. γ-Linolenic acid also causes an increase in the expression of the nm-23 metastasis-suppressor gene in cancer cells, which favors the inhibition of angiogenesis, cancer cell migration, and consequently, cancer metastasis [38,39]. The formation of these changes is also associated with a reduction in the expression of the vascular endothelial growth factor (VEGF), which plays a significant role in cancer (e.g., in the process of tumor angiogenesis) [40]. The above data suggest that evening primrose oil, as a rich source of gamma linolenic acid, supports anti-cancer therapy.

Moreover, it has been found that the oral supplementation of evening primrose oil (EPO) containing both linoleic acid and γ-linolenic acid reduces the inflammatory reaction and eases eye problems such as burning, dryness, and light sensitivity in people with Sjögren's syndrome [41]. Moreover, GLA reduces the levels of triacylglycerols and low-density lipoprotein (LDL) cholesterol in plasma [42]. It has been suggested that phytosterols, which are present in large quantities in EPO, also contribute to the above action [43].

Table 5. Biological effect and occurrence of selective eicosanoids [18,44,45].

	Metabolite	Biological Activity	Occurrence
anti-inflammatory	PGE$_1$	- anti-inflammatory - anti-proliferatory	keratinocytes fibroblasts sebocyte
	15-HETrE	- anti-inflammatory - anti-proliferatory	keratinocytes fibroblasts
	13-HODE	- anti-inflammatory - anti-proliferatory	keratinocytes fibroblasts
	15-HETE	- anti-inflammatory (lipoxin precursor) - anti-proliferatory - counteracts 12-HETE and LTB$_4$ effects - induces leukocyte chemotaxis	keratinocytes fibroblasts

Table 5. *Cont.*

	Metabolite	Biological Activity	Occurrence
inflammatory	LXA$_4$ LXB$_4$	- anti-inflammatory - LXA$_4$ inhibits expression of interleukin 6 (IL-6) and interleukin 8 (IL-8) - LXA$_4$ inhibits proliferation	neutrophils
	PGE$_2$	- proliferatory - chemotaxis - immunosuppression	keratinocytes fibroblasts
	5-HETE	- chemotaxis	keratinocytes
	LTB$_4$	- chemotaxis	leukocytes keratinocytes in chronic dermatitis (psoriasis, atopic dermatitis)
	Cys-LT (LTC$_4$ LTD$_4$ LTE$_4$)	- leukocyte activators - chemotaxis	leukocytes in chronic dermatitis (psoriasis, atopic dermatitis)
	12-HETE	- proliferatory - chemotaxis	keratinocytes fibroblasts Langerhans cells in chronic dermatitis (psoriasis)

Cys-LT: cysteinyl leukotrienes; LTC$_4$: leukotriene C4, LTD$_4$: leukotriene D4; LTE$_4$—leukotriene E4.

4. Conclusions

After analyzing the chemical composition of evening primrose (*Oenothera biennis*), especially the oil from its seeds and the biological activity of its components, it can be stated that it is a natural preparation supplementing the deficiency of essential fatty acids in the body. Therefore, it is beneficial in the treatment of chronic inflammation. It supports the metabolism of the body at various levels, especially in situations leading to the development of pathological conditions.

Author Contributions: M.T.: description of the biological activity of the evening primrose; K.B.: description of the chemical composition of the evening primrose; E.S.: major contribution in writing the manuscript. All authors read and approved the final manuscript.

Funding: This research received no external funding.

Conflicts of Interest: The authors declare no conflicts of interest.

References

1. Mihulka, S.; Pysek, P. Invasion history of Oenothera congeners in Europe: A comparative study of spreading rates in the last 200 years. *J. Biogeogr.* **2001**, *28*, 597–609. [CrossRef]
2. Granica, S.; Czerwińska, M.E.; Piwowarski, J.P.; Ziaja, M.; Kiss, A.K. Chemical composition, antioxidative and anti-inflammatory activity of extracts prepared from aerial parts of *Oenothera biennis* L. and Oenothera paradoxa Hudziok obtained after seeds cultivation. *J. Agric. Food Chem.* **2013**, *61*, 801–810. [CrossRef] [PubMed]

3.   Johnson, M.T.J.; Agrawal, A.A.; Maron, J.L.; Salminen, J.P. Heritability, covariation and natural selection on 24 traits of common evening primrose (*Oenothera biennis*) from a field experiment. *J. Evol. Biol.* **2009**, *22*, 1295–1307. [CrossRef] [PubMed]

4.   Singh, S.; Kaur, R.; Sharma, S.K. An updated review on the *Oenothera genus*. *J. Chin. Integr. Med.* **2012**, *10*, 717–725. [CrossRef]

5.   Ahmad, A.; Singh, D.K.; Fatima, K.; Tandon, S.; Luqman, S. New constituents from the roots of *Oenothera biennis* and their free radical scavenging and ferric reducing activity. *Ind. Crops Prod.* **2014**, *58*, 125–132. [CrossRef]

6.   Christie, W.W. The analysis of evening primrose oil. *Ind. Crops Prod.* **1999**, *10*, 73–83. [CrossRef]

7.   Zadernowski, R.; Polakowska-Nowak, H.; Rashed, A.A.; Kowalska, M. Lipids from evening primrose and borage seeds. *Oilseed Crops* **1999**, *20*, 581–589.

8.   Montserrat-de la Paz, S.; Fernandez-Arche, M.A.; Angel-Martin, M.; Garcia-Gimenez, M.D. Phytochemical characterization of potential nutraceutical ingredients from Evening Primrose oil (*Oenothera biennis* L.). *Phytochem. Lett.* **2014**, *8*, 158–162. [CrossRef]

9.   Zadernowski, R.; Naczk, M.; Nowak-Polakowska, H. Phenolic Acids of Borage (*Borago officinalis* L.) and Evening Primrose (*Oenothera biennis* L.). *J. Am. Oil Chem. Soc.* **2002**, *79*, 335–338. [CrossRef]

10.   Hudson, B.J.F. Evening primrose (*Oenothera* spp.) oil and seed. *J. Am. Oil Chem. Soc.* **1984**, *61*, 540–543. [CrossRef]

11.   Białek, M.; Rutkowska, J. The importance of γ-linolenic acid in the prevention and treatment. *Adv. Hyg. Exp. Med.* **2015**, *69*, 892–904. [CrossRef]

12.   Muggli, R. Systemic evening primrose oil improves the biophysical skin parameters of healthy adults. *Int. J. Cosmet. Sci.* **2005**, *27*, 243–249. [CrossRef] [PubMed]

13.   Kendall, A.C.; Kiezel-Tsugunova, M.; Brownbridge, L.C.; Harwood, J.L.; Nicolaou, A. Lipid functions in skin: Differential effects of n-3 polyunsaturated fatty acids on cutaneous ceramides, in a human skin organ culture model. *Biochim. Biophys. Acta* **2017**, *1859*, 1679–1689. [CrossRef] [PubMed]

14.   Mahfouz, M.M.; Kummerow, F.A. Effect of magnesium deficiency on delta 6 desaturase activity and fatty acid composition of rat liver microsomes. *Lipids* **1989**, *24*, 727–732. [CrossRef] [PubMed]

15.   Zietemann, V.; Kröger, J.; Enzenbach, C.; Jansen, E.; Fritche, A.; Weiker, C.; Boeing, H.; Schylze, M.B. Genetic variation of the FADS1 FADS2 gene cluster and n-6 PUFA composition in erythrocyte membranes in the European Prospective Investigation into Cancer and Nutrition-Potsdam study. *Br. J. Nutr.* **2010**, *104*, 1748–1759. [CrossRef] [PubMed]

16.   Ge, L.; Gordon, J.S.; Hsuan, C.; Stenn, K.; Prouty, S.M. Identification of the Δ-6 desaturase of human sebaceous glands: Expression and enzyme activity. *J. Investig. Dermatol.* **2003**, *120*, 707–714. [CrossRef] [PubMed]

17.   Sampath, H.; Ntambi, J.M. The role of fatty acid desaturases in epidermal metabolism. *Dermatoendocrinol* **2011**, *3*, 62–64. [CrossRef] [PubMed]

18.   Nicolaou, A. Eicosanoids in skin inflammation. *Prostaglandins Leukot. Essent. Fat. Acids* **2013**, *88*, 131–138. [CrossRef] [PubMed]

19.   Williard, D.E.; Nwankwo, J.O.; Kaduce, T.L.; Harmon, S.D.; Irons, M.; Moser, H.W.; Raymond, G.V.; Spector, A.R. Identification of a fatty acid Δ⁶-desaturase deficiency in human skin fibroblasts. *J. Lipid Res.* **2001**, *42*, 501–508. [PubMed]

20.   Huang, S.; Liu, R.; Niu, Y.; Hasi, A. Cloning and functional characterization of a fatty acid Δ6-desaturase from *Oenothera biennis*: Production of γ-linolenic acid by heterologous expression in *Saccharomyces cerevisiae*. *Russ. J. Plant Phys.* **2010**, *57*, 568–573. [CrossRef]

21.   Cho, H.P.; Nakamura, M.T.; Clarke, S.D. Cloning, expression, and nutritional regulation of the mammalian Delta-6 desaturase. *J. Biol. Chem.* **1999**, *274*, 471–477. [CrossRef] [PubMed]

22.   Senapati, S.; Sabyasachi, B.; Gangopadhyay, D.N. Evening primrose oil is effective in atopic dermatitis: A randomized placebo-controlled trial. *Indian J. Dermatol. Venereol. Leprol.* **2008**, *74*, 447–452. [CrossRef] [PubMed]

23.   Schlichte, M.J.; Vandersall, A.; Katta, R. Diet and eczema: A review of dietary supplements for the treatment of atopic dermatitis. *Dermatol. Pract. Concept* **2016**, *6*, 23–29. [CrossRef] [PubMed]

24.   Wang, W.; Lin, H.; Gu, Y. Multiple roles of dihomo-γ-linolenic acid against proliferation diseases. *Lipids Health Dis.* **2012**, *14*, 11–25. [CrossRef] [PubMed]

25. Fujiyama-Fujiwara, Y.; Ohmori, C.; Igarashi, O. Metabolism of γ-linolenic acid in primary cultures of rat hepatocytes and in Hep G2 cells. *J. Nutr. Sci. Vitaminol.* **1989**, *35*, 597–611. [CrossRef] [PubMed]

26. Ziboh, V.A.; Naguwa, S.; Vang, K.; Wineinger, J.; Morrissey, B.M.; Watnik, M.; Gershwin, M.E. Suppression of leukotriene B4 generation by ex-vivo neutrophils isolated from asthma patients on dietary supplementation with gammalinolenic acid-containing borage oil: Possible implication in asthma. *Clin. Dev. Immunol.* **2004**, *11*, 13–21. [CrossRef] [PubMed]

27. Belch, J.J.; Hill, A. Evening primrose oil and borage oil in rheumatologic conditions. *Am. J. Clin. Nutr.* **2000**, *71*, 352–356. [CrossRef] [PubMed]

28. Cao, D.; Luo, J.; Zang, W.; Chen, D.; Xu, H.; Shi, H.; Jing, H. Gamma-Linolenic Acid Suppresses NF-κB Signaling via CD36 in the Lipopolysaccharide-Induced Inflammatory Response in Primary Goat Mammary Gland Epithelial Cells. *Inflammation* **2016**, *39*, 1225–1237. [CrossRef] [PubMed]

29. Surette, M.E.; Koumenis, I.L.; Edens, M.B.; Tramposch, K.M.; Clayton, B.; Bowton, D.; Chilton, F.H. Inhibition of leukotriene biosynthesis by a novel dietary fatty acid formulation in patients with atopic asthma: A randomized, placebo-controlled, parallel-group, prospective trial. *Clin. Ther.* **2003**, *25*, 972–979. [CrossRef]

30. Khajeh, M.; Rahbarghazi, R.; Nouri, M.; Darabi, M. Potential role of polyunsaturated fatty acids, with particular regard to the signaling pathways of arachidonic acid and its derivatives in the process of maturation of the oocytes: Contemporary review. *Biomed. Pharmacother.* **2017**, *94*, 458–467. [CrossRef] [PubMed]

31. Larsson, S.C.; Kumlin, M.; Ingelman-Sunberg, M.; Wolk, A. Dietary long-chain n-3 fatty acids for the prevention of cancer: A review of potential mechanisms. *Am. J. Clin. Nutr.* **2004**, *79*, 935–945. [CrossRef] [PubMed]

32. Nilsen, D.W.T.; Aarsetoey, H.; Ponitz, V.; Brugger-Andersen, T.; Staines, H.; Harris, W.S.; Grundt, H. The prognostic utility of dihomo-gamma-linolenic acid (DGLA) in patients with acute coronary heart disease. *Int. J. Cardiol.* **2017**, *249*, 12–17. [CrossRef] [PubMed]

33. Łuczaj, W.; Gęgotek, A.; Skrzydlewska, E. Antioxidants and HNE in redox homeostasis. *Free Radic. Biol. Med.* **2017**, *111*, 87–101. [CrossRef]

34. Park, K.Y.; Ko, E.J.; Kim, I.S.; Li, K.; Kim, B.J.; Seo, S.J.; Kim, M.N.; Hong, C.K. The effect of evening primrose oil for the prevention of xerotic cheilitis in acne patients being treated with isotretinoin: A pilot study. *Ann. Dermatol.* **2014**, *26*, 706–712. [CrossRef] [PubMed]

35. Antal, O.; Peter, M.; Hackler, L., Jr.; Man, I.; Szebeni, G.; Ayaydin, F.; Hideghety, K.; Vigh, L.; Kitajka, K.; Balogh, G.; et al. Lipidomic analysis reveals a radiosensitizing role of gamma-linolenic acid in glioma cells. *Biochim. Biophys. Acta* **2015**, *1851*, 1271–1282. [CrossRef] [PubMed]

36. Das, U.N.; Rao, K.P. Effect of γ-linolenic acid and prostaglandins E1 on gamma-radiation and chemical-induced genetic damage to the bone marrow cells of mice. *Prostaglandins Leukot. Essent. Fat. Acids* **2006**, *74*, 165–173. [CrossRef] [PubMed]

37. Menendez, J.A.; Vellon, L.; Colomer, R.; Lupu, R. Effect of γ-Linolenic Acid on the Transcriptional Activity of the Her-2/neu (erbB-2) Oncogene. *J. Natl. Cancer Inst.* **2005**, *2*, 1611–1615. [CrossRef] [PubMed]

38. Marshall, J.C.; Lee, J.H.; Steeg, P.S. Clinical-translational strategies for the elevation of Nm23-H1 metastasis suppressor gene expression. *Mol. Cell Biochem.* **2009**, *329*, 115–120. [CrossRef] [PubMed]

39. Jiang, W.G.; Hiscox, S.; Bryce, R.P.; Horrobin, D.F.; Mansel, R.E. The effects of n-6 polyunsaturated fatty acids on the expression of nm-23 in human cancer cells. *Br. J. Cancer* **1998**, *77*, 731–738. [CrossRef] [PubMed]

40. Miyake, J.A.; Benadiba, M.; Colquhoun, A. Gamma-linolenic acid inhibits both tumour cell cycle progression and angiogenesis in the orthotopic C6 glioma model through changes in VEGF, Flt1, ERK1/2, MMP2, cyclin D1, pRb, p53 and p27 protein expression. *Lipids Health Dis.* **2009**, *17*. [CrossRef] [PubMed]

41. Aragona, P.; Bucolo, S.; Spinella, R.; Giuffrida, S.; Ferreri, G. Systemic Omega-6 Essential Fatty Acid Treatment and PGE₁ Tear Content in Sjögren's Syndrome Patients, Investigative Ophthalmology & Visual Science. *Investig. Ophthalmol. Vis. Sci.* **2005**, *46*, 4474–4479. [CrossRef]

42. Dasgupta, S.; Bhattacharyya, D.K. Dietary effect of γ-linolenic acid on the lipid profile of rat fed erucic acid rich oil. *J. Oleo Sci.* **2007**, *56*, 569–577. [CrossRef] [PubMed]

43. Ras, R.T.; Geleijnse, J.M.; Trautwein, E.A. LDL-cholesterol-lowering effect of plant sterols and stanols across different dose ranges: A metaanalysis of randomized controlled studies. *Br. J. Nutr.* **2014**, *112*, 214–219. [CrossRef] [PubMed]

44. Sivamani, R.K. Eicosanoids and Keratinocytes in Wound Healing. *Adv. Wound Care* **2014**, *3*, 476–481. [CrossRef] [PubMed]

45. Guimaraes, R.F.; Sales-Campos, H.; Nardini, V.; da Costa, T.A.; Fonseca, M.T.C.; Júnior, V.R.; Sorgi, C.A.; da Silva, J.S.; Chica, J.E.L.; Faccioli, L.H.; et al. The inhibition of 5-Lipoxygenase (5-LO) products leukotriene B4 (LTB$_4$) and cysteinyl leukotrienes (cysLTs) modulates the inflammatory response and improves cutaneous wound healing. *Clin. Immunol.* **2018**, *190*, 74–83. [CrossRef] [PubMed]

![antioxidants logo] *antioxidants*

MDPI

Article

Cell-Type-Specific Modulation of Hydrogen Peroxide Cytotoxicity and 4-Hydroxynonenal Binding to Human Cellular Proteins In Vitro by Antioxidant *Aloe vera* Extract

Vera Cesar [1,2], Iva Jozić [1], Lidija Begović [1], Tea Vuković [3], Selma Mlinarić [1] (ORCID), Hrvoje Lepeduš [2,4], Suzana Borović Šunjić [3] and Neven Žarković [3,*]

[1] Department of Biology, Josip Juraj Strossmyer University of Osijek, Cara Hadrijana 8/A, 31000 Osijek, Croatia; vcesarus@yahoo.com (V.C.); ivajoo@gmail.hr (I.J.); lbegovic@biologija.unios.hr (L.B.); smlinaric@biologija.unios.hr (S.M.)

[2] Faculty of Dental Medicine and Health, Josip Juraj Strossmyer University of Osijek, Cara Hadrijana 10/E, 31000 Osijek, Croatia; hlepedus@yahoo.com

[3] Laboratory for Oxidative Stress (LabOS), Rudjer Boskovic Institute, Bijenicka 54, 10000 Zagreb, Croatia; Tea.Vukovic@irb.hr (T.V.); borovic@irb.hr (S.B.Š.)

[4] Faculty of Humanities and Social Sciences, Josip Juraj Strossmyer University of Osijek, L. Jägera 9, 31000 Osijek, Croatia

* Correspondence: zarkovic@irb.hr; Tel.: +385-1-457-1234

Received: 24 July 2018; Accepted: 17 September 2018; Published: 21 September 2018

Abstract: Although *Aloe vera* contains numerous bioactive components, the activity principles of widely used *A. vera* extracts are uncertain. Therefore, we analyzed the effects of genuine *A. vera* aqueous extract (AV) on human cells with respect to the effects of hydrogen peroxide (H_2O_2) and 4-hydroxynonenal (HNE). Fully developed *A. vera* leaves were harvested and analyzed for vitamin C, carotenoids, total soluble phenolic content, and antioxidant capacity. Furthermore, human cervical cancer (HeLa), human microvascular endothelial cells (HMEC), human keratinocytes (HaCat), and human osteosarcoma (HOS) cell cultures were treated with AV extract for one hour after treatment with H_2O_2 or HNE. The cell number and viability were determined using Trypan Blue, and endogenous reactive oxygen species (ROS) production was determined by fluorescence, while intracellular HNE–protein adducts were measured for the first time ever by genuine cell-based HNE–His ELISA. The AV extract expressed strong antioxidant capacities (1.1 mmol of Trolox eq/g fresh weight) and cell-type-specific influence on the cytotoxicity of H_2O_2, as well as on endogenous production of ROS and HNE–protein adducts induced by HNE treatment, while AV itself did not induce production of ROS or HNE–protein adducts at all. This study, for the first time, revealed the importance of HNE for the activity principles of AV. Since HMEC cells were the most sensitive to AV, the effects of AV on microvascular endothelia could be of particular importance for the activity principles of *Aloe vera* extracts.

Keywords: *Aloe vera*; plant extract; antioxidants; cell growth; oxidative stress; reactive oxygen species (ROS); hydrogen peroxide; lipid peroxidation; 4-hydroxynonenal (HNE); cell-based ELISA; HNE–protein adducts; microvascular endothelium

---

1. Introduction

Aloe barbadensis Miller L. (trivially called *A. vera*) is one of more than 400 species of the *Aloe* genus belonging to family *Liliaceae* that originated in South Africa, but are indigenous to dry subtropical and tropical climates [1]. *Aloe vera* is widely used in different forms of medicinal remedies without a clear understanding of the activity principles that could make the basis for its therapeutic properties [2]. In addition to the medicinally most potent *A. barbadensis* Miller, at least three other species are known to have medicinal properties: *Aloe perryi* Baker, *Aloe ferox*, and *Aloe arborescens* [2].

The antioxidant composition of *A. vera* includes mostly α-tocopherol (vitamin E), carotenoids, ascorbic acid (vitamin C), flavonoids, and tannins. In vitro studies showed the scavenging potential of *A. vera* gel for various free radicals. Moreover, phytosterols purified from *A. vera*, namely lophenol and cycloartanol, can induce the downregulation of fatty-acid synthesis and show a tendency for the upregulation of fatty-acid oxidation in the liver, which favors the reduction in intra-abdominal fat and improvement of hyperlipidemia. It was claimed that the polysaccharides in *A. vera* gel have therapeutic properties such as immunostimulation, anti-inflammatory effects, wound healing, promotion of radiation damage repair, anti-bacterial, anti-viral, anti-diabetic, and anti-neoplastic activities, as well as stimulation of hematopoiesis and anti-oxidant effects [3]. *Lactobacillus brevis* strains isolated from naturally fermented *A. vera* gel inhibited the growth of many harmful enteropathogens without restraining most normal commensals in the gut. Moreover, aloin is metabolized by the colonic flora to reactive aloe emodin, which is responsible for purgative activity. Aloe emodin also inhibits colon cancer cell migration by downregulating matrix metalloproteinases 2 and 9 (MMP-2/9) [1–3].

Many of the medicinal effects of *A. vera* extracts were assigned to the polysaccharides found in the inner leaf parenchymatous tissue, while it is believed that these biological activities could mostly be due to synergistic action of the compounds contained therein rather than a single chemical substance [4]. The most investigated biomedical properties of *A. vera* gel involve the promotion of wound healing, including burns and frostbite, in addition to anti-inflammatory, antifungal, hypoglycemic, and gastroprotective properties. However, the healing properties of *A. vera* gel extracts were mostly tested using animal models. Hence, *A. vera* gel extract stimulated fibroblast growth in a synovial model, while also enhancing wound tensile strength and collagen turnover in wound tissue [5]. In another trial, *A. vera* gel increased levels of hyaluronic acid and dermatan sulfate in granulation tissue. *A. vera* treatment of wounded tissue also increased the blood supply, which is essential for the formation of new tissue. On the other hand, some reports mentioned inhibitory effects of *A. vera* gel on wound healing, which should not be a surprise, as the composition of *A. vera* gel varies even within the same species and depends on the source and climate of the region of plant growth, as well as on the processing method [5]. It was suggested that a standardized method could be necessary for the production of aloe gel products to avoid degradation of the polysaccharides, thereby preventing the removal of high-molecular-weight molecules in aloe gel extracts [3].

In vivo and in vitro studies demonstrated the potential of *A. vera* gel as an anti-hyperglycemic and anti-hyprecholesterolemic agent for type 2 diabetic patients without any significant effects on other normal blood lipid levels or liver/kidney function. *A. vera* also helps improve carbohydrate metabolism, with a recent report suggesting that it helps improve metabolic status in obese pre-diabetics and in early non-treated diabetic patients by reducing body weight, body fat mass, fasting blood glucose, and fasting serum insulin in obese individuals [3,6].

It was also shown that *A. vera* extracts can inhibit inflammatory processes via the reduction of leukocyte adhesion and the suppression of pro-inflammatory cytokines, thus attenuating lipid peroxidation and cerebral ischemia/reperfusion injury in rats [1].

The abovementioned effects of *A. vera* extracts, together with its content of different antioxidants, suggest that *A. vera* might influence biomedical effects of lipid peroxidation, and thus, of generated reactive aldehydes denoted as second messengers of free radicals, due to their high cytotoxic and mutagenic capacities combined with multiple regulatory activities [7–9]. Among these reactive aldehydes, of particular interest is 4-hydroxynonenal (HNE), generated from polyunsaturated fatty acids (PUFAs). In particular, HNE has high affinity for binding to proteins, consequently changing their structure and function, while still retaining toxic and regulatory activities of the aldehyde, including regulation of cell proliferation, differentiation, and apoptosis [10,11]. Therefore, HNE is currently considered to be major biomarker of lipid peroxidation, especially if bound to proteins, which already helped us better understand the pathophysiology of lipid peroxidation, as well as inflammatory and growth-regulating processes, and helped us revise modern concepts of major stress- and age-associated diseases [12–15].

Therefore, the aim of this study was to evaluate, using in vitro experiments, if *A. vera* extract could interfere with the cytotoxicity of reactive oxygen species (ROS), notably of hydrogen peroxide (H_2O_2), which is the most common (patho)physiological ROS, and with HNE, acting as a major second messenger of free radicals.

2. Materials and Methods

2.1. Plant Material and Extract Preparation

To avoid difficulties arising from the use of commercially available *A. vera* extracts, while also considering the inconvenience of gel extracts for in vitro studies, we prepared in-house aqueous extracts from the fresh plant, as can be done easily in any laboratory. *Aloe vera* (*Aloe barbadensis* Miller) plants were subjected to vegetative propagation. Young shoots were removed from the mother plant, and were planted in sand until the roots were developed. After that, plantlets were transferred in 0.5-kg plastic pots, two plants per pot, containing commercial soil. Plants were irrigated once every two weeks with tap water and grown under ambient irradiance and temperature (400–1400 μmol m^{-2} s^{-1} and 25 $\pm$ 1 °C, respectively). After one year of growth, the first fully developed leaves, fourth from the top, were used for crude-extract preparation, as well as for all biochemical analysis. *A. vera* leaves were grounded in liquid nitrogen using a pistil and mortar. The obtained leaf powder was used for further preparations and analyses.

2.2. Total Soluble Phenolic Content and Antioxidant Activity

2.2.1. Phenolic Content

For total soluble phenolic content estimation, approximately 600 mg of leaf powder was used and extraction was performed for 24 h at −20 °C in 2.5 mL of 96% ethanol [16]. The reaction mixture contained 100 μL of ethanol extract, 700 μL of distilled H_2O, 50 μL of Folin–Ciocalteu reagent, and 150 μL of sodium carbonate solution (200 g L^{-1}). Samples were incubated for 60 min at 37 °C in a water bath, and absorbance was measured spectrophotometrically at 765 nm using gallic acid (GA) as a standard. Total soluble phenolic content was expressed as gallic acid equivalent (GAEq) per g of fresh weight.

2.2.2. Ascorbic Acid Content

Approximately 600 mg of leaf powder was extracted in 10 mL of distilled water. The homogenates were centrifuged for 15 min at 3000× *g* and 4 °C. The reaction mixture contained 300 μL of aqueous extract, 100 μL of 13.3% trichloroacetic acid, 25 μL of deionized water, and 75 μL of 2,4-dinitrophenylhydrazine (DNPH) reagent. The DNPH reagent was prepared by dissolving 2 g of DNPH in 230 mg of thiourea and 270 mg of $CuSO_4$ in 100 mL of 5 M H_2SO_4 [17]. Blanks were made in parallel for each sample as described above without addition of DNPH reagent. Samples were incubated in a water bath for 60 min at 37 °C. After incubation and addition of DNPH reagent to the blanks, 500 μL of 65% H_2SO_4 was added to all reaction mixtures. The absorbance was measured at 520 nm. The concentration of ascorbic acid was obtained from a standard curve with known concentrations of ascorbic acid (2.5–20 μg mL^{-1}). The ascorbic acid content was expressed in mg per 100 g of fresh weight.

2.2.3. Measurement of Total Carotenoids

Approximately of 0.1 g of leaf tissue was ground in liquid nitrogen with the addition of $Mg(HCO_3)_2$. Fine powder was extracted in absolute acetone for 24 h at −20 °C. Samples were centrifuged at $18,000 \times g$ for 10 min and 4 °C; the absorbance was measured at 470 nm, 645 nm, and 662 nm. Absolute acetone was used as a blank test. The content of total carotenoids was estimated according to the method described by Lichtenthaler and Buschmann [18].

2.2.4. Antioxidant Activity

Antioxidant activity was determined using the Brand-Williams method [19]. The supernatant obtained by extraction with 96% ethanol for 24 h at −20 °C was used. The reaction mixture contained 20 µL of leaf extract and 980 µL of 0.094 mM 2,2-diphenyl-1-picrylhydrazyl (DPPH) previously dissolved in methanol. The reaction was carried out in the dark at 22 °C for 15 min with occasional stirring. After 15 min, the absorbance at 515 nm was measured. Then, 6-hydroxy-2,5,7,8-tetramethylchroman-2-carboxylic acid (Trolox) dissolved in methanol was used as a calibration standard, as Trolox is a water-soluble synthetic analog of α-tocopherol widely used as an antioxidant standard when plant extracts are analyzed [20,21]. Hence, the total antioxidant activity of *A. vera* extract was expressed as the equivalent (Eq) of Trolox per g of fresh weight of *A. vera* leaves (FW).

2.3. Cell Cultures and Treatments

Aiming to evaluate whether the activity principles of *A. vera* extract include interference with the bioactivities of ROS and HNE, we used four different human cell lines and three complementary analytical methods. Each experiment and analysis was done using triplicates of identical cultures, while statistical evaluation was done using a *t*-test, with values of $p < 0.05$ considered as significant. For these experimental treatments, obtained leaf powder was dissolved with ice-cold physiological saline solution at a 1:1 *w/v* ratio, vortexed, and left at 4 °C overnight. After centrifugation ($5000 \times g$, 4 °C, 10 min), supernatants were transferred to and combined in a new tube. Thus, the produced crude extract was filtered through a 0.45-µm filter (Millipore, Merck, Germany) and stored in a refrigerator in sterile plastic tubes as 1-mL aliquots before being used further in experimental work. This in-house prepared *A. vera* extract is denoted as AV in the study presented.

2.3.1. Cell Cultures

The human uterine cervical carcinoma cell line (HeLa), human dermal microvascular endothelial cells (HMEC), the human spontaneously transformed aneuploid immortal keratinocyte cell line from adult human skin (HaCaT), and the human osteosarcoma cell line (HOS), which grows resembling osteoblast cells in vitro, were purchased from the American Type Culture Collection (ATCC). The cells were cultivated in T75 cell culture flasks (TPP, Switzerland) in Dulbecco's modified Eagle's medium (DMEM) with 10% (*v/v*) fetal calf serum (FCS) at 37 °C in humidified atmosphere with 5% CO_2.

Prior to the experiments, the cells were harvested with 0.25% (*w/v*) Trypsin/0.53 mM ethylenediaminetetraacetic acid (EDTA) solution and counted with a Trypan Blue Exclusion Assay in a Bürker-Türk hemocytometer (Brand). The cells were seeded into 96-microwell plates (TPP, Trasadingen, Switzerland) at a specific density different for each cell line to acquire optimal short-term culturing conditions of 4×10^4 cells/well (HMEC and HaCaT), or 1×10^5 cells/well (HeLa and HOS), and left for 2 h to attach before further treatment.

2.3.2. Experimental Treatments with AV, H_2O_2, and HNE

In the first set of experiments, the potential influence of AV extract on the acute cytotoxicity of H_2O_2, which is the most common (patho)physiological non-radical ROS, was evaluated. To do that, AV extract was added either as 1% or as 10% final *v/v* concentration one hour after treating the cells with either 0.0025% or 0.05% *v/v* concentration, which should cover the range of median lethal dose (LD$_{50}$) for the majority of the cell lines. The respective control cell cultures were either treated only with H_2O_2 or with AV extract, or were not treated at all. After 24 h, the cells were harvested and analyzed using the Trypan Blue Exclusion Assay in a Bürker-Türk hemocytometer, counting not only total cells per culture, but also the incidence of live vs. dead cells.

According to the obtained data, the second set of experiments in which HNE was used to treat the cells was performed. The treatment protocol was almost identical to the first one, except that, instead of hydrogen peroxide, HNE was used at 50 μM concentration resembling the LD_{50} for the majority of the cells, while the AV extract was used only at 1% dose one hour after HNE. During the one-hour period, the aldehyde should mostly be metabolized, bound to the cellular proteins, or eliminated from the cells, thereby gaining its major immediate effects [22]. Two hours later, cell cultures were used for determination of the levels of HNE–protein adducts in the cells or for analysis of the endogenous (i.e., intracellular) production of ROS, notably of H_2O_2.

2.3.3. Determination of Intracellular HNE–Protein Adducts Using Cell-Based HNE–His ELISA

After the above-described treatment with HNE and/or AV extract, the cells were treated for 5 min with 90% ethanol and were fixed with 10% buffered formalin to be processed immunocytochemically for determination of the intracellular content of HNE–protein adducts. For determination of the HNE–protein adducts, the genuine monoclonal antibody obtained from the culture medium of the clone derived from a fusion of Sp2/Ag8 myeloma cells with B-cells of a BALBc mouse immunized with HNE-modified keyhole limpet hemocyanine specific for the HNE–His adducts (courtesy of Prof. G. Waeg from KF-University in Graz, Austria) were used [23]. Therefore, the well-known genuine HNE–His ELISA designed for in vitro research was combined for the first time as a cell-based ELISA with a standardized immunocytochemical procedure [22,24]. Shortly after the cells were fixed for 24 h, formalin was removed, and possible endogenous peroxidase activity of the samples was blocked with 1.5% H_2O_2, 0.1% NaN_3, and 2% bovine serum albumin (BSA), and the primary antibody against HNE–histidine conjugates was added. For detection of the HNE adducts, the immunoperoxidase technique was used, with secondary rabbit anti-mouse antibody (Dako, Glostrup, Denmark) applying 3,3′-diaminobenzidine tetrahydrocloride (DAB) as a chromogen. After the remaining reagents were removed, 100 μL of sterile saline was added to each microculture well, which was then analyzed at 620 nm using an ELISA plate reader with a 405-nm reference filter (Multiskan EX; Thermo Fisher Scientific, Waltham, MA, USA). To allow easier evaluation of the obtained data, the results are presented as a percentage of respective control values (100%).

2.3.4. Measurement of Intracellular ROS Production

The ROS measurement based on the intracellular oxidation of 2′,7′-dichlorodihydrofluorescein diacetate (DCFH-DA; Sigma-Aldrich, St. Louis, MO, USA) to fluorescent 2′,7′-dichlorofluorescein (DCF) was used to determine ROS generation inside the cells [25]. The cells were incubated with 10 μM DCFH-DA at 37 °C for 30 min. The medium was replaced with a fresh one and the zero point was measured with a Cary Eclipse Fluorescence Spectrophotometer (Varian, Agilent, Santa Clara, CA, USA) with an excitation wavelength of 500 nm and an emission detection wavelength of 530 nm, while fluorescence was measured after 30 min. To allow easier evaluation of the obtained data, the results obtained as relative fluorescence units (RFU) measured are presented as a percentage of respective control values (100%).

3. Results

3.1. The Levels of Antioxidants in A. vera Leaves

The amounts of major antioxidants present in *A. vera* leaves with respect to antioxidant capacity expressed in comparison to Trolox are shown in Figure 1.

The total antioxidant activity of *A. vera* leaves was 1102.42 ± 56.7 mg Trolox Eq/g of FW. Composition analysis revealed the following amounts of the known antioxidants in the *A. vera* leaves:

ascorbic acid (0.172 ± 0.03 mg/g of FW), total carotenoids (0.055 ± 0.002 mg/g of FW), and total soluble polyphenols (355.9 ± 10.2 mg GAEq/g of FW).

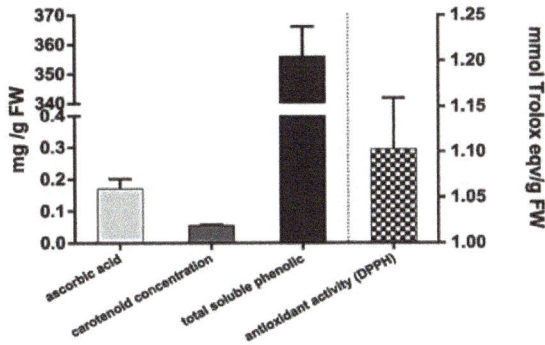

Figure 1. Partial characterization of the antioxidant levels of *Aloe vera* leaves. Values are given as mean values for triplicates.

3.2. The Effects of H_2O_2 and AV Extract on Cell Viability

The effects of different concentrations of AV extract added one hour after different doses of H_2O_2 are shown in Figures 2–5.

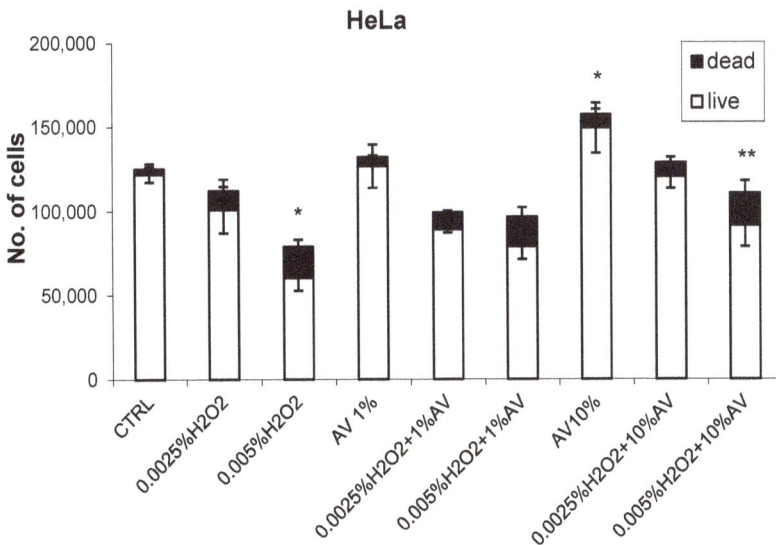

Figure 2. The effects of *Aloe vera* (AV) extract (1 or 10%) on the human cervical cancer (HeLa) cells with respect to H_2O_2 treatment (0.0025% or 0.05%). The cell count values (viability determined using Trypan blue) were obtained 24 h after treatment (H_2O_2 followed by AV extract 1 h later) and are given as mean values for triplicates. * significant difference to untreated control; ** significant difference to H_2O_2 treatment alone.

The HeLa cells did not show sensitivity to a lower dose of H_2O_2, while a higher concentration reduced the cell count and increased the incidence of dead cells, thus indicating a further decay of the H_2O_2-treated cells. The AV extract did not show any prominent effects if used at 1% concentration;

however, at the higher 10% concentration, the growth of the HeLa cells increased slightly and the cytotoxicity was reduced.

In contrast, in the case of HMEC cells, AV used at the 10% dose reduced the growth of the cells, while, if used at 1%, it caused a slight enhancement in the growth of these cells, as can be seen in Figure 3. Despite such concentration-dependent effects of AV on the HMEC cells if given alone, the plant extract did not influence the concentration-dependent cytotoxicity of H_2O_2 as could be expected. This was because 1% AV enhanced the growth of the HMEC cells; as such, its combined effect with a lower dose of H_2O_2 resulted in relatively (in comparison to 1% AV alone) more pronounced cytotoxicity of H_2O_2, while, in the case of the 10% AV extract, its influence on the cytotoxicity of H_2O_2 was the opposite. However, in comparison to the cells treated by H_2O_2 alone, these difference were not significant ($p > 0.05$).

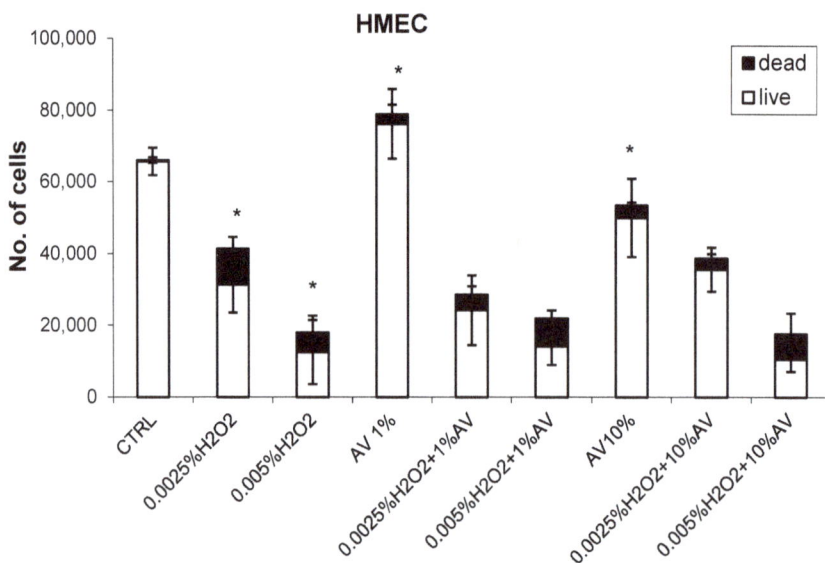

Figure 3. The effects of AV extract (1 or 10%) on the human microvascular endothelial cells (HMEC) with respect to H_2O_2 treatment (0.0025% or 0.05%). The cell count values (viability determined using Trypan blue) were obtained 24 h after treatment (H_2O_2 followed by AV extract 1 h later) and are given as mean values for triplicates. * significant difference to untreated control.

The concentration-dependent growth-inhibiting effects of AV extract were observed for the HaCaT cells (Figure 4), although the cytotoxicity of H_2O_2 for this cell line did not depend on the used dose of H_2O_2. The combined treatment with 1% AV did not influence the cytotoxicity of H_2O_2, while 10% concentration of the plant extract showed obviously additive suppressing (i.e., toxic) effects with H_2O_2 for the HaCaT cells.

Finally, it should be said that, for the HOS cells (Figure 5), the AV extract did not show any prominent effect, although these cells expressed relatively high sensitivity to the cytotoxic effects of H_2O_2 (more than 60% inhibition).

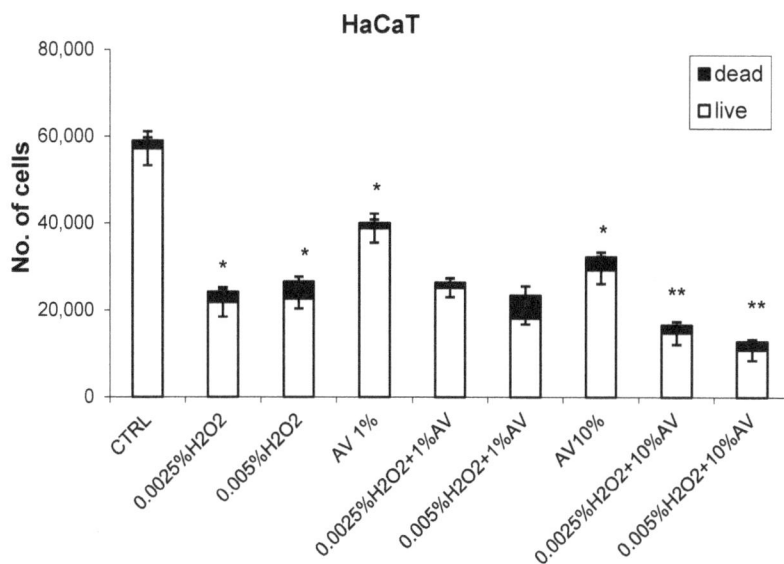

Figure 4. The effects of AV extract (1 or 10%) on the human keratinocyte (HaCaT) cells with respect to H_2O_2 treatment (0.0025% or 0.05%). The cell count values (viability determined using Trypan blue) were obtained 24 h after treatment (H_2O_2 followed by AV extract 1 h later) and are given as mean values for triplicates. * significant difference to untreated control; ** significant difference to respective H_2O_2 treatment alone.

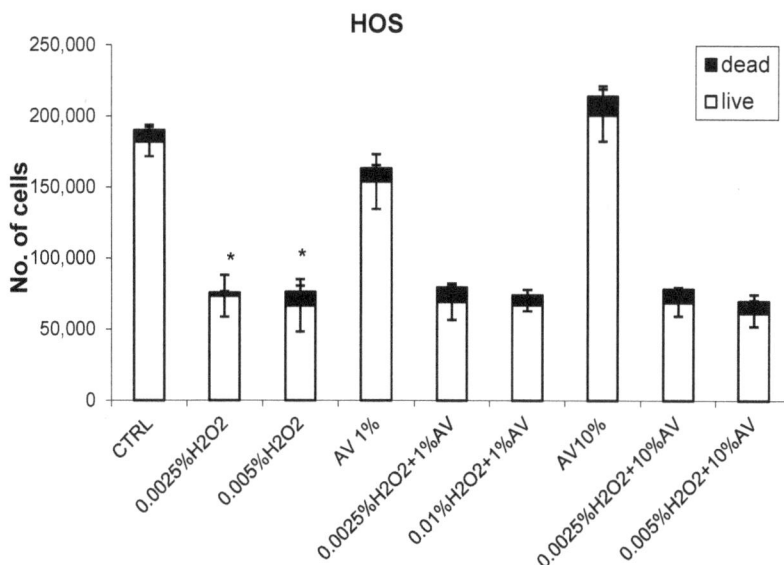

Figure 5. The effects of AV extract (1 or 10%) on the human osteosarcoma (HOS) cells with respect to H_2O_2 treatment (0.0025% or 0.05%). The cell count values (viability determined using Trypan blue) were obtained 24 h after treatment (H_2O_2 followed by AV extract 1 h later) and are given as mean values for triplicates. * significant difference to untreated control.

The toxic effects of H_2O_2 for the HOS cells did not depend on the dose of peroxide used.

3.3. The Effects of AV Extract on HNE-Pretreated Cells

In the next set of experiments, the cells were exposed to 50 µM HNE concentration, which was followed by AV extract after 1 h, applied at 1% concentration. The results of these treatments are shown in Figures 6 and 7.

3.3.1. The Effects AV on HNE Binding to Cellular Proteins

The amounts of HNE–protein adducts developed in the cells after treatment with HNE and AV 1% extract, or without the plant extract are presented in Figure 6.

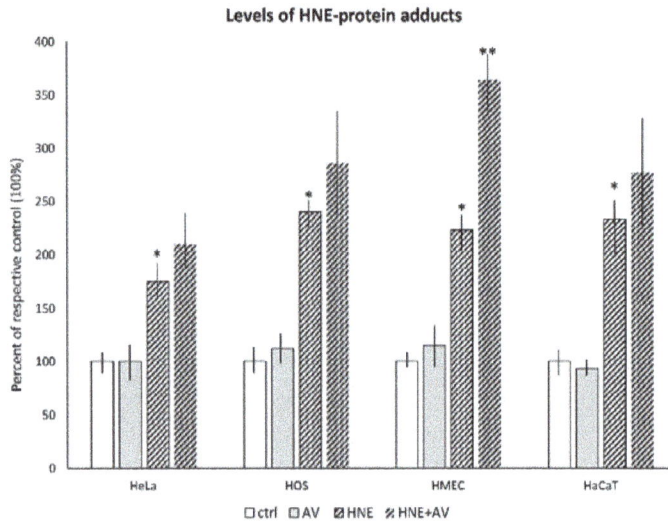

Figure 6. The effects of AV extract (1%) on the cellular generation of 4-hydroxynonenal (HNE)–protein adducts induced by HNE treatment (50 µM). The amounts of HNE–protein adducts were determined by cell-based HNE–His ELISA and are given as mean values for triplicates. * significant difference to respective untreated control (ctrl); ** significant difference to respective HNE-treated control.

The treatment with AV extract itself did not induce the production of HNE–protein adducts in any type of cell used, while increased amounts of HNE–protein adducts developed in the cells after treatment with HNE were observed for all cell lines (significant for all with respect to the plain controls, $p < 0.05$). If the cells were treated with the AV extract one hour after treatment with HNE, a tendency of enhanced accumulation of the cellular proteins modified by HNE was observed for all cell lines. However, it was significant only for the HMEC cells.

3.3.2. The Effects of AV on Cellular ROS Production Induced by HNE

The levels of ROS developed in the cells after treatment with HNE and AV 1% extract, or without the plant extract are presented in Figure 7.

The HMEC cells were the only cell line that responded to HNE treatment in terms of change in endogenous production of ROS. However, while treatment with HNE slightly increased (by 29%) intracellular production of ROS in the HMEC cells, the AV extract had no influence, as it did not induce ROS production in any cell line tested.

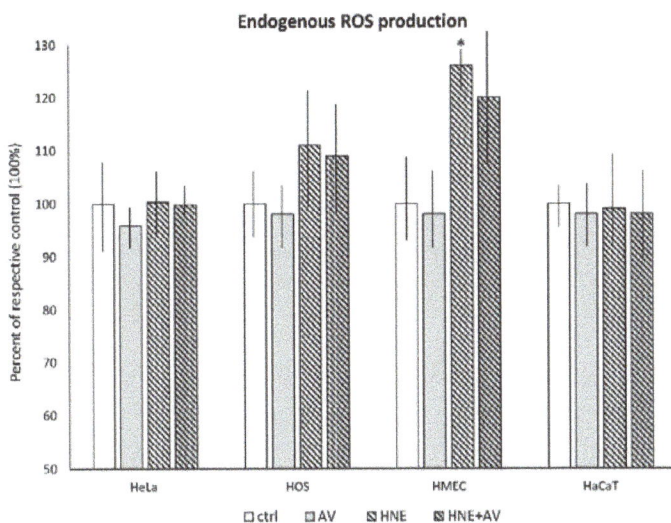

Figure 7. The effects of AV extract (1%) on the cellular generation of reactive oxygen species (ROS; mostly H_2O_2) induced by HNE treatment (50 µM). The amounts of the HNE–protein adducts were determined using luminescence and are given as mean values for triplicates. * significant difference to respective untreated control.

4. Discussion

The results obtained in this study show that *A. vera* extract prepared from plant leaves has prominent antioxidant capacity, most likely reflecting the activities of various antioxidants produced by the plant. Because the AV extract itself did not induce ROS or HNE production in any cell line used, its observed bioactivities probably reflect complex interactions of different plant substances with cellular redox homeostasis challenged by ROS- or HNE-induced oxidative stress. Since the cell lines used expressed differential sensitivity to the H_2O_2 toxicity, as well as reacting differently to the AV treatment, we assume that such complex cell differences might reflect not only redox alterations differently expressed by different types of cells upon H_2O_2 treatment interfering with antioxidants present in the AV extract, but also might be, at least in a part, due to the lipid peroxidation chain reactions that might generate HNE acting as a second messenger of free radicals. That should not be surprising since plants exposed to oxidative stress also experience lipid peroxidation generating reactive aldehydes, indicating that HNE and related aldehydes have important biological roles not only in animals and humans, but also in plants; these roles are not only toxic, but also regulatory, most likely related to the activity of antioxidants and regulatory proteins [26]. A possibility that such bioactive substances of plant origin could also affect the human cells is supported by our findings of enhancing effects of AV on accumulation of HNE–protein adducts upon HNE treatment, noticed for all cell lines used, especially for the HMEC cells, which also showed enhanced production of ROS upon HNE treatment associated with rapid accumulation of the advanced (aldehydic) lipoxidation end products (ALEs). Since microvascular endothelial cells (such as those used to establish the HMEC cell line) have crucial roles in various inflammatory and degenerative diseases, and above all, in tissue growth, either in wound regeneration or cancer development, the findings detailing the highest sensitivity of the HMEC cells to treatment with AV extract and HNE, in comparison to the other cell lines used, might be important for better understanding the bioactivity principles of AV extracts. Since this is the first study which reveals the possible relevance of HNE for the activity principles of AV extracts, we hope it will encourage further research in the field.

HNE acts in a concentration- and cell-type-dependent manner, regulating the majority of cellular processes interfering with lipids, especially PUFAs, and carbohydrate metabolism, crucial for cellular stress response and adaptation to stress, occurring even in yeast cells [27,28]. In the case of mammalian cells, HNE may interfere with cellular, as well as with extracellular, factors, eventually acting as a growth-regulating factor suppressing the growth of cancer, and enhancing the growth of non-malignant cells [29–32]. Among such interactions of HNE, the most important are its effects on enzymes involved in cellular metabolism and redox homeostasis, as well as on cytokines and their signaling pathways, which might result either in negative (co-carcinogenic) or positive (anti-cancer) effects [33,34]. Here, it should be stressed that such bioactivities of HNE occur not only in vitro, but also in vivo, and might represent an anti-cancer defense mechanism of the non-malignant cells [35–37]. Eventually, that might be of high importance for a better understanding of the interference of various antioxidants with carcinogens and anti-cancer therapies [38–41].

Since *Aloe vera* is a very popular medicinal plant, over 4000 studies were performed on the effectiveness of AV extracts in medical treatments, out of which many addressed the usefulness and activity principles of AV for cancer patients. Thus, aloe anthraquinones were quite extensively studied for their anticancer properties. In fact, the anthraquinones, aloin A and B, as well as aloe emodin, are structurally similar to DNA-binding drugs such as anthracyclines. The antitumor effect of aloe is also based on known mechanisms, including the induction of apoptosis and a significant elevation of key antioxidant enzymes, such as superoxide dismutase (SOD) and glutathione peroxidase (GPx). Pecere et al. [42] reported on selective in vitro and in vivo killing of neuroectodermal tumor cells by aloe emodin both in tissue cultures and in animal models. Grimaudo et al. investigated the effect of purified anthraquinines on sensitive and multidrug-resistant leukemia cells, and showed that only aloe emodin had reproducible cytotoxic activity, but at concentrations much higher than those of common anticancer agents such as daunorubicin and etoposide [43]. Lee et al. demonstrated that the time- and dose-dependent treatment of human lung squamous carcinoma CH27 cells by aloe emodin resulted in apoptosis, while combined effect of aloe emodin with cisplatinum confirmed that the inhibitory effect of aloe emodin acted in a dose-dependent manner [44]. Similar to AV extracts, HNE also has a strong pro-apoptotic capacity, which is related to its protein-binding capacity, while it can act also in cell-type-specific manner, being selectively toxic for cancer, but not for non-malignant cells [10,11,14,32]. Moreover, HNE can affect tumor–host relationships, acting as an effector of anti-cancer activities of leukocytes, stromal cells, and non-malignant cells bordering invading cancer, and might even result in the spontaneous regression of cancer, such as W256 [35,36,45].

Furthermore, concomitant administration of the potent antioxidant pineal indole melatonin (MLT) and *A. vera* extract had better effects than those obtained by MLT used alone in patients suffering either from lung cancer, gastrointestinal tract tumors, breast cancer, or glioblastoma, all of which are otherwise known to be associated with the synthesis of HNE–protein adducts [37,38,46,47]. Treatment with MLT plus *A. vera* extracts produced therapeutic benefits, at least in terms of stabilization of disease and survival, in patients with advanced solid tumors for whom no other standard effective therapy was available [48]. Since HNE has an important role in defense activities of normal cells against primary and metastatic cancer, and can reflect the overall tumor–host relationship, especially on a metabolic level, we believe that further studies on AV should include immunohistochemical evaluation of the HNE–protein adducts in cancer and in surrounding tissue, complemented by their determination using the HNE–His ELISA in the blood. Such an analytical approach might not only help better understand the biomedical effects of AV and the pathophysiology of HNE, but could also further enhance the development of modern integrative biomedicine [49–52].

5. Conclusions

Aloe vera leaves used to prepare the extract (AV) had prominent antioxidant capacity, reflecting the overall activities of various antioxidants. AV on its own did not at all induce ROS or HNE production in the cells treated, while its observed bioactivities might reflect a complex interaction of different plant substances with cellular redox homeostasis for cells challenged by ROS- or HNE-induced oxidative stress. The complexity of the biological effects of HNE (regulation of proliferation, differentiation, and apoptosis), particularly if bound to proteins, plays an important role in the pathogenesis of various diseases, including cancer, but also in the cellular and systemic defense against stress- and age-associated diseases. In particular, the effects of AV on microvascular endothelia could be an important activity principle of AV; thus, we suggest further studies on AV to include an immunohistochemical evaluation of HNE–protein adducts in cancer and surrounding tissue, complemented with their determination using HNE–His ELISA in the blood.

Author Contributions: Formal analysis, V.C., I.J. and S.M. Funding acquisition, S.B.Š. and N.Z. Methodology, T.V. Resources, L.B. Writing—Original draft, H.L.

Funding: This research received no external funding.

Acknowledgments: This study is dedicated to all the brave people fighting terror and to the family of Ogunmola Julius Femi, who inspired us to do the study.

Conflicts of Interest: The authors declare no conflicts of interest.

References

1. Radha, M.H.; Laxmipriya, N.P. Evaluation of biological properties and clinical effectiveness of *Aloe vera*: A systematic review. *J. Tradit. Complement. Med.* **2015**, *5*, 21–26. [CrossRef] [PubMed]
2. Eshun, K.; He, Q. *Aloe vera*: A valuable ingredient for the food, pharmaceutical and cosmetic industries—A review. *Crit. Rev. Food Sci. Nutr.* **2004**, *44*, 91–96. [CrossRef] [PubMed]
3. Hamman, J.H. Composition and applications of *Aloe vera* leaf gel. *Molecules* **2008**, *13*, 1599–1616. [CrossRef] [PubMed]
4. Lucini, L.; Pellizzoni, M.; Molinari, G.P.; Franchi, F. Aloe anthraquinones against cancer. *Med. Aromat. Plant Sci. Biotechnol.* **2012**, *6*, 20–24.
5. Choi, S.; Chung, M.H. A review on the relationship between *Aloe vera* components and their biologic effects. *Semin. Integr. Med.* **2003**, *1*, 53–62. [CrossRef]
6. Bourdeau, M.D.; Beland, F.A. An evaluation of the biological and toxicological properties of Aloe barbadensis (Miller), *Aloe vera*. *J. Environ. Sci. Health C* **2006**, *24*, 103–154.
7. Zarkovic, N. 4-Hydroxynonenal as a bioactive marker of pathophysiological processes. *Mol. Asp. Med.* **2003**, *24*, 281–291. [CrossRef]
8. Vistoli, G.; Maddis, D.D.; Cipak, A.; Zarkovic, N.; Carini, M.; Aldini, G. Advanced glycoxidation and lipoxidation end products (AGEs and ALEs): An overview of their mechanisms of formation. *Free Radic. Res.* **2013**, *47*, 3–27. [CrossRef] [PubMed]
9. Poli, G.; Zarkovic, N. Editorial Introduction to the Special Issue on 4-Hydroxynonenal and Related Lipid Oxidation Products. *Free Radic. Biol. Med.* **2017**, *111*, 2–5. [CrossRef] [PubMed]
10. Sovic, A.; Borović, S.; Lončarić, I.; Kreuzer, T.; Zarkovic, K.; Vukovic, T.; Wäg, G.; Hrascan, R.; Wintersteiger, R.; Klinger, R.; et al. The carcinostatic and proapoptotic potential of 4-Hydroxynonenal in HeLa cells is associated with its conjugation to cellular proteins. *Anticancer Res.* **2001**, *21*, 1997–2004. [PubMed]
11. Borovic Sunjic, S.; Cipak, A.; Rabuzin, F.; Wildburger, R.; Zarkovic, N. The influence of 4-hydroxy-2-nonenal on proliferation, differentiation and apoptosis of human osteosarcoma cells. *Biofactors* **2005**, *24*, 141–148. [CrossRef] [PubMed]
12. Zarkovic, N.; Cipak, A.; Jaganjac, M.; Borovic, S.; Zarkovic, K. Pathophysiological relevance of aldehydic protein modifications. *J. Proteom.* **2013**, *92*, 239–247. [CrossRef] [PubMed]
13. Milkovic, L.; Hoppe, A.; Detsch, R.; Boccaccini, A.R.; Zarkovic, N. Effects of Cu-doped 45S5 bioactive glass on the lipid peroxidation-associated growth of human osteoblast-like cells in vitro. *J. Biomed. Mater Res. Part A* **2014**, *102*, 3556–3561. [CrossRef] [PubMed]
14. Milkovic, L.; Cipak Gasparovic, A.; Zarkovic, N. Overview on major lipid peroxidation bioactive factor 4-hydroxynonenal as pluripotent growth regulating factor. *Free Radic. Res.* **2015**, *49*, 850–860. [CrossRef] [PubMed]

15. Zarkovic, K.; Jakovcevic, A.; Zarkovic, N. Contribution of the HNE-immunohistochemistry to modern pathological concepts of major human diseases. *Free Radic. Biol. Med.* **2017**, *111*, 110–125. [CrossRef] [PubMed]

16. Randhir, R.; Shetty, K. Developmental stimulation of total phenolics and related antioxidant activity in light-and dark-germinated corn by natural elicitors. *Process Biochem.* **2005**, *40*, 1721–1732. [CrossRef]

17. Bessey, O.A.; Lowky, O.H.; Brock, M.J. A method for the rapid determination of alkaline phosphatase with five cubic millimeters of serum. *J. Biol. Chem.* **1946**, *164*, 321–329. [PubMed]

18. Lichtenthaler, H.K.; Buschmann, C. Chlorophylls and Carotenoids: Measurement and Characterization by UV-VIS Spectroscopy. In *Current Protocols in Food Analytical Chemistry (CPFA)*; Wrolstad, R.E., Acree, T.E., An, H., Decker, E.A., Penner, M.H., Reid, D.S., Schwartz, S.J., Shoemaker, C.F., Sporns, P., Eds.; John Wiley and Sons: New York, NY, USA, 2001; pp. F4.3.1–F4.3.8.

19. Brand-Williams, W.; Cuvelier, M.E.; Berset, C.L.W.T. Use of a free radical method to evaluate antioxidant activity. *LWT-Food Sci. Tech.* **1995**, *28*, 25–30. [CrossRef]

20. Pisoschi, A.M.; Cheregi, M.C.; Danet, A.F. Total antioxidant capacity of some commercial fruit juices: Electrochemical and spectrophotometrical approaches. *Molecules* **2009**, *14*, 480–493. [CrossRef] [PubMed]

21. Tiveron, A.P.; Melo, P.S.; Bergamaschi, K.B.; Vieira, T.M.; Regitano-d'Arce, M.A.; Alencar, S.M. Antioxidant activity of Brazilian vegetables and its relation with phenolic composition. *Int. J. Mol. Sci.* **2012**, *13*, 8943–8957. [CrossRef] [PubMed]

22. Borović, S.; Rabuzin, F.; Waeg, G.; Žarković, N. Enzyme-linked immunosorbent assay for 4-hydroxynonenal-histidine conjugates. *Free Radic. Res.* **2006**, *40*, 809–820. [CrossRef] [PubMed]

23. Živković, M.; Žarković, K.; Škrinjar, L.; Georg, W.; Poljak-Blaži, M.; Šunjić, B.S.; Schaur, R.J.; Žarković, N. A new method for detection of HNE-histidine conjugates in rat inflammatory cells. *Croat Chem. Acta* **2005**, *78*, 91–98.

24. Spickett, C.M.; Wiswedel, I.; Siems, W.; Zarkovic, K.; Zarkovic, N. Advances in methods for the determination of biologically relevant lipid peroxidation products. *Free Radic. Res.* **2010**, *44*, 1172–1202. [CrossRef] [PubMed]

25. Jaganjac, M.; Almuraikhy, S.; Al-Khelaifi, F.; Al-Jaber, M.; Bashah, M.; Mazloum, N.A.; Zarkovic, K.; Zarkovic, N.; Waeg, G.; Kafienah, W.; et al. Combined metformin and insulin treatment reverses metabolically impaired omental adipogenesis and accumulation of 4-hydroxynonenal in obese diabetic patients. *Redox Biol.* **2017**, *12*, 483–490. [CrossRef] [PubMed]

26. Teklić, T.; Engler, M.; Cesar, V.; Lepeduš, H.; Parađiković, N.; Lončarić, Z.; Štolfa, I.; Marotti, T.; Mikac, N.; Žarković, N. Copper excess influence on lettuce (*Lactuca sativa* L.) grown in the soil and nutrient solution. *J. Food Agric. Environ.* **2008**, *6*, 439–444.

27. Wonisch, W.; Kohlwein, S.D.; Schaur, J.; Tatzber, F.; Guttenberger, H.; Zarkovic, N.; Winkler, R.; Esterbauer, H. Treatment of the budding yeast (Saccharomyces cerevisiae) with the lipid peroxidation product 4-HNE provokes a temporary cell cycle arrest in G1 phase. *Free Radic. Biol. Med.* **1998**, *25*, 682–687. [CrossRef]

28. Čipak, A.; Jaganjac, M.; Tehlivets, O.; Kohlwein, S.D.; Žarković, N. Adaptation to oxidative stress induced by polyunsaturated fatty acids in yeast. *BBA-Mol. Cell Biol. Lipids* **2008**, *178*, 283–287. [CrossRef] [PubMed]

29. Žarković, N.; Schaur, R.J.; Puhl, H.; Jurin, M.; Esterbauer, H. Mutual dependence of growth modifying effects of 4-hydroxy-nonenal and fetal calf serum in vitro. *Free Radic. Biol. Med.* **1994**, *16*, 877–884.

30. Žarković, N.; Žarković, K.; Schaur, R.J.; Štolc, S.; Schlag, G.; Redl, H.; Waeg, G.; Borović, S.; Lončarić, L.; Jurić, G.; et al. 4-Hydroxynonenal as a second messenger of free radicals and growth modifying factor. *Life Sci.* **1999**, *65*, 1901–1904. [CrossRef]

31. Semlitsch, T.; Tillian, M.H.; Žarković, N.; Borović, S.; Purtscher, M.; Hohenwarter, O.; Schaur, J.R. Differential Influence of the Lipid Peroxidation Product 4-Hydroxynonenal on the Growth of Human Lymphatic Leukaemia Cells and Human Peripheral Blood Lymphocytes. *Anticancer Res.* **2002**, *22*, 1689–1697. [PubMed]

32. Borović, S.; Čipak, A.; Meinitzer, A.; Kejla, Z.; Perovic, D.; Waeg, G.; Zarkovic, N. Differential effect of 4-hydroxynonenal on normal and malignant mesenchimal cells. *Redox Rep.* **2007**, *207*, 50–54. [CrossRef] [PubMed]

33. Mouthuy, P.A.; Snelling, S.J.B.; Dakin, S.G.; Milković, L.; Gašparović, A.C.; Carr, A.J.; Žarković, N. Biocompatibility of implantable materials: An oxidative stress viewpoint. *Biomaterials* **2016**, *109*, 55–68. [CrossRef] [PubMed]

34. Cipak-Gasparovic, A.; Milkovic, L.; Borovic-Sunjic, S.; Zarkovic, N. Cancer Growth Regulation by 4-Hydroxynonenal Article Type. *Free Radic. Biol. Med.* **2017**, *111*, 226–234. [CrossRef] [PubMed]

35. Bauer, G.; Zarkovic, N. Revealing mechanisms of selective, concentration-dependent potentials of 4-hydroxy-2-nonenal to induce apoptosis in cancer cells through inactivation of membrane-associated catalase. *Free Radic. Biol. Med.* **2015**, *81*, 128–144. [CrossRef] [PubMed]

36. Zhong, H.; Xiao, M.; Zarkovic, K.; Zhu, M.; Sa, R.; Lu, J.; Tao, Y.; Chen, Q.; Xia, L.; Cheng, S.; et al. Mitochondrial Control of Apoptosis through Modulation of Cardiolipin Oxidation in Hepatocellular Carcinoma: A Novel Link between Oxidative Stress and Cancer. *Free Radic. Biol. Med.* **2017**, *176*, 67–76. [CrossRef] [PubMed]

37. Piskač Živković, N.; Petrovečki, M.; Lončarić, T.Č.; Nikolić, I.; Waeg, G.; Jaganjac, M.; Žarković, K.; Žarković, N. Positron Emission Tomography-Computed Tomography and 4-Hydroxynonenal-histidine Immunohistochemistry Reveal Differential Onset of Lipid Peroxidation in Primary Lung Cancer and in Pulmonary Metastasis of Remote Malignancies. *Redox Biol.* **2017**, *11*, 600–605. [CrossRef] [PubMed]

38. Negre-Salvayre, A.; Auge, N.; Ayala, V.; Basaga, H.; Boada, J.; Brenke, R.; Chapple, S.; Cohen, G.; Feher, J.; Grune, T.; et al. Pathological aspects of lipid peroxidation. *Free Radic. Res.* **2010**, *44*, 1125–1171. [CrossRef] [PubMed]

39. Kujundžić, R.N.; Žarković, N.; Trošelj, K.G. Pyridine nucleotides in regulation of cell death and survival by redox and non-redox reactions. *Crit. Rev. Eukar. Gene Express.* **2014**, *24*, 287–309. [CrossRef]

40. Milkovic, L.; Siems, W.; Siems, R.; Zarkovic, N. Oxidative stress and antioxidants in carcinogenesis and integrative therapy of cancer. *Curr. Pharm. Des.* **2014**, *20*, 6529–6542. [CrossRef] [PubMed]

41. Milkovic, L.; Zarkovic, N.; Saso, L. Controversy about pharmacological modulation of Nrf2 for cancer therapy. *Redox Biol.* **2017**, *12*, 727–732. [CrossRef] [PubMed]

42. Pecere, T.; Gazzola, M.V.; Mucignat, C.; Parolin, C.; Vecchia, F.D.; Cavaggioni, A.; Basso, G.; Diaspro, A.; Salvato, B.; Carli, M.; et al. Aloe-emodin is a new type of anticancer agent with selective activity against neuroectodermal tumors. *Cancer Res.* **2000**, *60*, 2800–2804. [PubMed]

43. Grimaudo, S.; Tolomeo, M.; Gancitano, R.; Dalessandro, N.; Aiello, E. Effects of highly purified anthraquinoid compounds from *Aloe vera* on sensitive and multidrug resistant leukemia cells. *Oncol. Rep.* **1997**, *4*, 341–343. [CrossRef] [PubMed]

44. Lee, H.Z.; Hsu, S.L.; Liu, M.C.; Wu, C.H. Effects and mechanisms of aloe-emodin on cell death in human lung squamous cell carcinoma. *Eur. J. Pharmacol.* **2001**, *431*, 287–295. [CrossRef]

45. Jaganjac, M.; Poljak-Blazi, M.; Schaur, R.J.; Zarkovic, K.; Borović, S.; Čipak, A.; Cindrić, M.; Uchida, K.; Waeg, G.; Žarković, N. Elevated neutrophil elastase and acrolein-protein adducts are associated with W256 regression. *Clin. Exp. Immunol.* **2012**, *170*, 178–185. [CrossRef] [PubMed]

46. Biasi, F.; Tessitore, L.; Zanetti, D.; Citrin, J.C.; Zingaro, B.; Chiarpotto, E.; Zarkovic, N.; Serviddio, G.; Poli, G. Associated changes of lipid peroxidation and TGF 1 levels in human cancer during tumor progression. *Gut* **2002**, *50*, 361–367. [CrossRef] [PubMed]

47. Žarković, K.; Juric, G.; Waeg, G.; Kolenc, D.; Žarković, N. Immunohistochemical appearance of HNE-protein conjugates in human astrocytomas. *Biofactors* **2005**, *24*, 33–40. [CrossRef] [PubMed]

48. Harlev, E.; Nevo, E.; Lansky, E.P.; Ofir, R.; Bishayee, A. Anticancer potential of Aloes: Antioxidant, antiproliferative, and immunostimulatory attributes. *Planta. Med.* **2012**, *78*, 843–852. [CrossRef] [PubMed]

49. Frijhoff, J.; Winyard, P.G.; Zarkovic, N.; Davies, S.S.; Stocker, R.; Cheng, D.; Knight, A.R.; Taylor, E.L.; Oettrich, J.; Ruskovska, T.; et al. Clinical relevance of biomarkers of oxidative stress. *Antioxid. Redox Signal.* **2015**, *23*, 1144–1170. [CrossRef] [PubMed]

50. Gęgotek, A.; Nikliński, J.; Žarković, N.; Žarković, K.; Waeg, G.; Łuczaj, W.; Charkiewicz, R.; Skrzydlewska, E. Lipid mediators involved in the oxidative stress and antioxidant defence of human lung cancer cells. *Redox Biol.* **2016**, *9*, 210–219. [CrossRef] [PubMed]

51. Fedorova, M.; Zarkovic, N. Preface to the special issue on 4-hydroxynonenal and related lipid oxidation products. *Free Radic. Biol. Med.* **2017**, *111*, 1. [CrossRef] [PubMed]

52. Egea, J.; Fabregat, I.; Frapart, Y.M.; Ghezzi, P.; Görlach, A.; Kietzmann, T.; Kubaichuk, K.; Knaus, U.G.; Lopez, M.G.; Olaso-Gonzalez, G.; et al. European Contribution to the study of ROS: A Summary of the Findings and Prospects for the Future from the COST Action BM1203 (EU-ROS). *Redox Biol.* **2017**, *13*, 94–162. [CrossRef] [PubMed]

antioxidants

MDPI

Article

The Effect of Sea Buckthorn (*Hippophae rhamnoides* L.) Seed Oil on UV-Induced Changes in Lipid Metabolism of Human Skin Cells

Agnieszka Gęgotek, Anna Jastrząb, Iwona Jarocka-Karpowicz, Marta Muszyńska and Elżbieta Skrzydlewska * [ID]

Department of Inorganic and Analytical Chemistry, Medical University of Bialystok, Bialystok 15-089, Poland; agnieszka.gegotek@umb.edu.pl (A.G.); anna.jastrzab@umb.edu.pl (A.J.); iwona.jarocka-karpowicz@umb.edu.pl (I.J.-K.); marta.muszynska@umb.edu.pl (M.M.)
* Correspondence: elzbieta.skrzydlewska@umb.edu.pl; Tel./Fax: +48-857-485-882

Received: 9 July 2018; Accepted: 20 August 2018; Published: 23 August 2018

Abstract: Lipids and proteins of skin cells are the most exposed to harmful ultraviolet (UV) radiation contained in sunlight. There is a growing need for natural compounds that will protect these sensitive molecules from damage, without harmful side effects. The aim of this study was to investigate the effect of sea buckthorn seed oil on the redox balance and lipid metabolism in UV irradiated cells formed different skin layers to examine whether it had a protective effect. Human keratinocytes and fibroblasts were subjected to UVA (ultraviolet type A; 30 J/cm^2 and 20 J/cm^2) or UVB (ultraviolet type B; 60 mJ/cm^2 and 200 mJ/cm^2, respectively) radiation and treated with sea buckthorn seed oil (500 ng/mL), and the redox activity was estimated by reactive oxygen species (ROS) generation and enzymatic/non-enzymatic antioxidants activity/level (using electron spin resonance (ESR), high-performance liquid chromatography (HPLC), and spectrophotometry). Lipid metabolism was measured by the level of fatty acids, lipid peroxidation products, endocannabinoids and phospholipase A2 activity (GC/MS (gas chromatography/mass spectrometry), LC/MS (liquid chromatography/mass spectrometry), and spectrophotometry). Also, transcription factor Nrf2 (nuclear erythroid 2-related factor) and its activators/inhibitors, peroxisome proliferator-activated receptors (PPAR) and cannabinoid receptor levels were measured (Western blot). Sea buckthorn oil partially prevents UV-induced ROS generation and enhances the level of non-enzymatic antioxidants such as glutathione (GSH), thioredoxin (Trx) and vitamins E and A. Moreover, it stimulates the activity of Nrf2 leading to enhanced antioxidant enzyme activity. As a result, decreases in lipid peroxidation products (4-hydroxynonenal, 8-isoprostaglandin) and increases in the endocannabinoid receptor levels were observed. Moreover, sea buckthorn oil treatment enhanced the level of phospholipid and free fatty acids, while simultaneously decreasing the cannabinoid receptor expression in UV irradiated keratinocytes and fibroblasts. The main differences in sea buckthorn oil on various skin cell types was observed in the case of PPARs—in keratinocytes following UV radiation PPAR expression was decreased by sea buckthorn oil treatment, while in fibroblasts the reverse effect was observed, indicating an anti-inflammatory effect. With these results, sea buckthorn seed oil exhibited prevention of UV-induced disturbances in redox balance as well as lipid metabolism in skin fibroblasts and keratinocytes, which indicates it is a promising natural compound in skin photo-protection.

Keywords: fibroblasts; keratinocytes; sea buckthorn seeds oil; UV radiation; lipid metabolism

---

1. Introduction

Skin has been recognized as an organ that performs synthesis and processing of an astounding range of structural proteins, lipids, and signaling molecules. Moreover, skin is an integral component of the immune, nervous and endocrine systems, and therefore is also responsible for the first line of information and defense in the immunity process [1]. However, skin also protects the organism against the influence of the environment and is well-known as crucial for the maintenance of temperature, electrolyte and fluid balance [2]. As a natural biological barrier, skin is constantly exposed to numerous external factors that can impair the functioning of this barrier, including ultraviolet (UV) radiation contained in sunlight. High energy UVB (ultraviolet type B; 280–320 nm) and UVA (ultraviolet type A; 320–400 nm) radiation absorption by the skin may trigger mechanisms that defend skin integrity and also induces skin pathology, such as cancer [3]. These effects occurs as a result of the UV electromagnetic energy transduction into chemical, hormonal, and neural signals, defined by the nature of the chromophores and tissue compartments receiving specific UV wavelength. However, despite of the wavelength differences, both of them are characterized by high penetration into the deep layers of the epidermis and dermis [4]. UV radiation can upregulate local cytokines, corticotropin-releasing hormone, urocortins, proopiomelanocortin peptides expression, however, they can be released into circulation to exert systemic effects, including activation of the central hypothalamic-pituitary-adrenal axis, opioidogenic effects, and immunosuppression [5]. Similar effects of UV radiation on human organism are seen after exposure of the eyes and skin to UV, through which UV exerts very rapid stimulatory effects on the brain and central neuroendocrine system and impairs body homeostasis [5]. Moreover, UV radiation promotes the generation of reactive oxygen species (ROS) as well as impairs the antioxidant capacity of skin cells [6]. As a result, the redox imbalance leading to oxidative stress is increased. UV-induced oxidative stress in skin enhances the collagen degradation and elastin modification causes premature skin aging [7,8]. Moreover, increased oxidative phospholipid metabolism disturbs the function of cellular membranes, and promotes the generaton of reactive electrophiles, such as 4-hydroxyalkenals [9]. These reactive electrophiles interact with membrane phospholipids and proteins, including receptors, causing irreparable skin damage. Additionally, UV-induced oxidative stress affects phospholipid metabolism which influences the endocannabinoid system [10]. The main endocannabinoids, such as anandamide and 2-arachidonoylglycerol (2-AG), participate in cell signalling and are ligands for transmembrane receptors-CB1 and CB2 (cannabinoid receptor type 1 and 2) [11]. Activation of CB1 is responsible for ROS generation, whereas CB2 prevents ROS generation and inflammation, while also stimulating the MAP kinase (mitogen-activated protein kinase) pathway [12].

In connection with the permanent exposure to harmful UV radiation, cells from different skin layers-epidermal keratinocytes and dermal fibroblasts create a number of defense mechanisms that protect against UV-induced oxidative stress. Examples of such mechanisms are the high activity of repair and antioxidant enzymes, a large pool of non-enzymatic antioxidants, and redox-sensitive transcription factors including Nrf2 that is responsible for cytoprotective protein expression [13].

Despite the above well-developed mechanisms, skin cells exposed to long-lasting UV radiation are vulnerable to the depletion of the natural antioxidants present and therefore there is still a need for compounds with cytoprotective properties against UV-induced changes. One potential source are seed oils, the main source of phospholipids and triacylglycerols, which are solvents for other lipophilic compounds, i.e., sterols, fat-soluble vitamins (vitamin A and E), carotenoids, phenolic compounds, and free fatty acids [14]. One good source of seed oil is sea buckthorn (*Hippophae rhamnoides* L.) [15], which is increasingly appearing in skin care preparations. Its antioxidant potential is based on high content of polyphenols, vitamins (C and E), carotenoids as well as sterols [16–18]. In addition to the antioxidant activity, sea buckthorn also shows antibacterial and antifungal activity [19]. Carotenoids contained in the oil stimulate collagen synthesis, while phytosterols regulate inflammatory processes and have anticancer effects [20]. Because of the high concentration of these compounds sea buckthorn seed oil has been found as a promising therapeutic agent in the treatment of dermatitis [21]. However, there is no research on the effect of sea buckthorn seed oil on the UV-induced oxidative lipid metabolism of skin cells.

The aim of this study was to estimate the effect of sea buckthorn seed oil on the correlation between redox balance, lipid metabolism and endocannabinoid system in in vitro cultured human keratinocytes and fibroblasts exposed to the UV-radiation.

2. Materials and Methods

2.1. Examination of Sea Buckthorn Seed Oil Composition

Sea buckthorn (*Hippophae rhamnoides* L.) seed oil fatty acids levels as well as vitamins A and E were described below in the paragraphs Determination of fatty acids levels and Determination of non-enzymatic antioxidant levels.

2.1.1. Determination of Squalene Levels

Squalene level was analyzed with separation on HP-5ms capillary column (0.25 mm; 0.25 µm, 30 m) of GC/MS system (Agilent Technologies, Palo Alto, CA, USA) with 7890A GC–7000 quadrupole MS/MS (Agilent Technologies, Palo Alto, CA, USA). Operating conditions were as follows: Oven programming $-50\,°C$ (10 min), rate 2 $°C/min$ to 310 $°C$ (30 min); ion source (EI) $-230\,°C$; electron energy -70 eV [22]. The mass spectrometer source was run in selective ion monitoring (SIM) mode for the following ions: 191 and 81 m/z.

2.1.2. Phytosterols Profile

The content of sterols in oils was determined by GC/MS system with 7890A GC–7000 quadrupole MS/MS (Agilent Technologies, Palo Alto, CA, USA) [23]. Samples were separated on HP-5ms capillary column (0.25 mm; 0.25 µm, 30 m). Operating conditions were as follows: oven programming $-50\,°C$ (2 min), rate 15 $°C/min$ to 230 $°C$, to 310 $°C$ at the rate of 3 $°C/min$ (10 min); ion source (EI) $-230\,°C$; electron energy -70 eV. The mass spectrometer source was run in selective ion monitoring (SIM) mode for the following ions: 372 and 217 m/z for 5-α-cholestane (IS), 458 and 368 m/z for cholesterol, 470 and 255 m/z for brassicasterol, 382 and 343 m/z for campesterol, 394 and 255 m/z for stigmasterol and 396 and 357 m/z for β-sitosterol. The quantifications were carried out using the internal standard method.

2.1.3. Determination of β-Carotene Levels

HPLC methods was used to detect the level β-carotene [24]. Samples were dissolved in 1 mL of 2-propanol and 20 µL was taken analysis. Separation was performed using C18 column (150 nm × 4.6 mm) with UV detection at 454 nm. The concentration was determined using a calibration curve range 0.5–50 mg/L.

2.2. Cell Culture

Human skin cell lines used in experiment were obtained from American Type Culture Collection and cultured cultured in a humidified atmosphere of 5% CO_2 at 37 $°C$. Human immortalized keratinocytes (CDD 1102 KERTr) were transformed with human papillomavirus 16 (HPV-16) E6/E7, while fibroblasts (CCD 1112Sk) were isolated from normal foreskin of Caucasian newborn male and used in passage 11. The growth media for each line were prepared as follows: for keratinocytes-keratinocyte-SFM medium with 1% Bovine Pituitary Extract (BPE) and human recombinant Epidermal Growth Factor (hEGF); for fibroblasts-Dulbecco's Modified Eagle Medium (DMEM) with 10% fetal bovine serum (FBS). Media were supplemented with 50 U/mL penicillin and 50 µg/mL streptomycin. All sterile and cell culture reagents were obtained from Gibco (Grand Island, NY, USA).

Cells were exposed to UV radiation after reaching the 70% confluence. Radiation doses for keratinocytes were 30 J/cm^2 for UVA (365 nm, power density at 4.2 mW/cm^2) and 60 mJ/cm^2 for UVB (312 nm, power density at 4.08 mW/cm^2), while for fibroblasts were 20 J/cm^2 for UVA and 200 mJ/cm^2 for UVB (Bio-Link Crosslinker BLX 312/365, Vilber Lourmat, Germany). The distance of the cells from lamps was 15 cm. Exposure doses were chosen corresponding to 70% cell viability measured by the MTT (3-(4,5-dimethylthiazol-2-yl)-2,5-diphenyltetrazolium bromide) assay [25], what was previously shown as a doses that lead to the activation of pro-oxidative conditions, response of cell antioxidants, as well as activation of Nrf2 pathway and increases endocannabinoid system action [6]. Control cells were incubated in parallel without irradiation.

To examine the effect of sea buckthorn (*Hippophae rhamnoides* L.) seed oil (produced by "Szarłat" M i W Lenkiewicz Sp. J., Poland) on UV radiated skin cells, after keratinocytes and fibroblasts were exposed to UV radiation they were incubated 24 h under standard conditions in medium containing 500 ng/mL plant oil in 0.1% DMSO. The oil concentration was chosen corresponding to 100% cell viability compared to control cells measured by the MTT assay [25]. Parallel cells were cultured in medium containing plant oil without irradiation.

2.3. Examination of Pro-Oxidative Activity

2.3.1. Determination of ROS Generation

The superoxide anions generation was detected using stable nitroxide CM-radicals formation and detection with EPR spectrometer (Noxygen GmbH/Bruker Biospin GmbH, Germany) [10,26]. The results were reported in micromolar concentration of ROS per minute and normalised per milligram of protein.

2.3.2. Determination of Pro-Oxidants Enzyme Activities

NADPH oxidase (NOX-EC 1.6.3.1) activity was analysed by luminescence measurement using lucigenin as luminophore [27]. Enzyme specific activity was described in relative luminescence units (RLU) per milligram protein.

Xanthine oxidase (XO-EC1.17.3.2) activity was estimated by uric acid formation from xanthine [28]. Enzyme specific activity was described in microunits per milligram of protein.

2.4. Examination of Antioxidant Defence System

2.4.1. Determination of Antioxidant Enzymes Activity

Glutathione peroxidase (GSH-Px-EC.1.11.1.6) activity was assessed spectro-photometrically [29]. Enzyme specific activity was described in microunits per milligram of protein.

Glutathione reductase (GSSG-R-EC.1.6.4.2) activity was measured by monitoring the oxidation of NADPH at 340 nm [30]. Enzyme specific activity was described in milliunits per milligram of protein.

Superoxide dismutase (Cu/Zn–SOD-EC.1.15.1.1) activity was determined according to the method of Misra and Fridovich [31] as modified by Sykes [32]. Enzyme specific activity was calculated in milliunits per milligram of protein.

The thioredoxin reductase (TrxR-EC. 1.8.1.9) activity was measured using a commercially available kit (Sigma-Aldrich, St. Louis, MO, USA) in accordance with the included instruction [33]. Enzyme activity was measured in units of microunits per milligram of protein.

2.4.2. Determination of Non-Enzymatic Antioxidant Levels

Thioredoxin level were quantified using the ELISA (enzyme-linked immunosorbent assay) [34]. The ELISA plates coated with samples were incubated at 4 °C overnight with anti-thioredoxin primary antibody per well (diluted in 1% bovine serum albumin in phosphate buffered saline (PBS)) (Abcam, Cambridge, MA, USA). After washing and blocking the plates were incubated with goat anti-rabbit secondary antibody solution (Dako, Carpinteria, CA, USA). As a chromogen substrate solution 0.1 mg/mL 3,3', 5,5'-tetramethylbenzidine was used. Spectral absorption was read at 450 nm with the reference filter at 620 nm. Thioredoxin levels were described in micrograms per milligram of protein.

Glutathione level was measured using the capillary electrophoresis (CE) [35] with separation on 47 cm capillary operated at 27 kV. UV detection was set at 200 ± 10 nm. Calibration curve range of 1–120 nmol/mL ($r^2 = 0.9985$) was prepared to determined GSH concentration that was subsequently normalized for milligrams of protein.

To detect the level of vitamins A and E HPLC method was used [36]. Vitamins were extracted using hexane. The concentration of vitamins detected at 294 nm was determined using a calibration curves range: 0.125–1 mg/L ($r^2 = 0.9998$) for vitamin A, and 5–25 mg/L ($r^2 = 0.9999$) for vitamin E.

2.4.3. Determination of Protein Expression

Protein expression level was analyzed using Western blot technique [37]. Cell lysates or nuclear fractions (in the case of phosphorylated protein Nrf2) were separated by 10% Tris-Glycine SDS-PAGE (sodium dodecyl sulfate polyacrylamide gel electrophoresis). Following electrophoresis, separated proteins were transferred into a membrane and blocked with 5% skim milk. Primary antibodies against phospho-Nrf2 (pSer40), HO-1, Keap1, p21, p62, and peroxisome proliferator-activated receptors (PPARα, γ, and δ) (Sigma-Aldrich, St. Louis, MO, USA) were diluted 1:1000. Protein bands were visualized calorimetrically using the BCIP/NBT (5-bromo-4-chloro-3-indolyl-phosphate/nitro blue tetrazolium) Liquid substrate system (Sigma-Aldrich, St. Louis, MO, USA). Protein level was expressed as a percentage of control cells.

2.5. Examination of Lipid Metabolism

2.5.1. Determination of Cellular Membrane Integrity by LDH Test

Keratinocyte and fibroblast cellular membrane integrity was monitored using the assay based on the determination of the release of lactate dehydrogenase (LDH) into the medium. Cells were seeded into a 96 well plates in 200 μL of growth media at 5×10^3 for 24 h and then subjected to UV radiation or oil treatment range 5 ng/mL–5 mg/mL. Following 24 h incubation the activity of LDH in medium and in cell lysates was measured after 24 h. Spectrophotometric detection was carried out at 340 nm. The final concentrations of NADH and pyruvate in the reaction mixture were 1 mM and 2 mM, respectively [25]. The rate of LDH released from the treated cells was calculated by comparing its activity in medium to cell lysates. The results were expressed as the percentage of LDH activity obtained for control cells.

2.5.2. Determination of Fatty Acids Levels

Fatty acids profiles were determined by gas chromatography [38]. Free fatty acids (FFA) and phospholipids (PL) were methylated to fatty acid methyl esters (FAMEs). FAMEs were analyzed by Clarus 500 Gas Chromatograph (Perkin Elmer, Shelton, CT, USA) following separation on capillary column (50 m $\times$ 0.25 mm, ID 0.2 μm, Varian, Walnut Creek, USA). Identification of FAMEs was made by comparison to their retention time with authentic standards. Quantitation was achieved using an internal standard method.

2.5.3. Phospholipase A2 Activity

The activity of cPLA₂ (cytosolic phospholipase A₂) activity was measured using a commercially available kit (Cayman, No. 765021). To detect cPLA₂ activity specific inhibitors of sPLA2 (thioetheramide-PC) and iPLA2 (bromoenol lactone) were added to samples [39]. Enzyme specific activity was presented in nanomol of free thiol released per minute normalized per milligram of protein.

2.5.4. Determination of Lipid Peroxidation Products

Lipid peroxidation was estimated by measuring the level of 4-hydroxynonenal (4-HNE) and 8-Isoprostaglandin F2α.

Aldehydes were measured by GC/MS in selected ion monitoring (SIM) mode, as the O-PFB-oxime or O-PFB-oxime-TMS derivatives using benzaldehyde-D6 as an internal standard [40]. Aldehydes were analyzed using a 7890A GC—7000 quadrupole MS/MS (Agilent Technologies, Palo Alto, CA, USA) equipped with a HP-5ms capillary column (30-m length, 0.25-mm internal diameter, 0.25-µm film thickness). Target ion with a m/z 333.0 and 181.0 for 4-HNE-PFB-TMS and m/z 307.0 for IS derivatives were selected. Obtained results were normalized for milligrams of protein.

8-Isoprostaglandin F2α (8-isoPGF2α) level was measured using LC-MS (Agilent 1290 LC with triple quadruple mass spectrometer in negative multiple reaction mode [41]. SEP-PAK C18 column were used to samples purification. Target ions with a m/z 353→193 were selected. Target ion with a m/z 353→193 was selected.

2.5.5. Determination of Endocannabinoid System

Anandamide (AEA) and 2-arachidonoylglycerol (2-AG) were quantified using UPLC-MS/MS system (Agilent 1290 UPLC system interfaced with an Agilent 6460 triple quadropole mass spectrometer) [42]. As internal standards AEA-d8 and 2-AG-d8 were used. The samples were analyzed in positive-ion mode using multiple reaction monitoring (MRM) and the transition of the precursor to the product ion were: m/z 348.3→62.1 for AEA, and m/z 379.3→287.2 for 2-AG. Endocannabinoids concentrations were expressed as femtomols per milligram of protein.

The levels of endocannabinoid receptors CB1/2 were measured by Western blotting described above (*Determination of protein expression*). Primary monoclonal antibodies against CB1 and CB2 (Santa Cruse Biotechnology, Santa Cruz, CA, USA) were used at a concentration of 1:1000. Receptor level was expressed as a percentage of control cells.

2.6. Statistical Analysis

Data were analyzed using standard statistical analyses, including *t*-test, and the results are expressed as the mean ± standard deviation (SD) for $n = 5$. Experiment was performed on cell lines and all replicates were technical repetitions and do not reflect biodiversity. *p*-values less than 0.05 were considered significant.

3. Results

3.1. The Sea Buckthorn Seed Oil Composition

The composition of sea buckthorn seed oil is presented in Table 1. In addition to the large amount of fatty acids in the free form as well as in the form of DG, TAG and PL, the seed oil of sea buckthorn also contains phytosterols, such as cholesterol, brassicrakol, campesterol, stigmasterol and β-sitosterol, as well as antioxidants such as vitamins A and E, squalene and β-carotene.

Table 1. The fatty acids composition, antioxidant components and phytosterols in sea buckthorn (*Hippophae rhamnoides* L.) seeds oil. Mean values ± SD of five independent experiments are presented.

	Fatty Acids Composition of Sea Buckthorn Needs Oil			
	PL	TG	DG	FFA
			mg/mL	
14:0	0.060 ± 0.003	4.58 ± 0.14	0.240 ± 0.007	0.070 ± 0.003
16:0	1.14 ± 0.03	308.60 ± 9.25	3.10 ± 0.09	1.65 ± 0.05
18:0	0.51 ± 0.02	8.05 ± 0.24	0.47 ± 0.01	0.100 ± 0.003
20:0	n.d.	2.13 ± 0.06	0.130 ± 0.003	0.060 ± 0.002
22:0	n.d.	0.95 ± 0.03	0.140 ± 0.003	n.d.
24:0	n.d.	1.08 ± 0.03	0.170 ± 0.004	0.070 ± 0.002
SFA	1.71 ± 0.05	325.40 ± 9.76	4.26 ± 0.13	1.95 ± 0.06

Table 1. *Cont.*

Fatty Acids Composition of Sea Buckthorn Needs Oil					
	PL	TG	DG	FFA	
			mg/mL		
16:1n7	0.48 ± 0.01	310.85 ± 9.33	4.56 ± 0.14	2.22 ± 0.07	
18:1n9c	0.150 ± 0.005	45.34 ± 1.36	0.89 ± 0.03	0.34 ± 0.01	
18:1n7	0.110 ± 0.003	45.25 ± 1.35	0.83 ± 0.02	0.48 ± 0.01	
18:2n6	0.230 ± 0.007	87.68 ± 2.63	7.15 ± 0.21	1.53 ± 0.05	
18:3γ	n.d.	1.79 ± 0.05	n.d.	n.d.	
18:3α	0.280 ± 0.008	7.05 ± 0.21	n.d.	0.210 ± 0.006	
MUFA	0.74 ± 0.02	401.44 ± 12.04	6.29 ± 0.19	3.04 ± 0.09	
PUFA	0.51 ± 0.02	96.51 ± 2.90	7.15 ± 0.21	1.74 ± 0.05	
USFA	1.25 ± 0.04	497.95 ± 14.94	13.44 ± 0.40	4.78 ± 0.14	
Sum	2.95 ± 0.09	823.35 ± 24.70	17.70 ± 0.53	6.73 ± 0.20	
Antioxidant Components of Sea Buckthorn Needs Oil [mg/100 g]					
Squalene	β-Carotene	Vitamin E	Vitamin A		
240±11	15.11 ± 0.46	98.92 ± 2.89	648.0 ± 19.7		
Phytosterols in Sea Buckthorn Needs Oil [mg/100 g]					
Cholesterol	Brassicasterol	Campesterol	Stigmasterol	β-Sitosterol	Total phytosterols
0.51 ± 0.01	4.90 ± 0.14	17.61 ± 0.48	4.95 ± 0.12	223.88 ± 6.76	251.87 ± 7.48

Abbreviations: PL: phospholipid; DG: diacylglycerol; TG: triacylglycerol; EFA: essential fatty acid; n.d.: not detected.

3.2. *The Effect of Sea Buckthorn Seed Oil on Skin Cells Redox Balance*

Sea buckthorn seed oil treatment was found as a factor that partially prevented the UV-induced disturbances in redox balance occurring in experimental skin cells. It was found that fibroblast incubation with sea buckthorn oil caused decrease in ROS generation by approximately 25% as well as a decrease in the activity of NOX and XO by approximately 15% following UVA and 25% following UVB radiation. Similar character of changes was observed in the case of keratinocytes only following UVB radiation, were sea buckthorn oil caused decrease in NOX and XO activity by 10% and 15%, respectively (Table 2).

Table 2. The reactive oxygen species (ROS) generation, xanthine oxidase and NADPH oxidase activity in keratinocytes and fibroblasts after exposure to UVA (30 J/cm^2 and 20 J/cm^2) and UVB radiation (60 mJ/cm^2 and 200 mJ/cm^2, respectively) and sea buckthorn seeds oil (500 ng/mL) treatment.

Prooxidative Parameters		Keratinocytes			Fibroblasts		
	Oil	Control	UVA	UVB	Control	UVA	UVB
ROS	-	32.8 ± 1.6	89.1 ± 4.3 ^{x}	98.9 ± 4.8 ^{x}	50.6 ±2.4	126.5 ± 6.1 ^{x}	132.1 ± 6.4 ^{x}
[nM/min/mg protein]	+	63.5 ± 3.1 ^{x}	80.8 ± 3.9 ^{xy}	96.5 ± 4.7 ^{xy}	54.7 ±2.6	94.6 ± 4.6 ^{xya}	96.5 ± 4.7 ^{xyb}
NOX	-	158 ± 7.7	229 ± 11.2 ^{x}	256 ± 12.5 ^{x}	179 ± 8.7	321 ± 15.7 ^{x}	398 ± 19.5 ^{x}
[RLU/mg protein]	+	173 ± 8.4	231 ± 11.3 ^{xy}	227 ± 14.0 ^{xyb}	216 ± 10.5 ^{x}	268 ± 13.1 ^{xya}	321 ± 15.7 ^{xyb}
XO	-	164 ± 8.1	267 ± 13.1 ^{x}	297 ± 14.5 ^{x}	114 ± 5.5	248 ± 12.1 ^{x}	487 ± 23.8 ^{x}
[mU/mg protein]	+	158 ± 7.7	248 ± 12.1 ^{xy}	254 ± 12.4 ^{xyb}	278 ± 13.6 ^{x}	215 ± 10.5 ^{xya}	348 ± 17.1 ^{xyb}

Mean values ± SD of five independent experiments are presented. ^{x} statistically significant differences vs. control group, $p < 0.05$; ^{y} statistically significant differences between UVA/UVB and oil treated groups vs. oil treated control group, $p < 0.05$; ^{a} statistically significant differences between UVA and oil treated group vs. UVA treated group, $p < 0.05$; ^{b} statistically significant differences between UVB and oil treated group vs. UVB treated group, $p < 0.05$. Abbreviations: UVA: ultraviolet type A; UVB: ultraviolet type B; ROS: reactive oxygen species; NOX: NADPH oxidase; XO: xanthine oxidase.

Simultaneously, sea buckthorn oil altered the antioxidant system (Table 3), what was primarily visible in the case of the non-enzymatic antioxidant thioredoxin, its level was significantly increased by sea buckthorn oil in the keratinocytes and fibroblasts exposed to UV radiation by about 60% and 37% respectively in the case of UVA, and by 37% and 175%, respectively in the case of UVB radiation.

Sea buckthorn oil also enhanced the level of glutathione in UV exposed cells; it was increased 2.5 fold in the UV irradiated keratinocytes, and a 25% and 80% increase was observed in the case of UVA and UVB irradiated fibroblasts, respectively. In the case of vitamin levels, sea buckthorn oil treatment caused an increase in vitamin A levels in UVA and UVB irradiated keratinocytes by approximately 50% and 60%, respectively; and in UVA and UVB irradiated fibroblasts by about 20% and 15%, respectively. Also, the level of vitamin E following UV radiation was partially restored; by 55% in keratinocytes and by 30% in fibroblasts following both UV types. Described changes were accompanied by enhanced GSH-Px activity in keratinocytes (by 15% following UVA and even 2 times following UVB radiation). Moreover, sea buckthorn oil cell treatment lead to 4- and 3-fold increase in TrxR activity in keratinocytes following UVA and UVB radiation, and by 20% and 30% increases in the enzyme activity in fibroblasts following UVA and UVB radiation, respectively. However, in the case of SOD activity sea buckthorn oil additionally decreased its activity by 15% in the case of both cell lines.

Table 3. The activity of antioxidant enzymes (GSH-Px, GSSG-R, SOD, TrxR) and the level of non-enzymatic antioxidants (GSH, Trx, vitamins A, E, and C) in keratinocytes and fibroblasts after exposure to UVA ($30\,J/cm^2$ and $20\,J/cm^2$) and UVB radiation ($60\,mJ/cm^2$ and $200\,mJ/cm^2$, respectively) and sea buckthorn seeds oil (500 ng/mL) treatment.

Antioxidant Parameters	Oil	Keratinocytes			Fibroblasts		
		Control	UVA	UVB	Control	UVA	UVB
GSH-Px	-	13.6 ± 0.6	10.8 ± 0.5 [x]	7.8 ± 0.4 [x]	11.8 ± 0.6	26.4 ± 1.2 [x]	29.5 ± 1.4 [x]
[mU/mg protein]	+	15.6 ± 0.7 [x]	12.4 ± 0.6 [ya]	14.8 ± 0.7 [b]	16.3 ± 0.8 [x]	26.9 ± 1.3 [xy]	31.2 ± 1.5 [xy]
GSSG-R	-	26.3 ± 1.2	23.8 ± 1.2 [x]	19.2 ± 0.9 [x]	22.3 ± 1.1	45.6 ± 2.2 [x]	56.2 ± 2.7 [x]
[mU/mg protein]	+	22.5 ± 1.1 [x]	21.8 ± 1.1 [x]	18.7 ± 0.9 [xy]	26.9 ± 1.3 [x]	34.6 ± 1.7 [xy]	49.3 ± 2.4 [xy]
SOD	-	28.4 ± 1.3	23.1 ± 1.1 [x]	21.6 ± 1.1 [x]	26.3 ± 1.2	21.3 ± 1.0 [x]	14.6 ± 0.7 [x]
[mU/mg protein]	+	30.5 ± 1.5	25.6 ± 1.2 [xy]	24.7 ± 1.2 [xyb]	20.6 ± 1.0 [x]	18.5 ± 0.9 [xya]	12.3 ± 0.6 [xyb]
TrxR	-	112 ± 5	93 ± 4 [x]	84 ± 4 [x]	331 ± 16	204 ± 10 [x]	141 ± 7 [x]
[μU/mg protein]	+	104 ± 5	421 ± 11 [xya]	300 ± 9 [xyb]	312 ± 15	243 ± 11 [xya]	182 ± 8 [xyb]
Trx	-	1.5 ± 0.07	1 ± 0.07 [x]	0.8 ± 0.07 [x]	0.9 ± 0.04	0.4 ± 0.01 [x]	0.4 ± 0.02 [x]
[μg/mg protein]	+	1.6 ± 0.08	1.6 ± 0.08 [xa]	1.1 ± 0.07 [xb]	1.5 ± 0.07 [x]	1 ± 0.04 [xya]	1.1 ± 0.05 [xyb]
GSH	-	26.2 ± 1.2	11.4 ± 0.5 [x]	22.3 ± 1.1 [xyb]	14.6 ± 0.7	8.2 ± 0.4 [x]	5.6 ± 0.3 [x]
[nmol/mg protein]	+	37.7 ± 1.8 [x]	27.7 ± 1.3 [ya]	22.3 ± 1.1 [xyb]	15.6 ± 0.8	10.4 ± 0.5 [xya]	10.1 ± 0.5 [xyb]
vitamin A	-	53 ± 2.5	39 ± 1.9 [x]	35 ± 1.7 [x]	41 ± 2.1	34 ± 1.6 [x]	33 ± 1.6 [x]
[nmol/mg protein]	+	63 ± 3.0 [x]	59 ± 2.8 [xa]	56 ± 2.7 [yb]	46 ± 2.2	41 ± 2.1 [a]	38 ± 1.8 [xyb]
vitamin E	-	453 ± 22.1	325 ± 15.9 [x]	303 ± 14.8 [x]	336 ± 16.5	287 ± 14.1 [x]	269 ± 13.1 [x]
[nmol/mg protein]	+	517 ± 25.3 [x]	491 ± 24.1 [ya]	478 ± 23.4 [yb]	401 ± 19.5 [x]	381 ± 18.6 [xa]	351 ± 17.2 [yb]

Mean values ± SD of five independent experiments are presented. [x] statistically significant differences vs. control group, $p < 0.05$; [y] statistically significant differences between UVA/UVB and oil treated groups vs. oil treated control group, $p < 0.05$; [a] statistically significant differences between UVA and oil treated group vs. UVA treated group, $p < 0.05$; [b] statistically significant differences between UVB and oil treated group vs. UVB treated group, $p < 0.05$. Abbreviations: UVA: ultraviolet type A; UVB: ultraviolet type B; GSH-Px: glutathione peroxidase; GSSG-R: glutathione reductase; SOD: superoxide dismutase; TrxR: thioredoxin reductase; Trx: thioredoxin; GSH: glutathione.

3.3. The Effect of Sea Buckthorn Seed Oil on Skin Cells Transcription Factors

Sea buckthorn oil treatment also stimulated the antioxidant system, increased expression of phosphorylated factor Nrf2 in skin cells following UV radiation was observed in both cell lines as well as in the case of expression of Nrf2 targeted protein—HO-1. On the other hand, sea buckthorn oil restored the decreased activity following UV radiation, Keap1 level in UVA irradiated skin cells, and in UVB irradiated keratinocytes. Simultaneously, sea buckthorn oil lead to the decrease in Nrf2 activators—p21 and p62, in skin cells following both UVA and UVB radiation (Figure 1).

Additionally, sea buckthorn oil treatment following UV radiation induced various effects in keratinocytes and fibroblasts and in the other transcription factors such as peroxisome proliferator-activated receptors (PPARα, γ, and δ). In keratinocytes sea buckthorn oil lead to a decrease by 8% in PPARα expression following UVA radiation and by 20% in PPARδ expression following UVB radiation, while in UVA and UVB irradiated fibroblasts increases by 10% in PPARα, by 50% in PPARδ, and 2-fold in PPARγ expression were observed following sea buckthorn oil treatment (Figure 2).

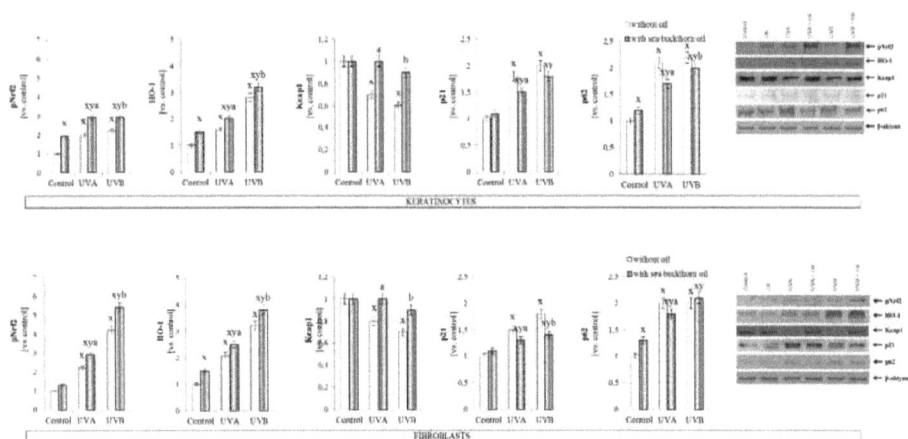

Figure 1. The expression of transcription factor phospho-Nrf2 (pSer40), its main target HO-1, and its inhibitor/activators-Keap1, p21, p62 in keratinocytes and fibroblasts after exposure to UVA (30 J/cm^2 and 20 J/cm^2) and UVB radiation (60 mJ/cm^2 and 200 mJ/cm^2, respectively) and sea buckthorn seeds oil (500 ng/mL) treatment. Mean values $\pm$ SD of five independent experiments are presented. [x] statistically significant differences vs. control group, $p < 0.05$; [y] statistically significant differences between UVA/UVB and oil treated groups vs. oil treated control group, $p < 0.05$; [a] statistically significant differences between UVA and oil treated group vs. UVA treated group, $p < 0.05$; [b] statistically significant differences between UVB and oil treated group vs. UVB treated group, $p < 0.05$.

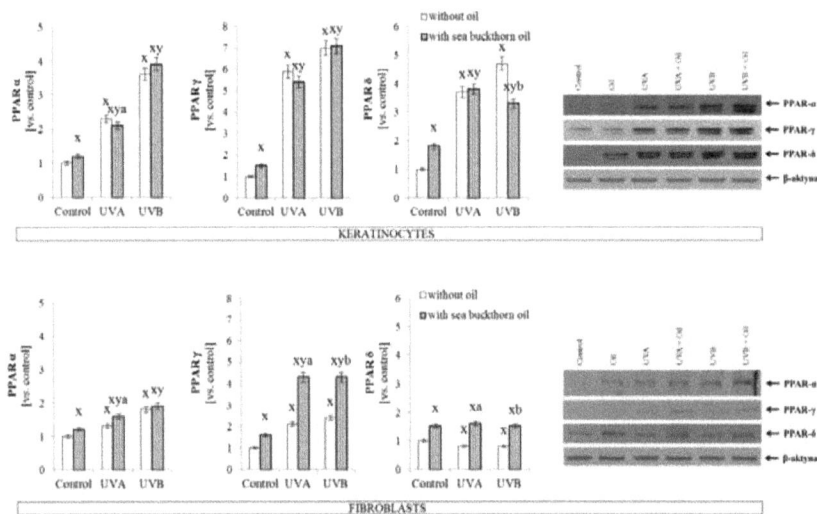

Figure 2. The expression of peroxisome proliferator-activated receptors (PPARα, γ, and δ) in keratinocytes and fibroblasts after exposure to UVA (30 J/cm^2 and 20 J/cm^2) and UVB radiation (60 mJ/cm^2 and 200 mJ/cm^2, respectively) and sea buckthorn seeds oil (500 ng/mL) treatment. Mean values $\pm$ SD of five independent experiments are presented. [x] statistically significant differences vs. control group, $p < 0.05$; [y] statistically significant differences between UVA/UVB and oil treated groups vs. oil treated control group, $p < 0.05$; [a] statistically significant differences between UVA and oil treated group vs. UVA treated group, $p < 0.05$; [b] statistically significant differences between UVB and oil treated group vs. UVB treated group, $p < 0.05$.

3.4. The Effect of Sea Buckthorn Seed Oil on Skin Cells Lipid Metabolism

The composition of sea buckthorn oil also had an impact on skin cells lipid metabolism. The greatest protective effect of this oil on membrane conditions was observed in the case of fibroblasts, where sea buckthorn oil not only reduced the permeability of the membrane compared to the control cells measured by LDH test (in the concentration rage 5 ng/mL–5 µg/mL), but also significantly prevented UVA and UVB-induced membrane damage (in the concentration rage 5 ng/mL–50 µg/mL). These effects were not observed in the case of keratinocytes (Figure 3).

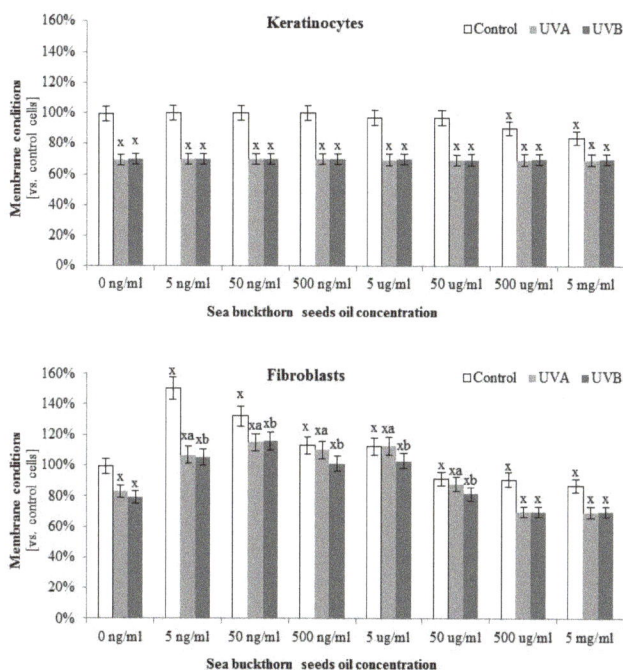

Figure 3. The effect of different concentrations ranging 50 ng/mL–5 µg/mL of sea buckthorn seed oil on the membrane conditions of control and UV irradiated keratinocytes and fibroblasts measured by LDH test. Total doses of UV irradiation for each cell line were: 30 J/cm^2 and 60 mJ/cm^2 for keratinocytes, and 20 J/cm^2 and 200 mJ/cm^2 for fibroblasts for UVA and UVB, respectively. Mean values $\pm$ SD of five independent experiments are presented. [x] statistically significant differences vs. control group, $p < 0.05$; [y] statistically significant differences between UVA/UVB and oil treated groups vs. oil treated control group, $p < 0.05$; [a] statistically significant differences between UVA and oil treated group vs. UVA treated group, $p < 0.05$; [b] statistically significant differences between UVB and oil treated group vs. UVB treated group, $p < 0.05$.

On the other hand, sea buckthorn oil treatment on skin cells significantly increased the level of all detected phospholipids as well as free fatty acids in both keratinocytes and fibroblasts (Table 4). Moreover, sea buckthorn oil influenced the UV-decreased level of numerous fatty acids, which was particularly evident in the case of keratinocytes and fibroblasts treated with oil following UVB radiation. Additionally, sea buckthorn oil treatment effects on short fatty acids (16:0; 16:1) in UVA irradiated cells was stronger in fibroblasts than in keratinocytes, while sea buckthorn oil treatment effect on the level of longer fatty acids (20:4) was visible only in the case of keratinocytes (Table 3). The Figure S1 presents the correlation between the level of free fatty acids in sea buckthorn seeds oil and changes in the level of free fatty acids in keratinocytes and fibroblasts.

Table 4. The level of phospholipid and free fatty acids in keratinocytes and fibroblasts after exposure to UVA (30 J/cm^2 and 20 J/cm^2) and UVB radiation (60 mJ/cm^2 and 200 mJ/cm^2, respectively) and sea buckthorn seeds oil (500 ng/mL) treatment.

	Oil		Keratinocytes			Fibroblasts	
		Control	UVA	UVB	Control	UVA	UVB
				Phospholipid fatty acids [ug/mg protein]			
14:0	-	5.3 ± 0.3	4.5 ± 0.2	4.5 ± 0.2	1.4 ± 0.1	1.2 ± 0.1	1.2 ± 0.1
	+	12.4 ± 0.6 [x]	11.3 ± 0.6 [xa]	11.8 ± 0.6 [xb]	6.3 ± 0.1 [x]	2.5 ± 0.1 [xya]	5.1 ± 0.3 [xyb]
16:0	-	152 ± 7.3	129 ± 6.5 [x]	128 ± 6.4 [x]	80 ± 4	68 ± 3.4	68 ± 3.4
	+	662 ± 33.1 [x]	145 ± 7.3 [y]	260 ± 13.0 [xyb]	190 ± 9.5 [x]	90 ± 4.5 [ya]	122 ± 6.1 [xyb]
16:1	-	15.6 ± 0.8	13.2 ± 0.7	13.1 ± 0.7	13.6 ± 0.7	11.6 ± 0.6	11.5 ± 0.6
	+	29.9 ± 1.5 [x]	27.5 ± 1.4 [xa]	28.5 ± 1.4 [xb]	23.2 ± 1.2 [x]	17.7 ± 0.9 [xya]	15.9 ± 0.8 [xyb]
18:0	-	123 ± 6.2	104 ± 5.2 [x]	103 ± 5.2 [x]	69 ± 3.5	58 ± 2.9 [x]	58 ± 2.9 [x]
	+	412 ± 20.6 [x]	137 ± 6.9 [xya]	187 ± 9.4 [xyb]	126 ± 6.3 [x]	109 ± 5.5 [xya]	110 ± 5.5 [xyb]
18:1nc	-	199 ± 10.1	169 ± 8.5 [x]	168 ± 8.4 [x]	112 ± 5.6	95 ± 4.8 [x]	94 ± 4.7 [x]
	+	580 ± 29.1 [x]	129 ± 6.5 [xya]	386 ± 19.3 [xyb]	181 ± 9.1 [x]	167 ± 8.4 [xa]	114 ± 5.7 [yb]
18:1nt	-	47 ± 2.4	39 ± 2.1 [x]	39 ± 2.1 [x]	18 ± 0.9	16 ± 0.8	15 ± 0.8 [x]
	+	170 ± 18.5 [x]	48 ± 2.4 [ya]	139 ± 7.1 [xyb]	58 ± 2.9 [x]	22 ± 1.1 [xya]	36 ± 1.8 [xyb]
18:2	-	116.2 ± 5.8	98.8 ± 4.9 [x]	98.1 ± 4.9 [x]	45.4 ± 2.3	38.6 ± 1.9 [x]	38.3 ± 1.9 [x]
	+	162.7 ± 8.1 [x]	111.4 ± 5.6 [ya]	137.2 ± 6.9 [xyb]	69.8 ± 3.5 [x]	60.4 ± 3.0 [xya]	64.5 ± 3.2 [xb]
18:3n3	-	24.6 ± 1.2	20.9 ± 1.1 [x]	20.8 ± 1.1 [x]	6.2 ± 0.3	5.2 ± 0.3	5.5 ± 0.3
	+	29.5 ± 1.5 [x]	30.2 ± 1.5 [xa]	31.2 ± 1.6 [xb]	12.5 ± 0.6 [x]	7.8 ± 0.4 [xya]	8.5 ± 0.4 [xyb]
20:4	-	38.1 ± 1.9	32.4 ± 1.6 [x]	32.2 ± 1.6 [x]	35.6 ± 1.8	30.2 ± 1.5	30.1 ± 1.5
	+	197.4 ± 9.9 [x]	36.4 ± 1.8 [ya]	104.7 ± 5.2 [xyb]	81.7 ± 4.1 [x]	30.3 ± 1.5 [y]	57.5 ± 2.9 [xyb]
22:6	-	11.1 ± 0.6	9.4 ± 0.5	9.3 ± 0.5	8.2 ± 0.4	6.9 ± 0.3	6.9 ± 0.3
	+	30.1 ± 1.5 [x]	18.1 ± 0.9 [xya]	23.1 ± 1.2 [xyb]	41.6 ± 2.1 [x]	16.2 ± 0.8 [xya]	33.8 ± 1.7 [xyb]
				Free fatty acids [μg/mg protein]			
16:0	-	8.8 ± 0.4	7.5 ± 0.4	7.4 ± 0.4	2.9 ± 0.1	2.4 ± 0.1	2.4 ± 0.1
	+	26 ± 1.3 [x]	14.1 ± 0.7 [xya]	19.8 ± 1 [xyb]	9.8 ± 0.5 [x]	3.9 ± 0.2 [xya]	7 ± 0.4 [xyb]
16:1	-	1.7 ± 0.1	1.4 ± 0.1	1.4 ± 0.1	0.7 ± 0.1	0.6 ± 0.1	0.6 ± 0.1
	+	5.5 ± 0.3 [x]	1.2 ± 0.1 [y]	4.4 ± 0.2 [xyb]	2.1 ± 0.1 [x]	1.3 ± 0.1 [xya]	1.6 ± 0.1 [xyb]
18:0	-	7.6 ± 0.4	6.4 ± 0.3	6.4 ± 0.3	3.4 ± 0.2	2.9 ± 0.1	2.9 ± 0.3
	+	16.8 ± 0.8 [x]	8.8 ± 0.4 [ya]	8.2 ± 0.4 [yb]	8.8 ± 0.4 [x]	3.2 ± 0.2 [y]	5.5 ± 0.3 [xyb]
18:1nc	-	13.3 ± 0.7	11.3 ± 0.6	11.2 ± 0.6	3.8 ± 0.2	3.3 ± 0.2	3.5 ± 0.2
	+	54.3 ± 2.7 [x]	51.1 ± 2.6 [xa]	45.1 ± 2.3 [xyb]	6.8 ± 0.3 [x]	4.1 ± 0.2 [ya]	5.5 ± 0.3 [xyb]
18:1nt	-	6.1 ± 0.3	5.2 ± 0.3	5.1 ± 0.3	1.1 ± 0.1	0.9 ± 0.1	0.9 ± 0.1
	+	9.8 ± 0.5 [x]	8.5 ± 0.4 [xya]	8.3 ± 0.4 [xyb]	4.5 ± 0.2 [x]	1.5 ± 0.1 [xya]	4.1 ± 0.2 [xb]
18:2	-	11.7 ± 0.6	9.9 ± 0.5	9.8 ± 0.5	3.2 ± 0.2	2.7 ± 0.1	2.7 ± 0.1
	+	15.1 ± 0.8 [x]	10.5 ± 0.5 [y]	12.7 ± 0.6 [xyb]	5.4 ± 0.3 [x]	3.3 ± 0.2 [ya]	3.3 ± 0.2 [y]
20:4	-	2.2 ± 0.1	1.8 ± 0.1	1.8 ± 0.1	0.7 ± 0.1	0.6 ± 0.1	0.6 ± 0.1
	+	9 ± 0.5 [x]	5.4 ± 0.3 [xya]	5.6 ± 0.3 [xyb]	3.4 ± 0.2 [x]	0.6 ± 0.1 [y]	2.7 ± 0.1 [xyb]
22:6	-	0.9 ± 0.1	0.8 ± 0.1	0.8 ± 0.1	0.3 ± 0.1	0.3 ± 0.1	0.3 ± 0.1
	+	11.3 ± 0.6 [x]	11.4 ± 0.6 [xa]	4.8 ± 0.2 [xyb]	2.3 ± 0.1 [x]	2.2 ± 0.1 [xa]	2.2 ± 0.1 [xb]

Mean values ± SD of five independent experiments are presented. [x] statistically significant differences vs. control group, $p < 0.05$; [y] statistically significant differences between UVA/UVB and oil treated groups vs. oil treated control group, $p < 0.05$; [a] statistically significant differences between UVA and oil treated group vs. UVA treated group, $p < 0.05$; [b] statistically significant differences between UVB and oil treated group vs. UVB treated group, $p < 0.05$.

Obtained results showed that sea buckthorn oil partially prevented UV-induced increase in phospholipase A2 activity by 50% in the case of UV irradiated keratinocytes, and by about 15% in the case of UV irradiated fibroblasts (Figure 4). As a result of UV-induced lipid oxidative metabolism a strong increase in the level of lipid peroxidation products was observed. Sea buckthorn oil treatment significantly prevented these changes in the case of 4-HNE levels, which was decreased by 15% in oil treated keratinocytes, and 30% in oil treated fibroblasts. Also, the level of isoprostaglandin in fibroblasts exposed to UV radiation was decreased by 25% following sea buckthorn oil treatment. However, in the case of isoprostaglandin levels in keratinocytes sea buckthorn oil lead to increases by approximately 20% under standard conditions and did not have a significant effect on irradiated cells (Figure 4).

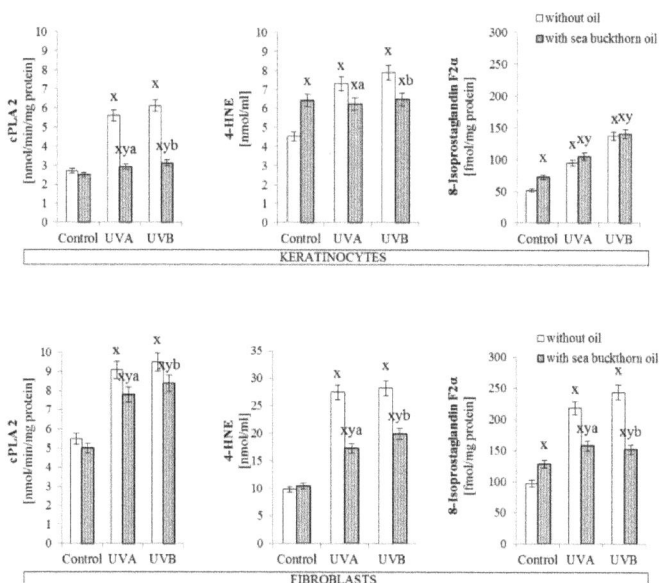

Figure 4. The activity of phospholipase A2 and level of lipid peroxidation products (4-HNE and 8-Isoprostaglandin F2α) in keratinocytes and fibroblasts after exposure to UVA (30 J/cm^2 and 20 J/cm^2) and UVB radiation (60 mJ/cm^2 and 200 mJ/cm^2, respectively) and sea buckthorn seeds oil (500 ng/mL) treatment. Mean values ± SD of five independent experiments are presented. ^{x} statistically significant differences vs. control group, $p < 0.05$; ^{y} statistically significant differences between UVA/UVB and oil treated groups vs. oil treated control group, $p < 0.05$; ^{a} statistically significant differences between UVA and oil treated group vs. UVA treated group, $p < 0.05$; ^{b} statistically significant differences between UVB and oil treated group vs. UVB treated group, $p < 0.05$.

Sea buckthorn oil treatment on UV irradiated skin cells was found as a factor influencing the endocannabinoid system. In both cell lines a significant increase in endocannabinoid levels were observed, however, keratinocytes exhibited greater susceptibility to oil treatment (approximately 3-fold and 7-fold increase in the case of AEA; and 2-fold and 3-fold increase in the case of 2-AG, compared to UVA and UVB irradiated cells, respectively, were observed). Moreover, endocannabinoid receptor expression was decreased by half in relation to UV irradiated keratinocytes. Parallel, in fibroblasts sea buckthorn oil following UV irradiation caused increases in the endocannabinoid levels by 25–30% compared to UVA or UVB irradiated cells. Also, the expression of endocannabinoid receptor CB1 was decreased by 25% compared to UVA or UVB irradiated cells. However, in the case of receptor CB2, sea buckthorn oil treatment decreased its expression only in fibroblasts exposed to UVB radiation (Figure 5).

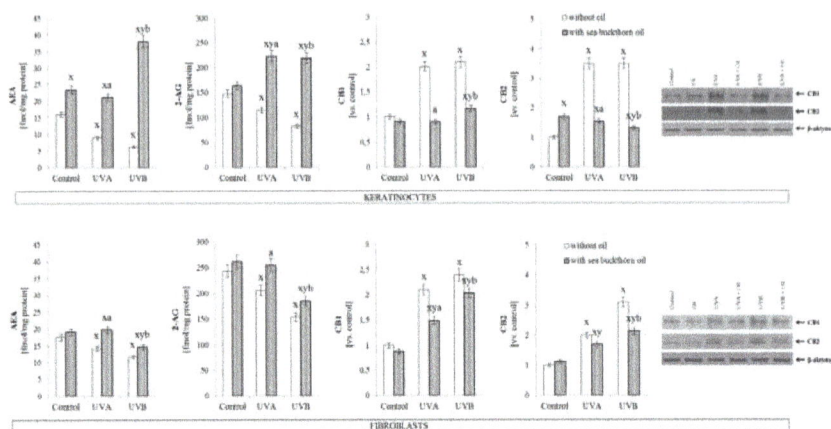

Figure 5. The level of endocannabinoids (AEA, 2-AG) and endocannabinoid receptors (CB1, CB2) in keratinocytes and fibroblasts after exposure to UVA (30 J/cm^2 and 20 J/cm^2) and UVB radiation (60 mJ/cm^2 and 200 mJ/cm^2, respectively) and sea buckthorn seeds oil (500 ng/mL) treatment. Mean values $\pm$ SD of five independent experiments are presented. [x] statistically significant differences vs. control group, $p < 0.05$; [y] statistically significant differences between UVA/UVB and oil treated groups vs. oil treated control group, $p < 0.05$; [a] statistically significant differences between UVA and oil treated group vs. UVA treated group, $p < 0.05$; [b] statistically significant differences between UVB and oil treated group vs. UVB treated group, $p < 0.05$.

4. Discussion

Since the major components of the outer skin layer are lipids, it can be assumed that the use of lipid derivatives will enhance skin conditions. Therefore, plant oils can be used as a source of compounds that may protect skin cells due to their multidirectional activities, including anti-inflammatory, antibacterial and antioxidants, as well as the source of fatty acids [43]. Sea buckthorn (*Hippophae rhamnoides*) seed oil has been traditionally used for accelerating wound healing [44] as well as treating skin diseases such as the first stage of atopic dermatitis [45], however it is not widely used in skin protection [46].

4.1. Sea Buckthorn Seeds Oil Effect on Antioxidant System

Due to the high content of polyphenolic compounds and vitamins, the sea buckthorn extract has been found as a antimicrobial factor, with strong antioxidant activity in in vitro examinations [47]. Moreover, it has been suggested that sea buckthorn seed oil may reduce ROS levels under oxidative conditions by its ability to scavenge free radicals [48].

This study indicates the effectiveness of the antioxidant actions of sea buckthorn oil in relation to the oxidative conditions found in skin cells exposed to UV radiation. However, in physiological conditions it is not so unambiguous, because by increasing the activity of cytosolic enzymes such as NADPH oxidase and xanthine oxidase this oil contributes to the intensification of ROS generation. After UV radiation, sea buckthorn oil promotes reduction of the above enzyme activity and consequently ROS generation in fibroblasts. This may be the result of a stronger fibroblast susceptibility to external factors including oxidative factors such as UVA and UVB radiation and for antioxidative factors in oils, which was also indicated for pomegranate seed oil [49]. It has been suggested that high content of unsaturated acids in plant oils are substrates for cellular oxygenases which can be a source of oxidation products in skin cells [50].

Independent of modifications of prooxidative conditions, sea buckthorn oil promotes increases in antioxidant enzyme activity and non-enzymatic antioxidant levels in control cells as well as cells treated with UV irradiation. Results of this study partially confirm earlier observations that sea buckthorn oil may lead to enhanced antioxidant abilities by promotion of glutathione accumulation in the whole animal body in the case of rats treated with this oil [51]. It was also shown that supplementation of rats with compounds of sea buckthorn extracts activate the enzymes superoxide dismutase, catalase, glutathione peroxidase, glutathione reductase and glutathione S-transferase in animal blood [52]. However, in this study sea buckthorn oil shows the tendency to decrease fibroblast Cu, Zn-SOD activity in control and UV-irradiated cells. This may be explained by the sea buckthorn oils ability to capture electrons and extinguish the singlet oxygen or superoxide anion [53] that is the main substrate of SOD in cell cytoplasm. Another important cellular redox regulator is the thioredoxin system consisting of thioredoxin and thioredoxin reductase (TrxR) [54]. This expression is diminished in skin cells after UV irradiation and enhanced after sea buckthorn oil treatment of keratinocytes and fibroblasts. Previously it has been shown that this system plays a significant role in the pathogenesis of a number of diseases, but it also participates in cellular protection against toxic compounds, which are reduced by thioredoxin or inhibited by thioredoxin reductase [55,56]. It has been indicated that sea buckthorn includes high levels of proteins containing thioredoxin domains and prevents the disulfide bond formation in oxidizing environments and stabilizing the tertiary and quaternary structure of proteins [57]. It has also been shown that Trx and TrxR synthesis is induced after exposure to prooxidative factors [58], this explains the strong positive response of this system to UV irradiation, particularly in fibroblasts. Therefore, enhanced effectiveness of this system after oil treatment may explain reduction of oxidative stress in these cells.

Cellular antioxidant responses dependent on proteins is promoted by the transcription factor Nrf2 which is responsible for cytoprotective gene transcription [59]. Under physiological conditions, cytoplasmic Nrf2 is bound to Keap1 [60], but oxidative stress caused by UV radiation was found to diminish Keap1 expression which prevents Nrf2–Keap1–Cul3 complex formation. Moreover, as observed in this paper, high levels of ROS and electrophiles such as 4-HNE may lead to the oxidation of Keap1 cysteine residues causing lack of binding and/or dissociation of Nrf2 from the complex resulting in its translocation into the nucleus, which was reported earlier [61]. Nrf2 activation, demonstrated also as its targeted genes-HO-1 expression, leads to cellular protection against pro-oxidative conditions. Sea buckthorn oil treatment of UV irradiated skin cells acts in two ways—it supports the cellular antioxidant capacity by the activation of Nrf2, and at the same time, affects the expression of proteins strongly related with Nrf2 activation. The results showing enhanced phospho-Nrf2 translocation to the nucleus despite of the increased Keap1 level in cytoplasm clearly suggest that sea buckthorn oil disrupts the Nrf2–Keap1 complex formation. Moreover, sea buckthorn seed oil decreases UV-induced levels of p21 and p62, reducing the possibility of these proteins to create adducts with Nrf2 or Keap1, thus encouraging Nrf2 inhibition [59,62]. Keap1 might also create adducts with small antioxidant molecules such as melatonin or GSH, what also prevents Nrf2-Keap1 binding, however, it does not lead to Keap1 degradation [63,64]. Melatonin has been previously shown to be downregulated in skin cells following UV radiation [65,66], while UV-induced decrease in GSH level, what is shown in this study is significantly enhanced as a result of sea buckthorn oil treatment. These modifications to the activators and inhibitors of the Nrf2 pathway are not able to cancel the protective effect of sea buckthorn oil to counteract prooxidative conditions.

4.2. Sea Buckthorn Seeds Oil Effect on Lipid Metabolism

The effects of sea buckthorn seed oil on redox balance can directly protect the metabolism of skin cells. Maintaining redox homeostasis is important for the physiological metabolism of lipids, which is important for the proper functioning of cells. Phospholipid protection may be associated with beneficial fatty acids present in sea buckthorn oil [67], which influences the skin cells fatty acid profile. The strongest correlation between oil fatty acids and increase in the cells fatty acid level is visible in the case of free fatty acids (FFA) regardless of whether that were control cells or treated with UV radiation. This relationship between FFA composition of the applied oil and cells might be associated with 24 h incubation with oil when cells were able only to take up and assimilate FFA from fatty rich oil composition. Sea buckthorn oil is rich in palmitoleic acid (16:1), which has been reported to play a role in many metabolic processes including intracellular lipid mediated signal transduction [68,69], which includes lipid metabolism, and is also responsible for maintaining the fluidity of biological membranes [70–72]. Palmitoleic acid levels are enhanced in the phospholipid profile of keratinocytes and fibroblasts after sea buckthorn oil treatment, this also causes the reduction of mRNA expression of proinflammatory genes, i.e., TNF-α [73], therefore this oil may affect intracellular signaling based on PI3K/Akt kinase cascades, and have anti-inflammatory activity, which was also confirmed by experiments on rats with diabetes [74]. Moreover, sea buckthorn oil contains a large amount of linoleic and α-linolenic acids, which can enhance the level of phospholipid α-linolenic acid in cells [75,76]. α-Linolenic acid contained in sea buckthorn oil has been found also as a source of eicosanoid prostaglandin E1, a signal precursor that produces antibacterial and cytoprotective extracellular fluids [77]. Shown in this study sea buckthorn oil added to skin cells enhances the level of these acids in phospholipid and free PUFAs fractions. Skin cells phospholipid PUFAs fraction is also enriched in γ-linolenic acid that is a precursor of anti-inflammatory eicosanoids, such as the 1-series prostaglandins and 15-hydroxyeicosatrienoic acid (15-HETrE) [78].

It has been previously shown that UV radiation significantly disturbs the metabolism of skin cell membrane phospholipids [6]. In this study, active compounds contained in sea buckthorn oil enhances the cell's antioxidant abilities and prevents lipid peroxidation [47] as well as enzymatic phospholipid metabolism. Sea buckthorn oil enhances, both phospholipid and free PUFA levels in keratinocytes and fibroblasts, as well as increases the ROS level favored by ROS-dependent lipid peroxidation manifested as oxidative fragmentation with enhanced 4-HNE levels and as oxidative cyclisation with enhanced 8-isoprostane levels. Independently of that, one of the main sources of peroxidation products-arachidonic acid (AA) is most enhanced in PUFAs of both skin cell lines, what indicates for huge tributary of this acid from used oil. As a result of AA peroxidation, increases in the 4-HNE levels can directly act as a signaling molecule or through protein adduct formation which significantly influences their structure and activity [79]. Phospholipid AA is also metabolized by enzymes among which the most important are phospholipases including PLA2 [80]. It has been previously shown that UV radiation significantly increases PLA2 activity [9,81], while PLA2 inhibition improves skin conditions [82]. Sea buckthorn oil reveals similar as other plant oils PLA2 inhibition properties, preventing UV-induced lipid metabolism [83]. Despite the observed decrease in PLA2 activity, the levels of endocannabinoids are increased after using sea buckthorn oil. It is believed that anandamide is a partial or full agonist of the CB1 receptor, depending on tissue and conditions, and is suggested to have low efficacy for CB2 receptors, whereas 2-AG is a full agonist of both CB1 and CB2 receptors [84]. However, an elevated level of endocannabinoids is accompanied by a down-regulation of cannabinoid receptors. Such response may indicate that redox and inflammatory regulation is independent from the cannabinoid receptors. Moreover, oil-induced changes in endocannabinoids level may influence the skin neuroendocrine capabilities regulated by UV radiation [85], what leads to disorders in steroid hormones, neuropeptides, and neurotransmitters biosynthesis [86]. Regardless of the above, it has been shown that a high level of AA delivered to skin cells from sea buckthorn oil may result in increased generation of the 4-series leukotrienes, which have a strong pro-inflammatory and hyperproliferative effect [87].

Endocannabinoid levels are enhanced by treating skin cells with sea buckthorn oil, which are agonists of peroxisome proliferator-activated receptors (PPAR). It is known that PPARs are activated by fatty acids and their derivatives, including lipid peroxidation products like 4-HNE, which act as PPAR-α agonists [88,89]. PPARs act as modulators of cellular processes including lipid metabolism, and thus create a lipid signaling network between the cell surface and the nucleus [89]. Enhanced expression of fibroblast PPARs indicates that sea buckthorn seed oil has anti-inflammatory activity. PPAR-α controls the expression of proteins that participate in inflammatory response [89], therefore enhanced activation of PPAR-α observed in fibroblasts indicates preventing NF-κB-dependent inflammation [90]. It has also been shown that anandamide as a ligand of PPAR-α can participate in its anti-inflammatory effect through impaired production of TNF-α [91]. Similarly, enhanced expression of PPAR-γ decreases the expression of TNF-α [92]. Moreover, sea buckthorn ethanolic extract has been beneficial in reducing fat pad mass and preventing weight gain in mice. The extract was effective in producing hypoglycemic effects in animals through up-regulating PPAR-γ and PPAR-α gene expression [92].

5. Conclusions

In summary, sea buckthorn oil significantly stimulates the antioxidant system in keratinocytes and fibroblasts. Therefore, sea-buckthorn seed oil prevents UV-induced impair in redox systems as well as lipid metabolism disorders in skin fibroblasts and keratinocytes, which makes it a promising natural substance in skin photo-protection. However, the influence of UV radiation on the stability and durability of oil components, their interactions and impact on the metabolism of skin cells has not been studied, therefore it is believed that current studies do not allow recommending the use of sea buckthorn seed oil in direct exposure to UV radiation.

Supplementary Materials: The following are available online at http://www.mdpi.com/2076-3921/7/9/110/s1, Figure S1: The correlation between the level of free fatty acids in sea buckthorn seeds oil and changes in the level of free fatty acids in keratinocytes and fibroblasts after exposure to UVA (30 J/cm^2 and 20 J/cm^2) and UVB radiation (60 mJ/cm^2 and 200 mJ/cm^2, respectively) and sea buckthorn seeds oil (500 ng/mL) treatment., Figure S2: The electrophorogram images of Western blot analyses of phospho-Nrf2 (pSer40) in keratinocytes and fibroblasts after exposure to UVA (30 J/cm^2 and 20 J/cm^2) and UVB radiation (60 mJ/cm^2 and 200 mJ/cm^2, respectively) and sea buckthorn seeds oil (500 ng/mL) treatment.

Author Contributions: Conceptualization, E.S.; Formal analysis, A.G., A.J., I.J.K. and M.M.; Investigation, A.G., A.J., I.J.K. and M.M.; Methodology, A.G., A.J., I.J.K. and M.M.; Validation, A.G., A.J., I.J.K. and M.M.; Visualization, A.G., A.J., I.J.K. and M.M.; Writing—Review & editing, E.S.

Funding: This research received no external funding.

Conflicts of Interest: The authors declare that they have no conflict of interests.

References

1. Chuong, C.M.; Nickoloff, B.J.; Elias, P.M.; Goldsmith, L.A.; Macher, E.; Maderson, P.A.; Sunberg, J.P.; Tagami, H.; Plonka, P.M.; Thestrup-Pederson, K.; et al. What is the 'ture' function of skin? *Exp. Dermatol.* **2002**, *11*, 159–187. [PubMed]

2. Agache, P.; Lihoreau, T.; Mac-Mary, S.; Fanian, F.; Humbert, P. The human skin: An overview. In *Agache's Measuring the Skin: Non-invasive Investigations*; Springer International Publishing AG: Cham, Switzerland, 2017; pp. 1–4.

3. Moan, J.; Grigalavicius, M.; Baturaite, Z.; Dahlback, A.; Juzeniene, A. The relationship between UV exposure and incidence of skin cancer. *Photodermatol. Photoimmunol. Photomed.* **2015**, *31*, 26–35. [CrossRef] [PubMed]

4. Natarajan, V.T.; Ganju, P.; Ramkumar, A.; Grover, R.; Gokhale, R.S. Multifaceted pathways protect human skin from UV radiation. *Nat. Chem. Biol.* **2014**, *10*, 542–551. [CrossRef] [PubMed]

5. Slominski, A.T.; Zmijewski, M.A.; Plonka, P.M.; Szaflarski, J.P.; Paus, R. How UV Light Touches the Brain and Endocrine System Through Skin, and Why. *Endocrinology* **2018**, *159*, 1992–2007. [CrossRef] [PubMed]

6. Gęgotek, A.; Biernacki, M.; Ambrożewicz, E.; Surażyński, A.; Wroński, A.; Skrzydlewska, E. The cross-talk between electrophiles, antioxidant defence and the endocannabinoid system in fibroblasts and keratinocytes after UVA and UVB irradiation. *J. Dermatol. Sci.* **2016**, *81*, 107–117. [CrossRef] [PubMed]

7. Mohamed, M.A.A.; Jung, M.; Lee, S.M.; Lee, T.H.; Kim, J. Protective effect of Disporum sessile D. Don extract against UVB-induced photoaging via suppressing MMP-1 expression and collagen degradation in human skin cells. *J. Photochem. Photobiol. B* **2014**, *133*, 73–79. [CrossRef] [PubMed]

8. Larroque-Cardoso, P.; Camaré, C.; Nadal-Wollbold, F.; Grazide, M.H.; Pucelle, M.; Garoby-Salom, S.; Bogdanowicz, P.; Josse, G.; Schmitt, A.M.; Uchida, K.; et al. Elastin modification by 4-hydroxynonenal in hairless mice exposed to UV-A. Role in photoaging and actinic elastosis. *J. Investig. Dermatol.* **2015**, *135*, 1873–1881. [CrossRef] [PubMed]

9.   Gęgotek, A.; Bielawska, K.; Biernacki, M.; Dobrzyńska, I.; Skrzydlewska, E. Time-dependent effect of rutin on skin fibroblasts membrane disruption following UV radiation. *Redox Biol.* **2017**, *12*, 733–744. [CrossRef] [PubMed]

10.   Gęgotek, A.; Bielawska, K.; Biernacki, M.; Zaręba, I.; Surażyński, A.; Skrzydlewska, E. Comparison of protective effect of ascorbic acid on redox and endocannabinoid systems interactions in in vitro cultured human skin fibroblasts exposed to UV radiation and hydrogen peroxide. *Arch. Dermatol. Res.* **2017**, *309*, 285–303. [CrossRef] [PubMed]

11.   Lutz, B.; Marsicano, G.; Maldonado, R.; Hillard, C.J. The endocannabinoid system in guarding against fear, anxiety and stress. *Nat. Rev. Neurosci.* **2015**, *16*, 705–718. [CrossRef] [PubMed]

12.   Gasperi, V.; Dainese, E.; Oddi, S.; Sabatucci, A.; Maccarrone, M. GPR55 and its interaction with membrane lipids: Comparison with other endocannabinoid-binding receptors. *Curr. Med. Chem.* **2013**, *20*, 64–78. [CrossRef] [PubMed]

13.   Masaki, H. Role of antioxidants in the skin: Anti-aging effects. *J. Dermatol. Sci.* **2010**, *58*, 85–90. [CrossRef] [PubMed]

14.   Kamal-Eldin, A. Effect of fatty acids and tocopherols on the oxidative stability of vegetable oils. *Eur. J. Lipid. Sci. Technol.* **2006**, *108*, 1051–1061. [CrossRef]

15.   Suryakumar, G.; Gupta, A. Medicinal and therapeutic potential of Sea buckthorn (*Hippophae rhamnoides* L.). *J. Ethnopharmacol.* **2011**, *138*, 268–278. [CrossRef] [PubMed]

16.   Ting, H.C.; Hsu, Y.W.; Tsai, C.F.; Lu, F.J.; Chou, M.C.; Chen, W.K. The in vitro and in vivo antioxidant properties of seabuckthorn (*Hippophae rhamnoides* L.) seed oil. *Food Chem.* **2011**, *125*, 652–659. [CrossRef]

17.   Zheng, L.; Shi, L.K.; Zhao, C.W.; Jin, Q.Z.; Wang, X.G. Fatty acid, phytochemical, oxidative stability and in vitro antioxidant property of sea buckthorn (*Hippophaë rhamnoides* L.) oils extracted by supercritical and subcritical technologies. *LWT-Food Sci. Technol.* **2017**, *86*, 507–513. [CrossRef]

18.   Teleszko, M.; Wojdyło, A.; Rudzińska, M.; Oszmiański, J.; Golis, T. Analysis of lipophilic and hydrophilic bioactive compounds content in sea buckthorn (*Hippophae rhamnoides* L.) berries. *J. Agric. Food Chem.* **2015**, *63*, 4120–4129. [CrossRef] [PubMed]

19.   Gupta, S.M.; Gupta, A.K.; Ahmed, Z.; Kumar, A. Antibacterial and antifungal activity in leaf, seed extract and seed oil of seabuckthorn (*Hippophae salicifolia* D. Don) plant. *J. Plant Pathol. Microbiol.* **2011**, *2*, 1–4.

20.   Punia, D.; Kumari, N. Potential health benefits of Sea buckthorn oil—A review. *Agric. Rev.* **2017**, *38*, 233–237.

21.   Hou, D.D.; Di, Z.H.; Qi, R.Q.; Wang, H.X.; Zheng, S.; Hong, Y.X.; Guo, H.; Chen, H.D.; Gao, X.H. Sea Buckthorn (*Hippophaë rhamnoides* L.) Oil Improves Atopic Dermatitis-Like Skin Lesions via Inhibition of NF-κB and STAT1 Activation. *Skin Pharmacol. Physiol.* **2017**, *30*, 268–276. [CrossRef] [PubMed]

22.   Nang Lau, H.L.; Puah, C.W.; Choo, Y.M.; Ma, A.N.; Chuah, C.H. Simultaneous quantification of free fatty acids, free sterols, squalene, and acylglycerol molecular species in palm oil by high-temperature gas chromatography-flame ionization detection. *Lipids* **2005**, *40*, 523–528. [CrossRef]

23.   Czaplicki, S.; Ogrodowska, D.; Derewiaka, D.; Tańska, M.; Zadernowski, R. Bioactive compounds in unsaponifiable fraction of oils from unconventional sources. *Eur. J. Lipid Sci. Technol.* **2011**, *113*, 1456–1464. [CrossRef]

24.   Zhao, B.; Tham, S.Y.; Lu, J.; Lai, M.H.; Lee, L.K.; Moochhala, S.M. Simultaneous determination of vitamins C, E and β-carotene in human plasma by high-performance liquid chromatography with photodiode-array detection. *J. Pharm. Sci.* **2004**, *7*, 200–204.

25.   Fotakis, G.; Timbrell, J.A. In vitro cytotoxicity assays: Comparison of LDH, neutral red, MTT and protein assay in hepatoma cell lines following exposure to cadmium chloride. *Toxicol. Lett.* **2006**, *160*, 171–177. [CrossRef] [PubMed]

26.   Kuzkaya, N.; Weissmann, N.; Harrison, D.G.; Dikalov, S. Interactions of peroxynitrite, tetrahydrobiopterin, ascorbic acid, and thiols: Implications for uncoupling endothelial nitricoxide synthase. *J. Biol. Chem.* **2003**, *278*, 22546–22554. [CrossRef] [PubMed]

27.   Griendling, K.K.; Minieri, C.A.; Ollerenshaw, J.D.; Alexander, R.W. Angiotensin II stimulates NADH and NADPH oxidase activity in cultured vascular smooth muscle cells. *Circ. Res.* **1994**, *74*, 1141–1148. [CrossRef] [PubMed]

28.   Prajda, N.; Weber, G. Malignant transformation-linked imbalance: Decreased xanthine oxidase activity in hepatomas. *FEBS. Lett.* **1975**, *59*, 245–259. [CrossRef]

29. Paglia, D.E.; Valentine, W.N. Studies on the quantitative and qualitative characterization of erythrocyte glutathione peroxidase. *J. Lab. Clin. Med.* **1967**, *70*, 158–169. [PubMed]

30. Mize, C.E.; Longdon, R.G. Hepatic glutathione reductase. Purification and general kinetic properties. *J. Biol. Chem.* **1962**, *237*, 1589–1595. [PubMed]

31. Misra, H.P.; Fridovich, I. The role of superoxide anion in the autoxidation of epinephrine and a simple assay for superoxide dismutase. *J. Biol. Chem.* **1972**, *247*, 3170–3175. [PubMed]

32. Sykes, J.A.; McCormac, F.X.; O'Breien, T.J. Preliminary study of the superoxide dismutase content of some human tumors. *Cancer Res.* **1978**, *38*, 2759–2762. [PubMed]

33. Holmgren, A. Thioredoxin and thioredoxin reductase. *Methods Enzymol.* **1995**, *252*, 199–208. [PubMed]

34. Lovell, M.A.; Xie, C.; Gabbita, S.P.; Markesbery, W.R. Decreased thioredoxin and increased thioredoxin reductase levels in Alzheimer's disease brain. *Free Radic. Biol. Med.* **2000**, *28*, 418–427. [CrossRef]

35. Maeso, N.; Garcia-Martinez, D.; Ruperez, F.J.; Cifuentes, A.; Barbas, C. Capillary electrophoresis of glutathione to monitor oxidative stress and response to antioxidant treatments in an animal model. *J. Chromatogr. B* **2005**, *822*, 61–69. [CrossRef] [PubMed]

36. Vatassery, G.T.; Brin, M.F.; Fahn, S.; Kayden, H.J.; Traber, M.G. Effect of high doses of dietary vitamin E on the concentrations of vitamin E in several brain regions, plasma, liver, and adipose tissue of rats. *J. Neurochem.* **1988**, *51*, 621–623. [CrossRef] [PubMed]

37. Eissa, S.; Seada, L.S. Quantitation of bcl-2 protein in bladder cancer tissue by enzyme immunoassay: Comparison with Western blot and immunohistochemistry. *Clin. Chem.* **1998**, *44*, 1423–1429. [PubMed]

38. Christie, W.W. Preparation of ester derivatives of fatty acids for chromatographic analysis. In *Advances in Lipid Methodology*; Oily Press: Dundee, UK, 1993; pp. 69–111.

39. Reynolds, L.J.; Hughes, L.L.; Yu, L.; Dennis, E.A. 1-Hexadecyl-2-arachidonoylthio-2-deoxy-sn-glycero-3-phosphorylcholine as a substrate for the microtiterplate assay of human cytosolic phospholipase A_2. *Anal. Biochem.* **1994**, *217*, 25–32. [CrossRef] [PubMed]

40. Luo, X.P.; Yazdanpanah, M.; Bhooi, N.; Lehotay, D.C. Determination of aldehydes and other lipid peroxidation products in biological samples by gas chromatography-mass spectrometry. *Anal. Biochem.* **1995**, *228*, 294–298. [CrossRef] [PubMed]

41. Coolen, S.A.; van Buuren, B.; Duchateau, G.; Upritchard, J.; Verhagen, H. Kinetics of biomarkers: Biological and technical validity of isoprostanes in plasma. *Amino Acids* **2005**, *29*, 429–436. [CrossRef] [PubMed]

42. Lam, P.M.; Marczylo, T.H.; El-Talatini, M.; Finney, M.; Nallendran, V.; Taylor, A.H.; Konje, J.C. Ultra-performance liquid chromatography tandem mass spectrometry method for the measurement of anandamide in human plasma. *Anal. Biochem.* **2008**, *380*, 195–201. [CrossRef] [PubMed]

43. Bakry, A.M.; Abbas, S.; Ali, B.; Majeed, H.; Abouelwafa, M.Y.; Mousa, A.; Liang, L. Microencapsulation of oils: A comprehensive review of benefits, techniques, and applications. *Compr. Rev. Food Sci. Food Saf.* **2016**, *15*, 143–182. [CrossRef]

44. Gendaszewska-Darmach, E.; Majewska, I. Proangiogenic activity of plant extract in accelerating wound healing—A new face of old phytomedicines. *Acta Biochim. Pol.* **2011**, *58*, 449–460.

45. Yang, B.; Kalimo, K.O.; Tahvonen, R.L.; Mattila, L.M.; Katajisto, J.K.; Kallio, H.P. Effect of dietary supplementation with sea buckthorn (*Hippophae rhamnoides*) seed and pulp oils on the fatty acid composition of skin glycerophospholipids of patients with atopic dermatitis. *J. Nutr. Biochem.* **2000**, *11*, 338–340. [CrossRef]

46. Zeb, A. Important therapeutic uses of sea buckthorn (Hippophae): A review. *J. Biol. Sci.* **2004**, *4*, 687–693.

47. Michel, T.; Destandau, E.; Le Flochc, G.; Lucchesi, M.E.; Elfakir, C. Antimicrobial, antioxidant and phytochemical investigations of sea buckthorn (*Hippophae rhamnoides* L.) leaf, stem, root and seed. *Food Chem.* **2012**, *131*, 754–760. [CrossRef]

48. Kumar, M.S.Y.; Dutta, R.; Prasad, D.; Mishra, K. Subcritical water extraction of antioxidant compounds from Seabuckthorn (*Hippophae rhamnoides* L.) leaves for the comparative evaluation of antioxidant activity. *Food Chem.* **2011**, *127*, 1309–1316. [CrossRef] [PubMed]

49. Baccarin, T.; Mitjans, M.; Ramos, D.; Lemos-Senna, E.; Vinardell, M.P. Photoprotection by Punica granatum seed oil nanoemulsion entrapping polyphenol-rich ethyl acetate fraction against UVB-induced DNA damage in human keratinocyte (HaCaT) cell line. *J. Photochem. Photobiol. B* **2015**, *153*, 127–136. [CrossRef] [PubMed]

50. Manku, M.S.; Horrobin, D.F.; Morse, N.L.; Wright, S.; Burton, J.L. Essential fatty acids in the plasma phospholipids of patients with atopic eczema. *Br. J. Dermatol.* **1984**, *110*, 643–648. [CrossRef] [PubMed]

51. Ganju, L.; Padwad, Y.; Singh, R.; Karan, D.; Chanda, S.; Chopra, M.K.; Bhatnagar, P.; Kashyap, R.; Sawhney, R.C. Anti-inflammatory activity of Seabuckthorn (*Hippophae rhamnoides*) leaves. *Int. Immunopharmacol.* **2005**, *5*, 1675–1684. [CrossRef] [PubMed]

52. Kumar, M.S.Y.; Maheshwari, D.T.; Verma, S.K.; Singh, V.K.; Singh, S.N. Antioxidant and hepatoprotectice activities of phenolic rich fraction of Seabuckthorn (*Hippophae rhamnoides* L.) leaves. *Food Chem. Toxicol.* **2011**, *49*, 2422–2428.

53. Myong-Jo, K.; Ju-Sung, K.; Yong-Soo, K.; Yeo-Jin, S. Isolation and identification of Sea buckthorn (*Hippophae rhamnoides*) phenolics with antioxidant activity and α-glucosidase inhibitory effect. *J. Agric. Food Chem.* **2011**, *59*, 138–144.

54. Schallreuter, K.U.; Wood, J.M. The role of thioredoxin reductase in the reduction of free radicals at the surface of the epidermis. *Biochem. Biophys. Res. Commun.* **1986**, *136*, 630–637. [CrossRef]

55. Arner, E.S.; Holmgren, A. The thioredoxin system in cancer. In *Seminars in Cancer Biology*; Academic Press: Orlando, FL, USA, 2006.

56. Carvalho, C.M.; Chew, E.H.; Hashemy, S.I.; Lu, J.; Holmgren, A. Inhibition of the human thioredoxin system a molecular mechanism of mercury toxicity. *J. Biol. Chem.* **2008**, *283*, 11913–11923. [CrossRef] [PubMed]

57. Sougrakpam, Y.; Deswal, R. Hippophae rhamnoides N-glycoproteome analysis: A small step towards sea buckthorn proteome mining. *Physiol. Mol. Biol. Plants* **2016**, *22*, 473–484. [CrossRef] [PubMed]

58. Gęgotek, A.; Rybałtowska-Kawałko, P.; Skrzydlewska, E. Rutin as a mediator of lipid metabolism and cellular signaling pathways interactions in fibroblasts altered by UVA and UVB radiation. *Oxid Med. Cell Longev.* **2017**, *2017*, 1–20. [CrossRef] [PubMed]

59. Gęgotek, A.; Skrzydlewska, E. The role of transcription factor Nrf2 in skin cells metabolism. *Arch. Dermatol. Res.* **2015**, *307*, 385–396. [CrossRef] [PubMed]

60. Konstantinopoulos, P.A.; Spentzos, D.; Fountzilas, E.; Francoeur, N.; Sanisetty, S.; Grammatikos, A.P.; Hecht, J.L.; Cannistra, S.A. Keap1 mutations and Nrf2 pathway activation in epithelial ovarian cancer. *Cancer Res.* **2011**, *71*, 5081–5089. [CrossRef] [PubMed]

61. Huang, Y.; Li, W.; Kong, A.N. Anti-oxidative stress regulator NF-E2-related factor 2 mediates the adaptive induction of antioxidant and detoxifying enzymes by lipid peroxidation metabolite 4-hydroxynonenal. *Cell Biosci.* **2012**, *2*, 40–47. [CrossRef] [PubMed]

62. Itoh, K.; Wakabayashi, N.; Katoh, Y.; Ishii, T.; Igarashi, K.; Engel, J.D.; Yamamoto, M. Keap1 represses nuclear activation of antioxidant responsive elements by Nrf2 through binding to the amino-terminal Neh2 domain. *Genes. Dev.* **1999**, *13*, 76–86. [CrossRef] [PubMed]

63. Abiko, Y.; Kumagai, Y. Interaction of Keap1 modified by 2-tert-butyl-1, 4-benzoquinone with GSH: Evidence for S-transarylation. *Chem. Res. Toxicol.* **2013**, *26*, 1080–1087. [CrossRef] [PubMed]

64. Ding, K.; Wang, H.; Xu, J.; Li, T.; Zhang, L.; Ding, Y.; Zhu, L.; He, J.; Zhou, M.L. Melatonin stimulates antioxidant enzymes and reduces oxidative stress in experimental traumatic brain injury: The Nrf2–ARE signaling pathway as a potential mechanism. *Free Radic. Biol. Med.* **2014**, *73*, 1–11. [CrossRef] [PubMed]

65. Fischer, T.W.; Sweatman, T.W.; Semak, I.; Sayre, R.M.; Wortsman, J.; Slominski, A. Constitutive and UV-induced metabolism of melatonin in keratinocytes and cell-free systems. *FASEB J.* **2006**, *20*, 1564–1566. [CrossRef] [PubMed]

66. Janjetovic, Z.; Jarrett, S.G.; Lee, E.F.; Duprey, C.; Reiter, R.J.; Slominski, A.T. Melatonin and its metabolites protect human melanocytes against UVB-induced damage: Involvement of NRF2-mediated pathways. *Sci. Rep.* **2014**, *7*, 1274. [CrossRef] [PubMed]

67. Ranjith, A.; Kumar, K.S.; Venugopalan, V.V.; Arumughan, C.; Sawhney, R.C.; Singh, V. Fatty acids, tocols, and carotenoids in pulp oil of three sea buckthorn species (*Hippophae rhamnoides*, H. *salicifolia*, and H. *tibetana*) grown in the Indian Himalayas. *J. Am. Oil Chem. Soc.* **2006**, *83*, 359–364. [CrossRef]

68. Erkkola, R.; Yang, B. Sea buckthorn oils: Towards healthy mucous membranes. *Agro Food Ind. Hi Tech* **2003**, *14*, 53–59.

69. Piłat, B.; Zadernowski, R. Rokitnik w produktach spożywczych. *Przemysł Spożywczy* **2016**, *70*, 35–38. [CrossRef]

70. Welters, H.J.; Diakogiannaki, E.; Mordue, J.M.; Tadayyon, M.; Smith, S.A.; Morgan, N.G. Differential protective effects of palmitoleic acid and cAMP on caspase activation and cell viability in pancreatic β-cells exposed to palmitate. *Apoptosis* **2006**, *11*, 1231–1238. [CrossRef] [PubMed]

71. Wu, Y.; Li, R.; Hildebrand, D.F. Biosynthesis and metabolic engineering of palmitoleate production, an important contributor to human health and sustainable industry. *Prog. Lipid Res.* **2012**, *51*, 340–349. [CrossRef] [PubMed]

72. Yang, Z.H.; Miyahara, H.; Hatanaka, A. Chronic administration of palmitoleic acid reduces insulin resistance and hepatic lipid accumulation in KK-A y Mice with genetic type 2 diabetes. *Lipids Health Dis.* **2011**, *10*, 120–128. [CrossRef] [PubMed]

73. Gao, S.; Guo, Q.; Qin, C.; Shang, R.; Zhang, Z. Sea buckthorn fruit oil extract alleviates insulin resistance through the PI3K/Akt signaling pathway in type 2 diabetes mellitus cells and rats. *J. Agric. Food Chem.* **2017**, *65*, 1328–1336. [CrossRef] [PubMed]

74. Bolsoni-Lopes, A.; Festuccia, W.T.; Chimin, P.; Farias, T.S.; Torres-Leal, F.L.; Cruz, M.M.; Andrade, P.B.; Hirabara, S.M.; Lima, F.B.; Alonso-Vale, M.I.C. Palmitoleic acid (n-7) increases white adipocytes GLUT4 content and glucose uptake in association with AMPK activation. *Lipids Health Dis.* **2014**, *13*, 199–209. [CrossRef] [PubMed]

75. Białek, M.; Rutkowska, J. The importance of γ-linolenic acid in the prevention and treatment. *Adv. Hyg. Exp. Med.* **2015**, *69*, 892–904. [CrossRef]

76. Larmo, P.S.; Järvinen, R.L.; Setälä, N.L.; Yang, B.; Viitanen, M.H.; Engblom, J.R.K.; Tahvonen, R.L.; Kallio, H.P. Oral Sea Buckthorn Oil Attenuates Tear Film Osmolarity and Symptoms in Individuals with Dry Eye–4. *J. Nutr.* **2010**, *140*, 1462–1468. [CrossRef] [PubMed]

77. Sergeant, S.; Rahbar, E.; Chilton, F.H. Gamma-linolenic acid, Dihommo-gamma linolenic, eicosanoids and inflammatory processes. *Eur. J. Pharmacol.* **2016**, *785*, 77–86. [CrossRef] [PubMed]

78. Luczaj, W.; Gęgotek, A.; Skrzydlewska, E. Antioxidants and HNE in redox homeostasis. *Free Radic. Biol. Med.* **2017**, *111*, 87–101. [CrossRef] [PubMed]

79. Korbecki, J.; Baranowska-Bosiacka, I.; Gutowska, I.; Chlubek, D. The effect of reactive oxygen species on the synthesis of prostanoids from arachidonic acid. *J. Physiol. Pharmacol.* **2013**, *64*, 409–421. [PubMed]

80. Davis, M.P. Cannabinoids in pain management: CB1, CB2 and non-classic receptor ligands. *Expert Opin. Investig. Drugs* **2014**, *23*, 1123–1140. [CrossRef] [PubMed]

81. Gresham, A.; Masferrer, J.; Chen, X.; Leal-Khouri, S.; Pentland, A.P. Increased synthesis of high-molecular-weight cPLA2 mediates early UV-induced PGE2 in human skin. *Am. J. Physiol.* **1996**, *270*, C1037–C1050. [CrossRef] [PubMed]

82. Lee, H.; Bae, S.K.; Pyo, M.; Heo, Y.; Kim, C.G.; Kang, C.; Kim, E. Anti-wrinkle effect of PLA2-free bee venom against UVB-irradiated human skin cells. *J. Agric. Life Sci.* **2015**, *49*, 125–135. [CrossRef]

83. Montserrat-De, L.P.S.; García-Giménez, M.D.; Ángel-Martín, M.; Pérez-Camino, M.C.; Arche, A.F. Long-chain fatty alcohols from evening primrose oil inhibit the inflammatory response in murine peritoneal macrophages. *J. Ethnopharmacol.* **2014**, *151*, 131–136. [CrossRef] [PubMed]

84. Yang, B.; Kalimo, K.O.; Mattila, L.M.; Kallio, S.E.; Katajisto, J.K.; Peltola, O.J.; Kallio, H.P. Effects of dietary supplementation with sea buckthorn (*Hippophae rhamnoides*) seed and pulp oils on atopic dermatitis. *J. Nutr. Biochem.* **1999**, *10*, 622–630. [CrossRef]

85. Paus, R. The Skin and endocrine disorders. In *Rook's Textbook of Dermatology*, 9th ed.; Wiley Online Library: Hoboken, NJ, USA, 2016; pp. 1–30.

86. Skobowiat, C.; Postlethwaite, A.E.; Slominski, A.T. Skin exposure to ultraviolet B rapidly activates systemic neuroendocrine and immunosuppressive responses. *Photochem. Photobiol.* **2017**, *93*, 1008–1015. [CrossRef] [PubMed]

87. Coleman, J.D.; Prabhu, K.S.; Thompson, J.T.; Reddy, P.S.; Peters, J.M.; Peterson, B.R.; Reddy, C.C.; Heuvel, J.P.V. The oxidative stress mediator 4-hydroxynonenal is an intracellular agonist of the nuclear receptor peroxisome proliferator-activated receptor-β/δ (PPARβ/δ). *Free Radic. Biol. Med.* **2007**, *42*, 1155–1164. [CrossRef] [PubMed]

88. Wahli, W.; Michalik, L. PPARs at the crossroads of lipid signaling and inflammation. *Trends Endocrinol. Metab.* **2012**, *23*, 351–363. [CrossRef] [PubMed]

89. Subramaniyan, S.A.; Kim, S.; Hwang, I. Cell-Cell Communication between fibroblast and 3T3-L1 cells under co-culturing in oxidative stress condition induced by H_2O_2. *Appl. Biochem. Biotechnol.* **2016**, *180*, 668–681. [CrossRef] [PubMed]

90. Benito, C.; Tolón, R.M.; Castillo, A.I.; Ruiz-Valdepeñas, L.; Martínez-Orgado, J.A.; Fernández-Sánchez, F.J.; Vázquez, C.; Cravatt, B.F.; Romero, J. β−Amyloid exacerbates inflammation in astrocytes lacking fatty acid amide hydrolase through a mechanism involving PPAR-α, PPAR-γ and TRPV1, but not CB1 or CB2 receptors. *Br. J. Pharmacol.* **2012**, *166*, 1474–1489. [CrossRef] [PubMed]
91. Lee, B.H.; Hsu, W.H.; Liao, T.H.; Pan, T.M. The monascus metabolite monascin against TNF-α-induced insulin resistance via suppressing PPAR-γ phosphorylation in C2C12 myotubes. *Food Chem. Toxicol.* **2011**, *49*, 2609–2617. [CrossRef] [PubMed]
92. Pichiah, P.T.; Moon, H.J.; Park, J.E.; Moon, Y.J.; Cha, Y.S. Ethanolic extract of seabuckthorn (*Hippophae rhamnoides* L.) prevents high-fat diet–Induced obesity in mice through down-regulation of adipogenic and lipogenic gene expression. *Nutr. Res.* **2012**, *32*, 856–864. [CrossRef] [PubMed]

antioxidants

MDPI

Article

Antioxidative 1,4-Dihydropyridine Derivatives Modulate Oxidative Stress and Growth of Human Osteoblast-Like Cells In Vitro

Lidija Milkovic [1] [ID], **Tea Vukovic** [1], **Neven Zarkovic** [1,*], **Franz Tatzber** [2], **Egils Bisenieks** [3], **Zenta Kalme** [3], **Imanta Bruvere** [3], **Zaiga Ogle** [3], **Janis Poikans** [3], **Astrida Velena** [3,*] **and Gunars Duburs** [3]

[1] Laboratory for Oxidative Stress, Rudjer Boskovic Institute, Bijenicka 54, 10000 Zagreb, Croatia; Lidija.Milkovic@irb.hr (L.M.); Tea.Vukovic@irb.hr (T.V.)
[2] Institute of Pathophysiology and Immunology, Medical University of Graz, A-8036 Graz, Austria; franz@tatzber.at
[3] Latvian Institute of Organic Synthesis, 21 Aizkraukles Str., LV-1006 Riga, Latvia; Egils.Bisenieks@osi.lv (E.B.); kalme@osi.lv (Z.K.); Imanta.Bruvere@gmail.com (I.B.); zaiga.ogle@osi.lv (Z.O.); japo@osi.lv (J.P.); gduburs@osi.lv (G.D.)
* Correspondence: zarkovic@irb.hr (N.Z.); astrida@osi.lv (A.V.); Tel.: +385-1-456-0937 (N.Z.); +371-67551822 (A.V.)

Received: 16 July 2018; Accepted: 15 September 2018; Published: 19 September 2018

Abstract: Oxidative stress has been implicated in pathophysiology of different human stress- and age-associated disorders, including osteoporosis for which antioxidants could be considered as therapeutic remedies as was suggested recently. The 1,4-dihydropyridine (DHP) derivatives are known for their pleiotropic activity, with some also acting as antioxidants. To find compounds with potential antioxidative activity, a group of 27 structurally diverse DHPs, as well as one pyridine compound, were studied. A group of 11 DHPs with 10-fold higher antioxidative potential than of uric acid, were further tested in cell model of human osteoblast-like cells. Short-term combined effects of DHPs and 50 µM H_2O_2 (1-h each), revealed better antioxidative potential of DHPs if administered before a stressor. Indirect 24-h effect of DHPs was evaluated in cells further exposed to mild oxidative stress conditions induced either by H_2O_2 or tert-butyl hydroperoxide (both 50 µM). Cell growth (viability and proliferation), generation of ROS and intracellular glutathione concentration were evaluated. The promotion of cell growth was highly dependent on the concentrations of DHPs used, type of stressor applied and treatment set-up. Thiocarbatone **III-1**, E2-134-1 **III-4**, Carbatone **II-1**, AV-153 **IV-1**, and Diethone **I** could be considered as therapeutic agents for osteoporosis although further research is needed to elucidate their bioactivity mechanisms, in particular in respect to signaling pathways involving 4-hydroxynoneal and related second messengers of free radicals.

Keywords: 1,4-dihydropyridine(s) (DHPs); oxidative stress; reactive oxygen species (ROS); antioxidant (AO); antioxidative activity (AOA); glutathione; cell viability and proliferation; 4-hydroxynonenal (4-HNE)

1. Introduction

Oxidative stress, defined as an imbalance between pro-oxidants and antioxidants in the favor of the former [1], has been implicated in the pathogenesis of numerous stress- and age-associated diseases, whether as a cause or as a consequence of respective illness progression [2]. The consensus among researchers in this field highlights the need of its better understanding as well as balancing oxidative homeostasis to the levels that promote health [3]. Indeed, antioxidants per se or drugs with antioxidative properties are important for reducing the detrimental levels of reactive oxygen species (ROS). Yet, the importance of ROS-related redox signaling in normal cellular maintenance should not be neglected, nor the fact that antioxidative and/or pro-oxidative activity are in the background of (un)desirable activities of many drugs and physiologically active compounds [3–5]. Thus, studies revealing bioactivities of natural antioxidants are complemented by scientific efforts aiming to synthetize new bioactive substances with antioxidant features that could help maintaining oxidative homeostasis of the living cells.

1,4-Dihydropyridine derivatives (DHPs) are a group of pleiotropic physiologically active compounds (reviewed in Swarnalatha et al. [6] and Velena et al. [7]), among which many 4-nitrophenyl- and differently substituted DHPs are known as effective antihypertensive agents. While condensed DHPs were suggested as agents that affect stem cell differentiation [8], several DHPs exert anti- or pro-oxidant effects in various systems both in vitro as well as in vivo [7,9].

Our present article extends the knowledge about antioxidative and potentially pro-oxidative activities of the well-known, water-insoluble antioxidant Diethone **I** (also known as Dietone, Diludine, Hantzsch ester, HEH) [7], in comparison to the water-soluble antimutagenic [10] and antimetastatic [11] compound Carbatone (disodium-2,6-dimethyl-1,4-dihydropyridine-3,5-bis(carbonyloxyacetate)) **II-1**, which prevents DNA lesions [12–15]. Moreover, comparison was done also with their analogues and with novel DHPs (see Table 1, compounds mentioned as MM) and their derivatives.

Therefore, a set of 27 structurally diverse (of monocyclic as well as of condensed ring structures, having symmetric as well as asymmetric alkyl-, alkoxyalkyl-, aryl-, aralkyl- or heteryl- substituents on positions 2, 3, 4, 5 and 6) synthetic 1,4-dihydropyridine compounds (as well as for comparison one pyridine type compound, an analogue (oxidized form) of diamide analogue J-12-25 **V**) (see Table 1 below) of which 14 are water soluble DHP compounds (including antimutagenic, DNA protecting, antimetastatic, antiischaemic) and 14 are water-insoluble (more lipid soluble): 13 DHP compounds (including antioxidant and radioprotector Hantzsch ester Diethone (Dietone) **I**) and one pyridine analogue were studied.

While possibly beneficial effects of antioxidants are mostly considered for patients suffering from malignant, cardiovascular, neurodegenerative and inflammatory diseases, osteoporosis is an age-related disease characterized by bone loss due to the impaired bone formation and increased bone resorption, which is less often considered as an oxidative stress-associated disease. However, requests for treatment strategies of this chronic disease that would promote bone formation [16] also point to oxidative stress as an important factor in pathogenesis of osteoporosis [17,18], consequently highlighting antioxidants as important factors for prevention and treatment of osteoporosis [17,19].

Therefore, in our study, we used human osteoblast-like cells (HOS cell line) which are well known as a model of human osteoblasts [20–22] aiming to reveal DHPs that can act as antioxidant(s) in cell-free systems (Total antioxidative capacity (TAC) and Total oxidative capacity (TOC) assays, respectively) but could also affect viability and growth of HOS cells under mild oxidative stress induced by two different stressors (hydrogen peroxide and tert-butyl hydroperoxide).

Table 1. Chemical structures and solubility of 27 1,4-dihydropyridine derivatives and one pyridine analogue investigated in this study.

Compounds	Trivial Name	R3	R4	R5	Solubility	Reference
I	Diethone				Ethanol with 10% DMSO	[23]
II-1	Carbatone	ONa	H	OCH$_2$COONa	Water	[24]
II-2	J-9-133-2	ONa	H	OCH$_3$	Water	[25]
II-3	J-9-46	OCH$_2$CH$_3$	H	OCH$_2$CH$_3$	Ethanol	[25]
II-4	Metcarbatone	ONa	CH$_3$	OCH$_2$COONa	Water	[26]
II-5	Etcarbatone	ONa	CH$_2$CH$_3$	OCH$_2$COONa	Water	[26]
II-6	J-9-117 (Styrylcarbatone)	ONa	CH=CHPh	OCH$_2$COONa	Water	*MM
III-1	Thiocarbatone	ONa	H	SCH$_2$COONa	Water	MM
III-2	TK-2	OCH$_3$	H	SCH$_2$COOCH$_3$	Ethanol with 10% DMSO	MM
III-3	E-170-4 (TK-1)	OCH$_2$CH$_3$	H	SCH$_2$COOCH$_2$CH$_3$	Ethanol with 10% DMSO	MM
III-4	E2-134-1	ONa	H	OCH$_2$CH$_3$	Water	MM
III-5	E2-135	OCH$_2$CH$_3$	H	OCH$_2$CH$_3$	Ethanol	MM
III-6	E2-136-2	ONa	H	OCH$_2$COONa	Water	MM
III-7	E2-131	ONa	CH$_2$CH$_3$	OCH$_2$CH$_3$	Water	MM
III-8	E2-130-3	OCH$_2$CH$_3$	CH$_2$CH$_3$	OCH$_2$CH$_3$	Ethanol	MM
III-9	E2-120 (ETK-2)	OCH$_3$	CH$_2$CH$_3$	SCH$_2$COOCH$_3$	Ethanol	MM
III-10	E2-113 (ETK-1)	OCH$_2$CH$_3$	CH$_2$CH$_3$	SCH$_2$COOCH$_2$CH$_3$	Ethanol	MM
III-11	E-163-1	ONa	3-Py	OCH$_2$CH$_3$	Water	MM
III-12	E-163-K	OCH$_2$CH$_3$	3-Py	OCH$_2$CH$_3$	Ethanol	MM
IV-1	AV-153	OCH$_2$CH$_3$	ONa	OCH$_2$CH$_3$	Water	[27]
IV-2	EE-126	OCH$_3$	ONa	OCH$_3$	Water	MM
IV-3	E3-46	OCH$_2$CH$_3$	OH	OCH$_2$COOCH$_2$CH$_3$	Ethanol	MM
IV-4	V-6-55-1	OCH$_2$COOCH$_2$CH$_3$	OH	OCH$_2$COOCH$_2$CH$_3$	Ethanol	MM
IV-5	AV-154-Na	CH$_3$	ONa	CH$_3$	Water	[27]
IV-6	J-11-61B	CH$_3$	OH	OCH$_2$CH$_3$	Ethanol	MM
IV-7	J-11-71-2	OCH$_2$CH$_3$	OH		Ethanol	MM
IV-8	Glutapyrone	OCH$_2$CH$_3$	NHCH(CH$_2$)$_2$COONa | COONa	OCH$_2$CH$_3$	Water	[28]
V	J-12-25				Ethanol	MM

*MM—see in Materials and Methods. DMSO (Dimethyl sulfoxide).

2. Materials and Methods

2.1. Compounds

The studied 27 DHP compounds belong to several (five, **I–V**, see Table 1) relative types. All tested DHPs were synthesized at the Latvian Institute of Organic Synthesis [24–30]. Already described compounds are referenced in Table 1; new compounds have been synthesized by making use of methods indicated below for each type and indexed as MM (*MM) in Table 1.

Dietone **I** is the basic most studied compound, all studied types are its derivatives. It could be used as standard. It has been elaborated and developed, used as radioprotector, antioxidant to preserve carotene, vitamins A and E, growth stimulant. Dietone **I** was synthesized according to modified Hantzsch syntesis: heterocyclization of acetoacetic ester and urotropine [23]. **Type II** compounds (**II-1–II-6**) or Carbatone **II-1** and its derivatives were prepared by making use of 1 or 2 equivalents of ethoxycarbonylmethyl ester of acetoacetic acid instead of or as addition to acetoacetic ester. For synthesis of soluble in water salts mild hydrolysis of distant ester group was performed. Compounds (**II-1–II-6**) were obtained according to procedure described in patents [24–26]. **Type III** compounds (**III-1–III-12**) or Thiocarbatone **III-1** and its derivatives were prepared by making use of one or two equivalents of ethoxycarbonylmethyl ester of acetothioacetic acid instead of or as addition to acetoacetic ester. Again, for synthesis of soluble in water salts mild hydrolysis of distant ester group was performed. Compounds (**III-1–III-12**) were obtained following the reported method [26].

Type IV compounds (derivatives of 1,4-dihydroisonicotinic acid **IV-1–IV-8**) were synthesized by making use of glyoxylic acid or its ester as aldehyde part, ethyl or methyl ester of acetoacetic acid (and/or additionally ethoxycarbonylmethyl ester of acetylacetic acid) and ammonia. AV-153 (**IV-1**) can be as intermediate for synthesis of Glutapyrone (**IV-8**) [28]. Compounds (**MM, IV-1–IV-6**) were obtained according to procedure described in [27]. Pyridine derivative J-12-25 (**V**) was obtained via oxidation of related DHP with sodium nitrite in acetic acid.

Major features of the substances tested are presented in Table 1.

Fresh stock solutions of DHPs were prepared in adequate solvent (listed in Table 1) in a concentration of 10 mg/mL prior to each experiment.

2.2. Total Antioxidative Capacity (TAC) Assay

The total antioxidative capacity assay relying on the ability of tested compounds to scavenge hydrogen peroxide, thus, competing with peroxidase and preventing oxidation of a chromogenic substrate tetramethylbenzidine (TMB), was performed as previously described [31], with a slight modification as described. The scavenging ability of DHP compounds was compared to ranging concentration of uric acid serving as standard antioxidant. Briefly, 25 μL of a standard (uric acid serial dilutions from 0–12 mM) or a sample (27 DHPs, as well one pyridine type compound) each 10 mg/mL) were pipetted into each well of the 96-well microplate and mixed with 100 μL Reagent A (0.1 M citric buffer containing 0.03% (*v/v*) hydrogen-peroxide). The first absorbance was measured at 450 nm (Multiscan Ex, Thermo Electron Corporation, Shanghai, China) followed by the addition of 50 μL Reagent B containing 1.25 mU horse radish peroxidase (HRP; Sigma Aldrich, St. Louis, MO, USA) and 0.416 mM TMB (Sigma Aldrich, St. Louis, MO, USA) in citric buffer. After 15-min incubation time, reaction was stopped with 2 M H_2SO_4 and the second absorbance was measured at 450 nm (Multiscan Ex, Thermo Electron Corporation, Shanghai, China). The difference between the second and the first absorbance values of the tested compounds was interpolated from the uric acid standard curve and results are presented as mM Uric acid equivalent.

2.3. Total Oxidative Capacity (TOC) Assay

A total oxidative capacity assay was performed as previously described [32]. The assay determines peroxides present in the sample by their reaction with peroxidase followed by a color reaction with chromogenic substrate TMB. Quantification is achieved by serial dilutions of a standard hydrogen peroxide solution. Briefly, 10 μL of a standard (hydrogen peroxide serial dilutions from 0–0.9791 mM) or a sample (27 DHPs, as well one pyridine type compound) was pipetted into each well of the 96-well microplate, mixed with 200 μL of reaction mixture (25 mU HRP, TMB, and substrate buffer in a proportion of 1:10:100 (*v:v:v*)) and initial absorbance at 450 nm was measured (Multiscan Ex, Thermo Electron Corporation, Shanghai, China). Plate was further incubated for 20 min and the second absorbance at 450 nm was measured after stopping the reaction with 2 M H_2SO_4. The results are expressed as μM H_2O_2 hydrogen peroxide equivalent.

2.4. Cell Culture and Treatments

The human osteosarcoma cell line HOS (ATCC® CRL-1543™), purchased from the American Type Culture Collection (ATCC, LGC Standards GmbH, Wesel, Germany) was used as an osteoblast model to test possible the in vitro effects of selected compounds. The cells were cultivated in T75 cell culture flasks (TPP, Trasadingen, Switzerland) in Dulbecco's Modified Eagle's Medium (DMEM) with 10% (*v/v*) fetal calf serum (FCS) at 37 °C in humidified atmosphere with 5% CO_2. Prior to experiments, cells were harvested with 0.25% (*w/v*) Trypsin-0.53 mM EDTA (Ethylenediaminetetraacetic acid) solution and counted with Trypan Blue Exclusion Assay in Bürker-Türk hemocytometer (Brand, Wertheim, Germany). Thus prepared cells were seeded at a density of 2×10^4 cells/well into 96-microwell plates (TPP, Trasadingen, Switzerland) and left for 24 h to attach. Afterwards, we evaluated the potential of selected compounds to influence intracellular ROS production upon administration of hydrogen peroxide. Cell cultures were either first exposed to 50 µM hydrogen peroxide for one hour before addition of selected compounds (100 or 1000 µM) or vice versa. The immediate effect on cellular viability was also evaluated in case of cells that were first exposed to selected compounds for one hour, followed by one-hour incubation with hydrogen peroxide, aiming to test possible beneficial effects of the tested substances that might prevent the onset of the cellular oxidative stress.

The second series of experiments were conducted with prolonged (24-h) exposure of cells to selected compounds before one-hour exposure to hydrogen-peroxide. The experiments are explained in more details in the following subsections.

2.5. Cellular Viability (Tetrazolium Reduction Assay)

Viability (metabolic activity) of the cultured HOS cells was evaluated with the EZ4U assay (Biomedica, Wien, Austria) according to manufacturer's instructions. The assay principle is based on reduction of tetrazolium salts to colored formazan derivatives by living cells, the intensity of which is measured spectrophotometrically as proportional to the number of viable cells in the culture sample.

Cells seeded at a density of 2×10^4 cells/well in 96-microwell plates (TPP, Trasadingen, Switzerland) were left to attach for 24 h prior to treatments. For the short-term treatment, cells were first treated with selected compounds (100 µM and 1000 µM concentration; except for Diethone **I** for which only 100 µM concentration was used because it precipitates at 1000 µM) for 1 h, after which hydrogen peroxide was added for an additional hour. In case of experiments with longer exposure to selected compounds, HOS cells were treated with different concentrations of selected compounds (ranging from 10 µM to 1000 µM, depending on the compound used) and left for additional 24 h, followed by the change of cell culture medium. Cellular viability was determined for cell culture either treated just by selected compounds or for those additionally treated with 50 µM hydrogen peroxide or with 50 µM tert-butyl hydroperoxide (tBHP; Sigma Aldrich, St. Louis, MO, USA) for one-hour, respectively. Afterwards, the medium was replaced with 200 µL Hanks' solution (pH 7.4) and 20 µL of dye solution was added to each well followed by two-hour incubation at 37 °C. The absorbance was measured at 450 nm with a reference wavelength of 620 nm using a microplate reader (Multiscan Ex, Thermo Electron Corporation, Shanghai, China). Thus obtained results were expressed as percentage of non-treated control.

2.6. Cell Proliferation (BrdU Assay)

Cell proliferation was determined in the experiments where cells were exposed to selected DHP compounds for 24-h used alone or in combination with 50 µM hydrogen peroxide or tBHP, as described in the previous section. The 5-bromo-2'-deoxyuridine (BrdU) colorimetric assay (Roche Applied Science, Mannheim, Germany) was used, according to the manufacturer's instruction. The assay relies on the ability of the BrdU to incorporate into cellular DNA during proliferation which is further detected by an anti-BrdU antibody. Briefly, after specified time points of treatments (24-h treatment with the compound alone or followed by 1-h treatment with hydrogen peroxide and tBHP), the BrdU was added to each well and left for two-hours at 37 °C. The culture medium was removed and the cells were washed twice with Hanks' solution before fixation with FixDenat solution for 30 min. Subsequently, the anti-BrdU-peroxidase (1:100) was added to each well and incubated at room temperature for 90 min. After removing the unbound antibody conjugate, 100 μL of the substrate solution was added and allowed to stand for 15 min. The reaction was quenched by adding 25 μL of a 1 M H_2SO_4 solution. The absorbance was measured at 450 nm with a reference wavelength of 620 nm using a microplate reader (Multiscan Ex, Thermo Electron Corporation, Shanghai, China).The results are presented as percentage of non-treated control.

2.7. Measurement of intracellular ROS production

The ROS measurement is based on the intracellular oxidation of 2′,7′-dichlorodihydrofluorescein diacetate (DCFH-DA; Sigma-Aldrich, St. Louis, MO, USA) to a fluorescent 2′,7′-dichlorofluorescein (DCF) that can be measured. Intracellular ROS levels were measured upon short-term (1-h) treatment and after 24-h treatment with selected DHPs, described hereafter. For all experiments, HOS cells were seeded at the density of 2×10^4 cells/well into white 96-microwell plates (Thermo Fisher Scientific, Nunc A/S, Roskilde, Denmark) and left for 24 h to attach.

In the short-term treatment, time-dependent ROS production was measured in cells treated at specific time points as follows: (a) hydrogen peroxide 1-h + DHPs, or (b) DHPs 1-h + hydrogen peroxide. Cells were first incubated with 10 μM DCFH-DA at 37 °C for 30 min, followed by fresh media exchange and a zero-point ROS measurement with a Cary Eclipse Fluorescence Spectrophotometer (Varian Australia Pty Ltd, Mulgrave, Victoria, Australia) with excitation at 500 nm and emission detection at 530 nm. Treatments were carried out with 1000 μM concentration of DHPs (exception Diethone I used as 100 μM because it precipitates at 1000 μM) and 50 μM hydrogen peroxide without the medium change. Further fluorescence measurements were done: immediately after addition of the first treatment (5-min point), 30- and 60-min points, immediately after addition of the second treatment (65-min point) and up to four hours afterwards (90-, 120-, 180-, and 240-min points). The results are expressed as relative fluorescence units (RFU) measured at specified time points.

Intracellular ROS production was also evaluated upon 24-h treatment with selected DHPs. Following 24-h treatment, media was replaced with the Hanks' solution containing 10 μM DCFH-DA and cells were further incubated for 30 min on 37 °C. Next, the medium was replaced with the fresh one and fluorescence was measured prior to the addition of 50 μM tBHP (zero-point) and after 1-h. Measurements were carried with a Cary Eclipse Fluorescence Spectrophotometer (Varian Australia Pty Ltd, Mulgrave, Victoria, Australia) as previously stated. Cells were further trypsinized with 0.25% (*w/v*) Trypsin-0.53 mM EDTA solution and counted with Trypan Blue Exclusion Assay in a Bürker–Türk hemocytometer (Brand, Wertheim, Germany). The results are expressed in arbitrary units which are a ratio of the difference in fluorescence (1-h point–zero-point) and cell number.

2.8. Determination of Glutathione (GSH) Levels

Cells were first trypsinized (0.25% (*w/v*) Trypsin-0.53 mM EDTA solution), washed twice with phosphate-buffer saline (PBS) and stored as dry pellets at −80 °C until analysis. Cell lysates were obtained by addition of 100 μL PBS and repeated thaw-freeze cycles, followed by centrifugation at 16,000× *g*/15 min. and collection of a supernatant containing proteins. Protein concentration in each sample was determined according to Bradford method [33], using bovine serum albumin as a standard.

The total intracellular GSH content (oxidized and reduced) was measured as previously described [34]. Briefly, 150 μL of each sample, containing 0.03 mg/mL of protein, or standard, reduced glutathione in serial dilutions (0–20 mg/mL) was pipetted into each well of 96-microwell plate. Reaction was started by addition of freshly prepared reaction mixture: 1.8 mM 5,5-dithio-bis-2-nitrobenzoic acid, 0.4 U GSH reductase, and 0.6 mM NADPH in phosphate buffer (100 mM NaH_2PO_4, 5 mM EDTA pH 7.4). The formation of 2-nitro-5-thiobenzoic acid was monitored spectrophotometrically in a plate reader at 405 nm (Multiscan Ex, Thermo Electron Corporation,

Shanghai, China). Total GSH concentration in cell lysates was calculated from the standard curve by linear regression and expressed as µM of total GSH.

2.9. Statistical Analysis

Cell-free colorimetric assays (TAC and TOC) were conducted in duplicates while cell-based methods (EZ4U assay, BrdU assay, ROS and GSH measurements) were carried out in at least triplicates. The respective numbers of biological replicates (n) is given in each figure legend. Data are expressed as mean values with standard deviations. Results were analyzed with Student's *t*-test and values of $p < 0.05$ were considered as statistically significant.

3. Results

3.1. Total Antioxidative Capacity

The results of the total antioxidative capacity assay for tested DHPs are listed in Table 2. In comparison to uric acid, used as a standard, the following compounds (at 1 mM concentration) were more effective: Diethone **I**, J-9-133-2 **II-2**, AV-153 **IV-1**, AV-154-Na **IV-5**, J-11-71-2 **IV-7**, Carbatone **II-1**, Thiocarbatone **III-1**, E-2-134-1 **III-4**, E-2-136-2 **III-6**, V-6-55-1 **IV-4**, and E-3-46 **IV-3** (about 10-fold or more); J-11-61B **IV-6** (8-fold); J-9-46 **II-3** and E-2-135 **III-5** (2-fold). J-9-117 **II-6** and E2-130-3 **III-8** were as effective as uric acid while other tested compounds did not have pronounced antioxidative potential. Accordingly, the DHPs with pronounced antioxidative capacities were selected for the treatment of human osteoblast-like cells (HOS).

Table 2. Total antioxidant capacity of selected compounds expressed as equivalent to mM uric acid.

Compounds	Trivial Name	Concentrations Tested		
		10 µM	100 µM	1000 µM
		Equivalent to mM Uric Acid		
I	Diethone	1.463 ± 1.016	1.465 ± 0.085	>10
II-1	Carbatone	0.438 ± 0.177	1.534 ± 0.064	9.692 ± 0.036
II-2	J-9-133-2	0.585 ± 0.074	3.756 ± 0.312	>10
II-3	J-9-46	0	0.148 ± 0.088	2.284 ± 0.504
II-4	Metcarbatone	0	0.526 ± 0.033	0.852 ± 0.013
II-5	Etcarbatone	0.027 ± 0.038	0.057 ± 0.019	0.206 ± 0.085
II-6	J-9-117 (Styrylcarbatone)	0.680 ± 0.051	0.421 ± 0.596	1.073 ± 0.04
III-1	Thiocarbatone	1.279 ± 0.046	2.735 ± 0.363	>10
III-2	TK-2	0	0	0
III-3	E-170-4 (TK-1)	0	0	0
III-4	E2-134-1	0.770 ± 0.075	0.541 ± 0.304	>10
III-5	E2-135	0.027 ± 0.302	0.564 ± 0.422	1.968 ± 0.294
III-6	E2-136-2	0.498 ± 0.115	1.082 ± 0.175	>10
III-7	E2-131	0.778 ± 0.085	0	0
III-8	E2-130-3	0.355 ± 0.234	0.285 ± 0.398	0.980 ± 0.129
III-9	E2-120 (ETK-2)	0	0.042 ± 0.007	0
III-10	E2-113 (ETK-1)	0.032 ± 0.195	0.077 ± 0.219	0
III-11	E-163-1	0	0	0
III-12	E-163-K	0	0	0
IV-1	AV-153	0.153 ± 0.216	2.005 ± 0.053	9.828 ± 0.081
IV-2	EE-126	0.856 ± 0.614	0.304 ± 0.078	0.292 ± 0.083
IV-3	E3-46	0.075 ± 0.163	1.542 ± 0.120	>10
IV-4	V-6-55-1	0.636 ± 0.220	4.013 ± 0.228	>10
IV-5	AV-154-Na	0.481 ± 0.028	2.774 ± 0.199	>10
IV-6	J-11-61B	0	0.948 ± 0.056	8.222 ± 0
IV-7	J-11-71-2	0	3.160 ± 0.094	>10
IV-8	Glutapyrone	0.190 ± 0.015	0.127 ± 0.096	0
V	J-12-25	0.016 ± 0.193	0.050 ± 0.040	0.012 ± 0.044

3.2. Total Oxidative Capacity

The results of the total oxidative capacity assay are listed in Table 3. Oxidative potential of the majority of the tested DHPs was either absent or, if present, not proportional to the concentration used. AV-153 **IV-1**, Carbatone **II-1** and E-170-4 **III-3** (TK1 **III-3**) were the only compounds showing oxidative potential somewhat proportional to the concentration used. While pro-oxidative activity of E-170-4 **III-3** (TK-1 **III-3**) was very low, but proportional to the concentration range, mild oxidative potential of lower concentrations (10 μM and 100 μM) and negligible of the highest (1000 μM) was observed for AV-153 **IV-1** and Carbatone **II-1**. Still, it was about 2-fold (AV-153 **IV-1**) and 5-fold (Carbatone **II-1**) lower than of hydrogen peroxide.

Table 3. The oxidative capacity of selected compounds expressed as equivalent to μM H_2O_2.

Compounds	Trivial Name	Concentrations Tested		
		10 μM	100 μM	1000 μM
		Equivalent to μM H_2O_2		
I	Diethone	1.346 ± 1.904	0	0
II-1	Carbatone	2.692 ± 0.000	18.846 ± 1.088	1.538 ± 0.544
II-2	J-9-133-2	0.250 ± 0.374	0	0
II-3	J-9-46	0	0	0
II-4	Metcarbatone	0	0	0
II-5	Etcarbatone	0	0	0
II-6	J-9-117 (Styrylcarbatone)	0.577 ± 0.816	0	0.192 ± 0.272
III-1	Thiocarbatone	0	0	0
III-2	TK-2	0	0	7.558 ± 1.791
III-3	E-170-4 (TK-1)	0	3.375±1.768	38.529 ± 1.663
III-4	E2-134-1	0	0	0
III-5	E2-135	0	0	0
III-6	E2-136-2	0	0	0
III-7	E2-131	0	5.411 ± 1.663	6.588 ± 3.328
III-8	E2-130-3	0	0	0
III-9	E2-120 (ETK-2)	0	0	13.235 ± 5.407
III-10	E2-113 (ETK-1)	0	0	0.147±0.208
III-11	E-163-1	0	4.5 ± 1.179	13.059 ± 1.664
III-12	E-163-K	0	0	0
IV-1	AV-153	6.923 ± 0.544	41.154 ± 1.088	5.00 ± 4.351
IV-2	EE-126	3.625 ± 0.884	3.938 ± 0.442	3.938 ± 0.442
IV-3	E3-46	0	7.243 ± 0.114	0
IV-4	V-6-55-1	0.481 ± 0.680	0	0
IV-5	AV-154-Na	3.938 ± 0.442	3.625 ± 0.884	0
IV-6	J-11-61B	0	0	0
IV-7	J-11-71-2	0.938 ± 0.000	1.875 ± 0.000	0
IV-8	Glutapyrone	0	0	0
V	J-12-25	0	0	0

3.3. Short-Term (1-h) Treatment of HOS Cells with DHPs and Hydrogen Peroxide

3.3.1. Measurement of Intracellular ROS Production

The ability of the tested DHPs to scavenge hydrogen peroxide revealed better reduction of intracellular ROS levels if DHPs were applied before the addition of the stressor. Such antioxidative effects were especially pronounced for water-soluble DHPs and Diethone **I**, observed as reduction of ROS levels almost to level of non-stressed control (Figure 1A,B). The E3-46 **IV-3** was not so effective, its effects were pronounced after 120 min and were not reaching the effects of the other effective DHPs (Figure 1B). On the other hand, if the cells were first exposed to hydrogen peroxide for one hour, the scavenging ability of the majority of tested DHPs was not so pronounced for, still being present (Figure 1C,D). In this type of treatment, E3-46 **IV-3** was the only compound unable to reduce but even increased ROS levels during 4-h measurement (Figure 1D). In both types of treatments, intracellular ROS levels of control cells (without stressor) remained low during the course of 4-h measurement.

Taken together results of both treatments (DHPs before or after stressor) revealed better antioxidative potential of the tested DHPs given before stressor, i.e., when cells were not already damaged by hydrogen peroxide. Moreover, ROS reduction indicated direct scavenging of hydrogen peroxide by the tested DHPs along with possible induction of the cellular protective mechanisms (endogenous antioxidants).

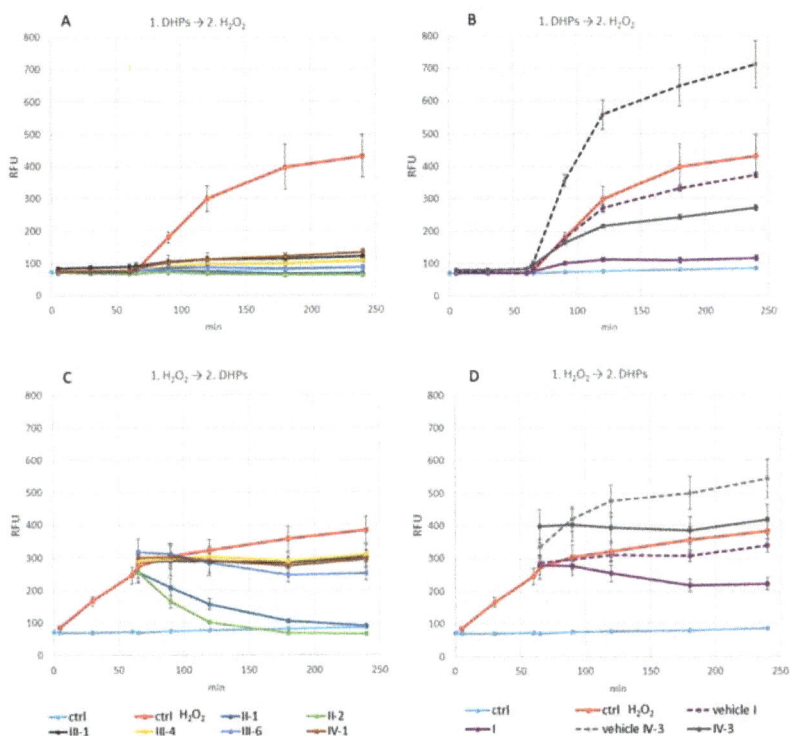

Figure 1. Intracellular ROS of HOS cells exposed to DHPs and H_2O_2.Cells were treated with DHPs and hydrogen peroxide, without the exchange of culture medium, as follows: (**A,B**) 1-h DHPs + hydrogen peroxide, or (**C,D**) 1-h hydrogen peroxide + DHPs. The fluorescence reflecting intracellular ROS levels was measured at different time points (0—no treatment, 5—immediately after the first treatment, 30, 60, 65—immediately after the second treatment, 90, 120, 180, 240 min). Results of DHPs (**A,C**—water-soluble; **B,D**—soluble in organic solvents) are presented as mean of relative fluorescence units (RFU) ± standard deviation (n = 2). Controls: ctrl (untreated cells), ctrl H_2O_2 (cells treated with hydrogen peroxide) and solvent-controls (vehicle Diethone **I** and vehicle E3-46 **IV-3**).

3.3.2. Cellular Viability

The effects of selected DHPs on cellular viability (Figure 2) were analyzed for cells treated with DHPs alone (non-stress condition) or in combination with 50 μM hydrogen peroxide (IC_{30}—corresponding to induction of mild oxidative stress). The E-2-134-1 **III-4** (1000 μM) significantly increased cellular viability both in case of non-stress and in mild-stress culturing conditions ($p < 0.05$ and $p < 0.005$, respectively) while for AV-153 **IV-1** (1000 μM) and Thiocarbatone **III-1** (1000 μM) cellular viability was significantly increased only under mild-stress conditions ($p < 0.005$ and $p < 0.0005$, respectively). Interestingly decreased cellular viability (metabolism) was observed by EZ4U assay for Carbatone **II-1** and J-9-133-2 **II-2**, which were previously (Figure 1) found to reduce ROS levels, even below non-treated control levels.

Therefore, evaluation of ROS production and cellular viability for short-term treatments have accentuated Thiocarbatone **III-1**, AV-153 **IV-1** and in particular E-2-134-1 **III-4** (at 1000 μM concentration) as likely the most potent and bioactive DHP antioxidants in the tested group due to their beneficial growth promoting and ROS scavenging effects under mild oxidative stress conditions.

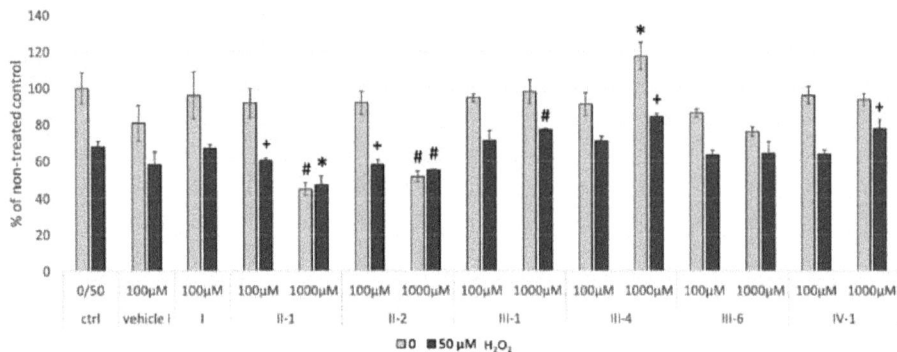

Figure 2. Cellular viability measured by EZ4U assay upon 1 h treatment with DHPs and H_2O_2. HOS cells were first treated with DHPs (100 μM and 1000 μM) for one-hour, followed by one-hour treatment with 50 μM hydrogen peroxide or with plain medium. Values are given as mean ± standard deviation; n = 3. Statistically significant differences to their respective controls are shown as follows: * $p < 0.05$; + $p < 0.005$; # $p < 0.0005$.

3.4. The Effects of 24-h Pre-Treatment of Cells with DHPs before Exposure to Hydrogen Peroxide or Tert-Butyl Hydroperoxide

3.4.1. Hydrogen Peroxide as a Stressor

The possible growth-promoting potential of DHPs was further evaluated in non-stress (without addition of stressor) as well as in mild oxidative stress conditions (addition of H_2O_2 as stressor). Stimulatory effects of 1000 μM AV-153 **IV-1** and AV-154-Na **IV-5** on cell viability (EZ4U assay, $p < 0.05$) or proliferation (BrdU assay, $p < 0.005$), respectively, were observed. Other DHPs did not enhance but have even inhibited cell growth, particularly 1000 μM Carbatone **II-1** and J-9-133-2 **II-2**. Treatment with hydrogen peroxide affected more cell proliferation than viability. Negative effect of hydrogen peroxide was attenuated with low concentration (10 μM) of AV-153 **IV-1**, AV-154-Na **IV-5**, J-9-133-2 **II-2**, Carbatone **II-1** and Diethone **I** (Figure 3A,B).

The results have shown attenuation of oxidative stress-induced suppression of cell growth for AV-153 **IV-1**, AV-154-Na **IV-5**, J-9-133-2 **II-2**, Carbatone **II-1**, and Diethone **I** (when present at low concentrations), thus suggesting potential beneficial effects as of these DHPs as bioactive antioxidants.

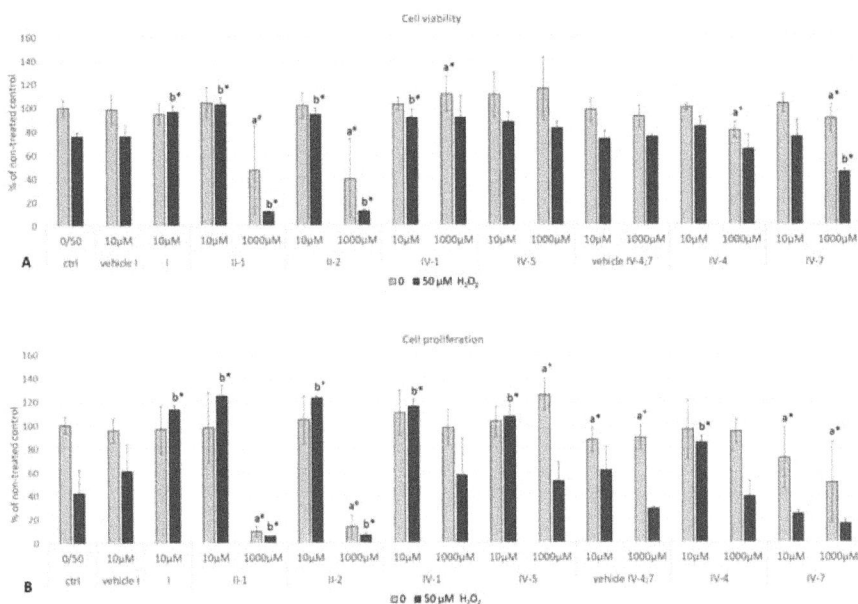

Figure 3. Cell viability (**A**) and proliferation (**B**) upon 24-h treatment with DHPs followed by 1-h exposure to H_2O_2. Results are expressed as a percentage of non-treated control. Statistically significant differences between—(a) DHPs alone vs. non-treated control (pale grey bars) or (b) DHPs treated vs. treated control (dark grey bars) are shown as follows: a*/b* $p < 0.05$; a+/b+ $p < 0.005$; a#/b# $p < 0.0005$. Values are given as mean ± standard deviation, n = 4.

3.4.2. Tert-Butyl Hydroperoxide(tBHP) as a Stressor

Cell Viability and Proliferation

Cell viability and proliferation were tested for a range of concentrations (10–1000 μM) of AV-153 **IV-1**, AV-154-Na **IV-5**, J-9-133-2 **II-2**, Carbatone **II-1**, and Diethone **I** (five of previously tested compounds), as well as for Thiocarbatone **III-1**, E2-134-1 **III-4**, E2-136-2 **III-6**, and E3-46 **IV-3** (four newly tested compounds). The majority of selected DHPs, per se, significantly stimulated cellular viability, according to the EZ4U assay, but did not affect cell proliferation according to the BrdU assay (Figure 4A). Hence, different concentrations of DHPs were selected for evaluation of respective DHPs' effects on HOS cells under mild oxidative stress conditions using 50 μM tBHP as a stressor. As in case of hydrogen peroxide, tBHP suppressed cell proliferation more than cellular viability. The E2-134-1 **III-4** stimulated both, cell viability (500 μM, $p < 0.05$) and proliferation (250 μM, $p < 0.05$) while Carbatone **II-1** (250 μM; $p < 0.05$) and Thiocarbatone **III-1** (100 μM; $p < 0.05$) stimulated either viability or proliferation (Figure 4B,C).

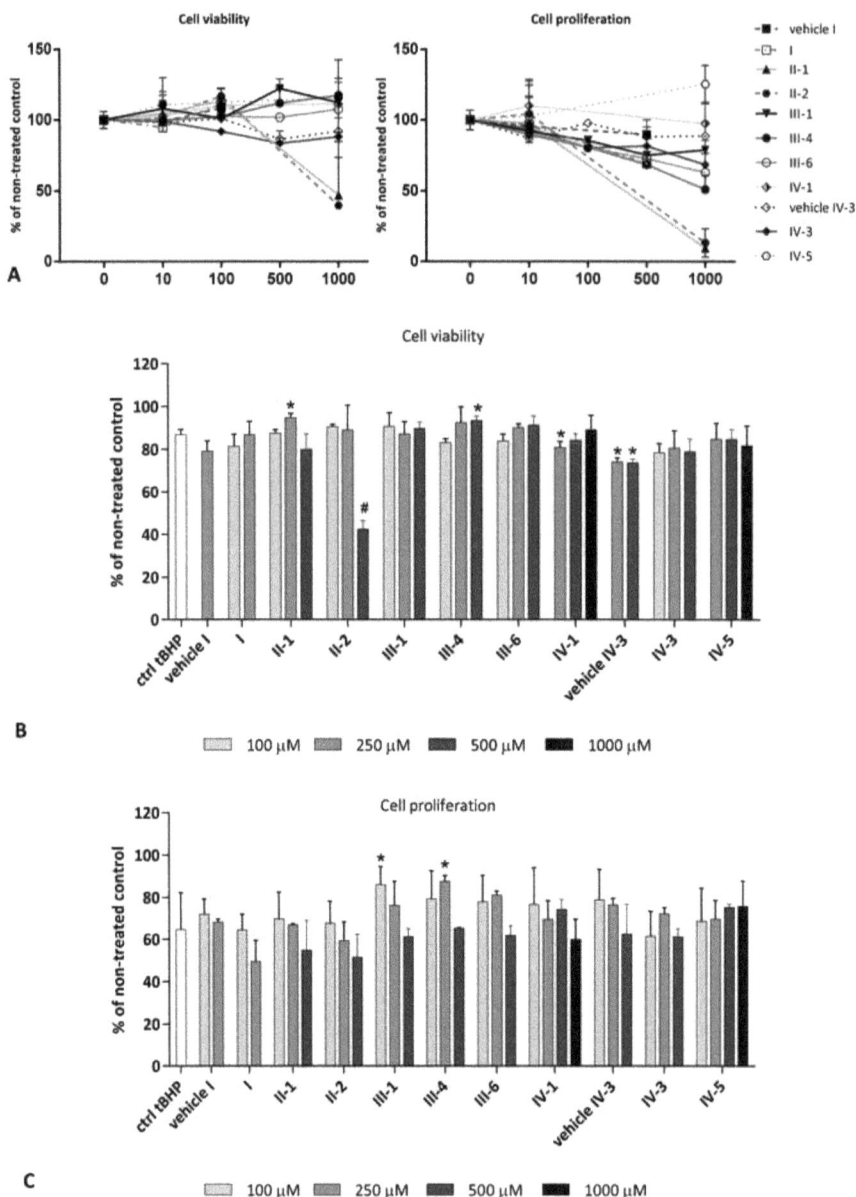

Figure 4. Cell viability (EZ4U assay) and proliferation (BrdU assay). (**A**) just DHPs without tBHP; (**B**) and (**C**) cell cultures treated with 50 μM tBHP. Results are expressed as a percentage of non-treated control. Values are given as mean ± standard deviation; n = 4. In comparison to the treatment control (ctrl tBHP), statistically significant differences are shown as follows: * $p < 0.05$; # $p < 0.0005$.

Intracellular ROS Production

As expected, pro-oxidant stressor tBHP increased the production of intracellular ROS while DHPs have shown diverse mostly concentration-dependent patterns (Figure 5). The highest reduction of ROS was observed with 250 μM Diethone **I**, which was the only DHP able to decrease ROS levels even below that of non-treated controls ($p < 0.005$). The highest applied concentration (500 μM) of Thiocarbatone **III-1** and of E2-134-1 **III-4**, decreased ROS levels to the level of non-treated controls ($p < 0.0005$). These DHPs were also effective at 250 μM ($p < 0.05$ and $p < 0.005$), while the lowest concentration applied (100 μM) increased intracellular ROS production ($p < 0.005$ and $p < 0.0005$). The AV-153 **IV-1** and AV-154-Na **IV-5** have shown a concentration-dependent bell-shaped curve, whereas in general, AV-154-Na **IV-5** increased ROS levels (suggesting its pro-oxidant role), while similar pro-oxidative effect of AV-153 **IV-1** was just observed in the middle concentration range of the substance used (250 μM; $p < 0.05$), to be followed by antioxidant effect of ROS reduction observed for the highest concentration (100 μM; $p < 0.05$). Carbatone **II-1** was effective in reducing the ROS levels at higher concentrations (250 μM, $p < 0.0005$; 500 μM, $p < 0.05$), while E2-136-2 **III-6** was effective as antioxidant at 250 μM ($p < 0.05$), not affecting ROS levels in other concentrations. Finally, E3-46 **IV-3** and J-9-133-2 **II-2** were mostly effective as pro-oxidants, particularly E3-46 **IV-3** if used at a lower concentration range (100 μM, $p < 0.05$; 250 μM, $p < 0.005$).

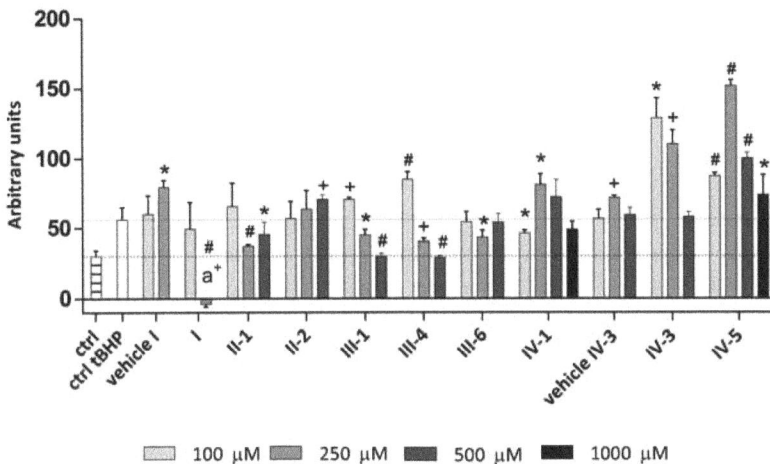

Figure 5. Intracellular ROS levels upon 24-h treatment with DHPs in mild oxidative stress conditions (1-h treatment with 50 μM t-BHP). Results are expressed in arbitrary units, adjusting the difference in fluorescence measurement after 1-h treatment with tBHP and zero-point (before addition of tBHP) with the cell number. Values are given as mean ± standard deviation ($n = 3$). Statistically significant differences are shown as follows: * $p < 0.05$; + $p < 0.005$; # $p < 0.0005$ (in comparison to the treatment control (ctrl tBHP; dotted line)); and a+ $p < 0.005$ (in comparison to the non-treated control (ctrl; dashed line)).

Determination of Total Glutathione Level in the Cells

Glutathione depletion was observed in HOS cells upon tBHP treatment (Figure 6). All tested DHPs were able to amend such tBHP-induced glutathione depletions at least in the case of some used concentrations. The only case of additional decrease of glutathione below the levels reduced by tBHP was observed for AV-153 **IV-1** (1000 μM; $p < 0.05$), E2-134-1 **III-4** (100 μM; $p < 0.05$), and J-9-133-2 **II-2** (500 μM; $p < 0.005$). Opposite to that some of tested DHPs were able to increase glutathione above the level of non-treated control. Such enhancing effect was observed for AV-154-Na **IV-5** (1000 μM; $p < 0.05$), E2-134-1 **III-4** (500 μM; $p < 0.05$), Diethone **I** (100 μM; $p < 0.05$), Carbatone **II-1** (250 μM; $p < 0.005$), E3-46 **IV-3** (250 μM; $p < 0.05$), and Thiocarbatone **III-1** (100 μM; $p < 0.005$ and 500 μM; $p < 0.05$). Surprisingly, Thiocarbatone **III-1**, while being able to increase glutathione more than the other tested DHPs (100 μM; $p < 0.005$), if applied at 250 μM concentration this DHP did not amend the tBHP-caused glutathione-depletion, thus producing the U-shaped concentration-dependence curve.

The glutathione levels for other DHPs were either proportional with their increasing concentrations (e.g., E2-134-1 **III-4**, E-2-136-2 **III-6**) or decreasing with their increasing concentrations (e.g., AV-153 **IV-1**, Diethone **I**) or forming a bell-shaped curve (e.g., Carbatone **II-1**, E3-46 **IV-3**).

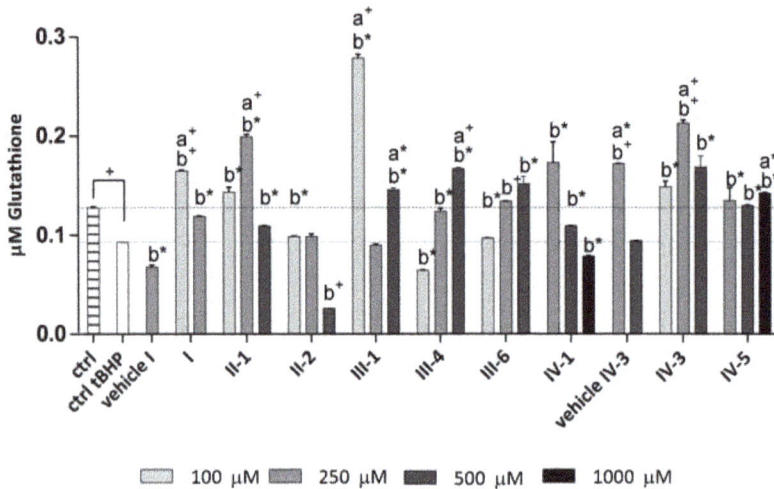

Figure 6. Concentration of total glutathione in cell lysates after 24-h treatment with DHPs followed by 1-h exposure to mild oxidative stress (50 μM tBHP). Statistically significant difference between DHPs and (a) non-treated control (ctrl (dashed line); only for the significant increase) or (b) treated control (ctrl tBHP (dotted line)) is given as follows: a*/b* $p < 0.05$; a+/b+ $p < 0.005$ with the difference between ctrl vs. ctrl tBHP as + $p < 0.005$. Values are given as mean ± standard deviation, n = 3.

In summary, the results of 24-h exposure of cells to DHPs prior the induction of mild oxidative stress (50 μM tBHP) revealed that Thiocarbatone **III-1**, Carbatone **II-1** and in particular E2-134-1 **III-4** were the only DHPs able to promote cell growth under oxidative stress conditions, acting in a concentration-dependent manner. While growth promotion was associated, as expected, with reduction of ROS and induction of the intracellular glutathione for E2-134-1 **III-4** (250 μM and 500 μM) and Carbatone **II-1** (250 μM), in case of the most pronounced increase of glutathione observed for Thiocarbatone **III-1** (100 μM), an increase of intracellular ROS was shown, too.

4. Discussion

The results of our study are in line with emerging evidence supporting pleiotropic, against oxidative stress-oriented, actions of DHPs. Thus, 4-aryl-2,6-dimethyl-3,5-bis-*N*-(aryl)-carbamoyl-1,4-dihydropyridines is already in use as novel skin-protecting agents, which inhibit elastase enzyme and protect against ROS [35]. In addition, the antioxidative ability of DHPs was used for creating more stable, light-sensitive DHP polythiophene derivatives (PTDHPs) as better alternatives to fluorescent probes used for cell imaging. These PTDHPs, via DHP groups, were found to regulate angiotensin-induced intracellular oxidative stress [36]. Recently, effective ROS radical scavenging was found also for some mitochondria-targeting DHPs [37]. These Mito-DHPs are promising antioxidants since they could protect both against radiation- and ROS-induced DNA strand breaks. These findings further confirm previous data on antioxidant and reductant activities of DHPs (Diethone **I** and its analogues) [38,39], pointing to mitochondria as targets for cell protective, bioactive DHP antioxidants [40,41]. This assumption is further supported by the current study since we observed differences in biological activities of DHPs in respect to biological parameter analyzed (cell proliferation vs. viability), which might be due to the fact that cellular viability was analyzed by EZ4U assay that reflects mitochondrial dehydrogenase activities. The same stands also for the observed differential effects of DHPs in respect to their direct pro- and anti-oxidative capacities (TOC/TAC assays based on hydrogen peroxide/peroxidase activity principle) vs. influence of DHPs on intracellular production of ROS, which is dependent on cellular oxidative homeostasis affecting viability and mitochondrial stability of the cells.

This study of pro- and anti-oxidant capacities of 27 structurally different DHPs (as well as one pyridine type compound) support previous findings, further indicating growth-regulating bioactivities of several DHPs acting on human osteoblast-like cells. The used DHPs, roughly divided into four subtypes comprising Diethone (Diludine) **I** and its analogues, Carbatone **II-1** and its analogues, Thiocartbatone (**III-1**) and its analogues, as well 1,4-dihydroisonicotinic acid derivatives (such as AV-153 **IV-1**) and its analogues, were primarily evaluated for their antioxidative ability (TAC assay) as well as for their potential pro-oxidative ability (TOC assay). Antioxidative potential of 1 mM concentration of Diethone **I**, AV-154-Na **IV-5**, Thiocarbatone **III-1**, E2-134-1 **III-4**, E2-136-2 **III-6**, J-9-133-2 **II-2**, E3-46 **IV-3**, J-11-71-2 **IV-7**, V-6-55-1 **IV-4**, Carbatone **II-1**, and AV-153 **IV-1** was about 10-fold higher than was antioxidant capacity of uric acid, the well-known natural antioxidant used as standard. There are main, node structures of DHPs possessing quite high antioxidative capacity known for Diethone **I**, Carbatone **II-1**, AV-153 **IV-1** and AV-154-Na **IV-5**, according to the structure: function analysis. Extension of substituents in positions 3 and 5 of Diethone **I** (ethoxycarbonyl to alkoxycarbonyl-methoxycarbonyl) and its thioderivative (E170-4, TK-2 **III-2**) leads to substantial diminution of antioxidative capacity. On the contrary, compounds possessing carboxylate anions in positions 3 (or 3 and 5) of Diethone **I** possess high antioxidative activity (Carbatone **II-1**, Thiocarbatone **III-1**, also 4-carboxycarbatone V-6-55-1 **IV-4**, hybrids of Carbatone **II-1** or Thiocarbatone **III-1** and Diethone **I** such as E2-134-1 **III-4**, J-9-133-2 **II-2**). Insertion of substituents in position 4 of Carbatone **II-1** (methyl, ethyl, styryl groups) diminishes antioxidative activity to the levels of uric acid for Styrylcarbatone **II-6** (J-9-117 **II-6**) and Metcarbatone **II-4** or almost completely suppresses in case of Etcarbatone **II-5**. However, a carboxylate anion in position 4 leads to the already mentioned antioxidative activity observed for AV-153 **IV-1** and AV-154-Na **IV-5**. Moreover, hybrids of Carbatone **II-1** ester and Diethone **I** possess high antioxidative capacity if compounds have 4-carboxylate anion (V-6-55-1 **IV-4** and E3-46 **IV-3**). If carboxylate groups in position 4 are distant from the DHP cycle (Glutapyrone **IV-8**), antioxidative activity is absent. The oxidized form (pyridine type compound J-12-25 **V**) was found to be inactive. Antioxidant activity mechanisms of DHPs studied may include direct scavenging of ROS (namely, hydrogen peroxide and free radicals derived from it) and decomposition of hydrogen peroxide in the manner, in which ROS will be produced in lesser degree. Namely, it was observed before that Carbatone **II-1** and its derivatives scavenge hydroxyl radicals (spectrophotometric and EPR detection) produced in Fenton reaction [7,14]. Furthermore, EPR spectroscopy showed that Metcarbatone **II-4** acts as effective scavenger of hydroxyl radicals produced in the Fenton reaction, while Etcarbatone **II-5**, and Propcarbatone (close analogue of Etcarbatone **II-5** having propyl group instead of ethyl group) are less effective, and Styrylcarbatone **II-6** is not effective at all. In addition, DHPs affecting TAC could be converted in their heteroaromatic oxidized form [42].

Pro-oxidative potential, measured by the TOC assay (Table 3), of the majority of the tested DHPs was either absent or minor. AV-153 **IV-1**, Carbatone **II-1** and E-170-4 **III-3** (TK1, **III-3**) were the only compounds showing oxidative potential somewhat proportional to the concentration used. While pro-oxidative activity of E-170-4 **III-3** (TK-1 **III-3**; diethyl ester of Thiocarbatone **III-1**) was very low, mild oxidative potential of lower concentrations (10 µM and 100 µM) and negligible of the highest (1000 µM) were observed for AV-153 **IV-1** and Carbatone **II-1**. Still, it was about 2-fold (AV-153 **IV-1**) and 5-fold (Carbatone **II-1**) lower than of hydrogen peroxide. Therefore, the oxidative ability tested by this assay was very moderate or absent suggesting that DHPs are less prone to oxidative activity. Concerning TOC effects, it should be mentioned that Tirzit et al. [43] showed possibility of free-radical reaction propagation by DHPs in vitro. If water-soluble DHPs (where R^4 = H, or CH_3, or COONa) have methyl group in the 4-th position, the possibility of that compounds reactivity against hydroxyl radical generation is lowered. However, introduction of strong electron donor group carboxylate-ion (COONa) causes sharp increase of reaction power. Even the introduction of carboxylate-ion in the positions 2 and 6 of the DHP ring increased this reactivity. The generation potency was observed at 0.5 mM and 1 mM, respectively 500 μM and 1000 μM concentrations, while at higher concentrations, the hydroxyl radical (HO•) generation is more pronounced. In addition, both assays, TAC and TOC, use HRP (horse radish peroxidase)-catalyzed oxidation of 3,3′,5,5′-tetramethylbenzidine (TMB) (reduced form of this is colorless) which includes either free radical (one-electron oxidation product) and/or charge transfer complex intermediates, both resulting in the formation of blue-colored reaction product [44]. However, it is difficult to discriminate between a one-electron and a two-electron mechanism for the initial enzymatic step in the reaction, thus the DHPs could react even as free radical quenchers or propagators (by one-electron or two-electron mechanisms) of free radical chain reactions (depending on assays used, DHPs chemical structure and concentration). Indeed, both, strong antioxidative activity and mild oxidative potential was observed for Carbatone **II-1** and AV-153 **IV-1**. Moreover, as being effectors of TAC/TOC systems in the presence of HRP, 1,4-DHPs themselves could also be enzymatically oxidized [45,46]. Thus, the elementary steps of oxidation of water-insoluble lipophilic calcium antagonist Nifedipine (NF) catalyzed by enzyme HRP have been described by analysis of kinetic magnetic field effects (MFEs). It has been shown that the first step of the catalytic cycle is single electron transfer resulting in formation of NF*(+) radical cation and ferroperoxidase (Per$^{(2+)}$) [46]. As a result, comparison with an earlier studied oxidation reaction of NADH catalyzed by HRP evidenced that the enzymatic oxidations of two substrates, native, NADH, and its synthetic DHP analogue, NF-catalyzed by HRP in the absence of hydrogen peroxide follow identical mechanisms. It is likely, that analogous reaction route could proceed in the case of the set of lipophilic DHPs (see Table 1) studied here. Besides, peroxide oxidation (without peroxidase, in Fenton system a.o.) of DHPs could occur [47].

Since previous studies have shown that various DHPs can achieve modulation of cellular responses to oxidative stress and thus influence cell growth and differentiation [30,48], the major aim of the current study was to find a DHP compound or a group of them that will act as potential antioxidants and could modify the growth of human–osteoblast cells indicating their possibly beneficial effects for treatment of osteoporosis and related bone disorders. Since oxidative stress was implicated in pathogenesis of osteoporosis (manifested as deterioration of bone) and antioxidants, such as lycopene and polyphenols have been suggested to benefit the therapy of osteoporosis [17,19,49], we studied the antioxidative potential of the selected DHPs which were 10-fold more effective than uric acid on HOS cells in mild oxidative stress conditions. Although normal human osteoblasts would be preferred model, due to their limited availability and donor-related influences, human osteoblast-like cancer cell lines have been widely used model systems for human osteoblasts, in particular HOS cells are known as such in vitro model [20–22].

To evaluate short-term, cell-protective (ROS decreasing) effects of DHPs, cells were treated with the selected DHPs and hydrogen peroxide, in general, for 1-h each. Thiocarbatone **III-1**, AV-153 **IV-1**, E2-134-1 **III-4**, E2-136-2 **III-6**, and Diethone **I** decreased ROS levels showing antioxidative potential, which was pronounced if the cells were not already damaged by hydrogen peroxide. Moreover, observed ROS reduction indicates direct scavenging of hydrogen peroxide by the mentioned DHPs along with possible induction of cellular-protective, antioxidant mechanisms. In addition, while Diethone **I** and E2-136-2 **III-6** did not affect cell growth, Thiocarbatone **III-1**, AV-153 **IV-1** and in particular E2-134-1 **III-4** did, thus emerging as the most potent antioxidants in the tested group, able to reduce oxidative stress-induced damage. Noteworthy, we cannot state with certainty whether the observed beneficial effects upon short-term treatment (DHPs and hydrogen peroxide being present in the medium at the same time for specified time-points) are just due to scavenging of hydrogen peroxide by DHPs (which is negligible without added catalysts) or their ability to induce cellular protective mechanism(s), or even both.

We further explored whether DHPs can induce cellular protective mechanism(s) under mild oxidative stress conditions and if longer exposure of cells to DHPs can potentiate possible effects. To do so, cells were treated with the selected DHPs for 24-h which were removed from the medium prior to exposure to stressors (1-h treatment). Hydrogen peroxide and tert-butyl hydroperoxide (tBHP) were used for induction of mild oxidative stress. In addition to the evaluation of cell growth, measurements of ROS generation along with intracellular glutathione content were carried out with the latter (tBHP) since it is known as more stable oxidative stress-inducing agent than hydrogen peroxide that provides more consistent effects [50]. Notably, for HepG2 cells tBHP, but not hydrogen peroxide, was found to be able to decrease reduced glutathione as wells as to increase malondialdehyde (MDA) levels and activity of antioxidative enzymes [50]. Moreover, tBHP was shown to induce caspase-dependent apoptosis in endothelial cells, which was mediated by ROS generation deriving from NADPH oxidase and mitochondria [51]. However, we should also stress the fact that hydrogen peroxide in very important natural ROS with multiple positive and negative bioactivities.

Both used stressors decreased cell growth, thus creating mild oxidative stress, while AV-153 **IV-1**, AV-154-Na **IV-5**, J-9-133-2 **II-2**, Carbatone **II-1**, and Diethone **I** (at low concentration) ameliorated negative effects of hydrogen peroxide, by promoting cell growth and acting as antioxidants.

The more in-depth analysis of the selected DHPs, considering the influence on cell growth, ROS production and glutathione content, was carried out with tBHP, known to be more stable and consistent in induction of oxidative stress, unlike hydrogen peroxide, which has multiple biological activities and is rapidly metabolized as natural ROS. Glutathione is one of the most important intracellular antioxidants able to scavenge ROS thus becoming oxidized. The increase in ratio oxidized/reduced glutathione indicates severity of oxidative stress. tBHP is known to increase this ratio thus causing depletion of reduced glutathione [50]. We measured total glutathione content (oxidized + reduced) which should be influenced only by de novo synthesis. tBHP decreased glutathione content, also observed in Rat clone-9 hepatocytes [52] which could be due to attenuated activation of the GCLc (glutamate cysteine ligase, catalytic subunit) promoter [53], resulting in reduced biosynthesis.

The selected DHPs were able to amend tBHP-induced glutathione depletions at least in some of the tested concentrations. Such effect was observed in rats exposed to benzo(a)pyrene-induced oxidative stress, where pre-treatment with Nitrendipine increased the glutathione but also superoxide dismutase (SOD) levels while decreasing TBARS levels [54].

Moreover, AV-154-Na **IV-5** (1000 µM), E2-134-1 **III-4** (500 µM), Diethone **I** (100 µM), Carbatone **II-1** (250 µM), E3-46 **IV-3** (250 µM), and Thiocarbatone **III-1** (100 µM and 500 µM) were able to induce glutathione biosynthesis. Indeed, the concentration-dependent diversity in patterns of intracellular glutathione and of ROS levels was observed in our study. Yet, Thiocarbatone **III-1**, Carbatone **II-1** and in particular E2-134-1 **III-4** were the only DHPs able to promote cell growth under such mild oxidative stress conditions acting in a concentration dependent manner. While growth promotion was related to the reduction of ROS and the induction of intracellular glutathione for E2-134-1 **III-4** (250 µM and 500 µM) and Carbatone **II-1** (250 µM), with the strongest observed increase of glutathione, the increase of intracellular ROS was also noticed for Thiocarbatone **III-1** (100 µM).The major pathway activated upon stress conditions is NRF2/KEAP1 [55]. NRF2 activates genes that encode phase II detoxifying enzymes and antioxidant enzymes thus influencing glutathione content. Recently, DHPs (Nifedipine and Amlodipine) were shown to activate NRF2 via a phosphoinositide 3-kinase (PI3K)-dependent mechanism [56]. At this point, we cannot fully explain observed concentration-dependent diversities in ROS and glutathione content, being sometimes reversal as expected, and the others not, or can we explain lack of linearity. However, we can assume that the onset of lipid peroxidation, generating reactive aldehydes, especially 4-hydronynoenal known to act as a second messenger of free radicals and signaling molecules regulating cell growth and inducing endogenous production of ROS, eventually affecting the activity and synthesis of glutathione might be, at least in part, responsible for that [57–59]. Interestingly, these DHPs did not affect cell growth, unlike growth-supporting DHPs E2-134-1 **III-4** and Carbatone **II-1**, which were found to increase glutathione and reduce ROS production, thus acting as potent antioxidants. On the other hand, Thiocarbatone **III-1** could not decrease ROS although it enhanced the growth of the HOS cells. These findings again suggest involvement of other survival pathway(s) that might include second messengers of free radicals, which are since recently also in focus of research as growth regulators [60–62]. In favor of this possibility are also pro-oxidative effects of Thiocarbatone **III-1**, resembling similar effects of HNE, while the other DHPs were also acting as pro-oxidants at certain concentration.

This dual phenomenon is characteristic also for some other compounds (e.g., ascorbic acid), acting either as antioxidants or as pro-oxidants, depending on given conditions [63]. For alpha-tocopherol (vitamin E) and even for polyphenols and thiols, the persistent rhetoric question is if vitamin E should be considered as a pro- or an anti-oxidant [64,65]. This question is actual similar for DHPs, in the case of which the likely answer could be both antioxidants and/or pro-oxidants, depending on the compound's individual structure determinants, like C. Winterburn stated: "Antioxidants are individuals" [65]. The same appears to be valid for bioactive DHP antioxidants, too.

DHPs, such as Thiocarbatone **III-1**, E2-134-1 **III-4**, Carbatone **II-1**, AV-153 **IV-1**, AV-154-Na **IV-5**, J-9-133-2 **II-2**, and Diethone **I** were able to attenuate negative effects of oxidative stress in human osteoblast-like cells. Their action was highly dependent on the concentration used, stressor applied and treatment set-up. Thus, the observed growth-supporting effects suggest that they might be eventually considered as promising therapeutic agents in the treatment of osteoporosis as was observed for Cilnidipine [66] and Amlodipine [67] in murine translation models.

5. Conclusions

Eleven out of the 28 structurally different compounds tested (27 DHP derivatives and one oxidized form pyridine-type heteroaromatic compound), in particular, Diethone **I**, AV-154-Na **IV-5**, Thiocarbatone **III-1**, E2-134-1 **III-4**, E2-136-2 **III-6**, J-9-133-2 **II-2**, E3-46 **IV-3**, J-11-71-2 **IV-7**, V-6-55-1 **IV-4**, Carbatone **II-1** and AV-153 **IV-1** were about 10-fold more effective antioxidants than uric acid and were further tested in the HOS model of human osteoblast-like cells. From the selected DHPs, Thiocarbatone **III-1**, E2-134-1 **III-4**, Carbatone **II-1**, AV-153 **IV-1**, AV-154-Na **IV-5**, J-9-133-2 **II-2**, and Diethone **I** were able to attenuate negative effects of oxidative stress in human osteoblast-like cells induced by pro-oxidants. These bioactive DHP antioxidants acted in a concentration-dependent manner, while their efficiency was also related to the pro-oxidant stressor applied and experimental treatment set-up. Thiocarbatone **III-1**, Carbatone **II-1** and E2-134-1 **III-4** were found to be undoubtedly the most potent compounds with growth-promoting effects and the only DHPs being able to act so in tBHP-induced oxidative stress conditions. The observed induction of de novo glutathione synthesis by HOS cells, observed upon treatment with these DHPs, implicates involvement of the NRF2 signaling pathway. While Carbatone **II-1** and E2-134-1 **III-4** have shown concordant reduction in ROS, for Thiocarbatone **III-1**, increased ROS contributing to cell growth suggested other survival mechanism(s) involved. Surprisingly, Diethone **I** (known antioxidant) did not affect cell growth even though it decreased ROS and mostly resulted in increasing glutathione in such conditions. Likewise, Diethone **I** and AV-153 **IV-1** were more protective if hydrogen peroxide was used as a stressor. In addition, AV-153 **IV-1** appeared to be more effective if directly acting as a scavenger of hydrogen-peroxide for shorter time, like Diethone **I**, while Thiocarbatone **III-1** and E2-134-1 **III-4** acted seemingly both as scavengers of ROS and as inducers of growth promoting signaling pathways. Therefore, we assume that HNE and related second messengers of free radicals generated by lipid peroxidation might be relevant factors for the bioactivity principles of bioactive DHPs, which should be further studied.

Finally, the results of our study together with supporting evidence for other DHPs [66,67] as well as with other antioxidants [17], suggest that DHPs could be considered as therapeutic agents for osteoporosis although further research is needed to elucidate mechanisms involved.

Author Contributions: Conceptualization, N.Z., A.V., G.D. and L.M.; Methodology, F.T., T.V., E.B., Z.K. and L.M.; Formal analysis, L.M.; Investigation, E.B., J.P., Z.O., I.B., Z.K., T.V. and L.M.; Resources, G.D. and N.Z.; Writing: Original Draft Preparation, A.V., G.D., N.Z. and L.M.; Supervision, G.D. and N.Z.; Funding Acquisition, G.D.

Funding: This research was funded by INNOVABALT (REGPOT—CT-2013-316149) project.

Acknowledgments: The authors acknowledge Latvian Council of Science, National Research Program PUBLIC HEALTH/Biomedicine of the Republic of Latvia and INNOVABALT (REGPOT—CT-2013-316149) project for financial support. The support of Cooperation in European System of Science and Technology (COST) Domain of Chemistry, Molecular Sciences and Technologies (CMST) (COST B35) and COST Action BM1203 (EU-ROS) was of highest importance for preparation of this paper.

Conflicts of Interest: The authors declare no conflict of interest.

Abbreviations

DHP(s)	1,4-dihydropyridine(s)
OS	oxidative stress
AO	Antioxidant
AOA	antioxidative activity
TAC	total antioxidative capacity
TOC	total oxidative capacity
ROS	reactive oxygen species
HRP	horse radish peroxidase
TMB	3,3′,5,5′-tetramethylbenzydine
DCFH-DA	2′,7′-dichlorodihydrofluorescein diacetate
tBHP	tert-butyl hydroperoxide
BrdU	5-bromo-2′-deoxyuridine
HOS	human osteosarcoma.

References

1. Sies, H. Oxidative stress: Introductory remarks. In *Oxidative Stress*; Sies, H., Ed.; Academic Press: London, UK, 1985; pp. 1–7.
2. Milkovic, L.; Siems, W.; Siems, R.; Zarkovic, N. Oxidative stress and antioxidants in carcinogenesis and integrative therapy of cancer. *Curr. Pharm. Des.* **2014**, *20*, 6529–6542. [CrossRef] [PubMed]
3. Egea, J.; Fabregat, I.; Frapart, Y.M.; Ghezzi, P.; Görlach, A.; Kietzmann, T.; Kubaichuk, K.; Knaus, U.G.; Lopez, M.G.; Olaso-Gonzalez, G.; et al. Corrigendum to "European contribution to the study of ROS: A summary of the findings and prospects for the future from the COST action BM1203 (EU-ROS)". *Redox Biol.* **2018**, *14*, 694–696. [CrossRef] [PubMed]
4. Long, M.J.C.; Poganik, J.R.; Ghosh, S.; Aye, Y. Subcellular redox targeting: Bridging in vitro and in vivo chemical biology. *ACS Chem. Biol.* **2017**, *12*, 586–600. [CrossRef] [PubMed]
5. Žarković, N.; Lončarić, I.; Čipak, A.; Jurić, G.; Wonisch, W.; Borović, S.; Waeg, G.; Vuković, T.; Žarković, K. [Pathophysiological characteristics of secondary messengers of free radicals and oxidative stress]. Patofiziološke značajke sekundarnih glasnika slobodnih radikala i oksidativni stress. In *Oksidativni Stres i Djelotvornost Antioksidansa (Second Messengers of Free Radicals; Oxidative Stres)*; Bradamante, V., Lacković, Z., Eds.; Medicinska Naklada: Zagreb, Croatia, 2001; pp. 13–32. ISBN 953-176-127-2.
6. Swarnalatha, G.; Prasanthi, G.; Sirisha, N.; Madhusudhana Chetty, C. 1,4-Dihydropyridines: A multtifunctional molecule—A review. *Int. J. ChemTech Res.* **2011**, *3*, 75–89.
7. Velena, A.; Zarkovic, N.; Gall Troselj, K.; Bisenieks, E.; Krauze, A.; Poikans, J.; Duburs, G. 1,4-Dihydropyridine derivatives: Dihydronicotinamide analogues—Model compounds targeting oxidative stress. *Oxid. Med. Cell. Longev.* **2016**, *2016*, 1892412. [CrossRef] [PubMed]
8. Mercola, M.; Cashman, J.; Lanier, M.; Willems, E.; Schade, D. Compounds for Stem Cell Differentiation. U.S. Patent OO9233926B2, 12 January 2016.
9. Velēna, A.; Zilbers, J.; Duburs, G. Derivatives of 1,4-dihydropyridines as modulators of ascorbate-induced lipid peroxidation and high-amplitude swelling of mitochondria, caused by ascorbate, sodium linoleate and sodium pyrophosphate. *Cell Biochem. Funct.* **1999**, *17*, 237–252. [CrossRef]
10. Ryabokon, N.I.; Goncharova, R.I.; Duburs, G.; Rzeszowska-Wolny, J. A 1,4-dihydropyridine derivative reduces DNA damage and stimulates DNA repair in human cells in vitro. *Mutat. Res. Genet. Toxicol. Environ. Mutagen.* **2005**, *587*, 52–58. [CrossRef] [PubMed]

11. Uldrikis, Y.R.; Zidermane, A.A.; Biseniex, E.A.; Preisa, I.E.; Dubur, G.Y.; Tirzit, G.D. Esters of 2,6-dimethyl-1,4-dihydropyridine-3,5-dicarboxylic Acid and Method of Obtaining. Thereof. Patent WO 8000345 A1, 8 August 1978.

12. Buraka, E.; Chen, C.Y.-C.; Gavare, M.; Grube, M.; Makarenkova, G.; Nikolajeva, V.; Bisenieks, I.; Brūvere, I.; Bisenieks, E.; Duburs, G.; et al. DNA-binding studies of AV-153, an antimutagenic and DNA repair-stimulating derivative of 1,4-dihydropiridine. *Chem. Biol. Interact.* **2014**, *220*, 200–207. [CrossRef] [PubMed]

13. Ošiņa, K.; Rostoka, E.; Sokolovska, J.; Paramonova, N.; Bisenieks, E.; Duburs, G.; Sjakste, N.; Sjakste, T. 1,4-Dihydropyridine derivatives without Ca^{2+}-antagonist activity up-regulate Psma6 mRNA expression in kidneys of intact and diabetic rats. *Cell Biochem. Funct.* **2016**, *34*, 3–6. [CrossRef] [PubMed]

14. Ošiņa, K.; Leonova, E.; Isajevs, S.; Baumane, L.; Rostoka, E.; Sjakste, T.; Bisenieks, E.; Duburs, G.; Vīgante, B.; Sjakste, N. Modifications of expression of genes and proteins involved in DNA repair and nitric oxide metabolism by carbatonides [disodium-2,6-dimethyl-1,4-dihydropyridine-3,5-bis(carbonyloxyacetate) derivatives] in intact and diabetic rats. *Arh. Hig. Rada Toksikol.* **2017**, *68*, 212–227. [CrossRef] [PubMed]

15. Rostoka, E.; Sokolovska, J.; Sjakste, N.; Sjakste, T.; Ošiņa, K. 1,4-Dihidropiridīna atvasinājums DNS pārrāvumu novēršanai cukura diabēta apstākļos. Patent of the Republic of Latvia, Nr. P-15-89 LV15181B, 20 August 2015.

16. Marie, P.J.; Kassem, M. Osteoblasts in osteoporosis: Past, emerging, and future anabolic targets. *Eur. J. Endocrinol.* **2011**, *165*, 1–10. [CrossRef] [PubMed]

17. Rao, L.G.; Rao, A.V. Oxidative stress and antioxidants in the risk of osteoporosis—Role of the antioxidants lycopene and and polyphenols. *Top Osteoporos.* **2013**, 117–161. [CrossRef]

18. Marinucci, L.; Balloni, S.; Fettucciari, K.; Bodo, M.; Talesa, V.N.; Antognelli, C. Nicotine induces apoptosis in human osteoblasts via a novel mechanism driven by H$_2$O$_2$ and entailing Glyoxalase 1-dependent MG-H1 accumulation leading to TG2-mediated NF-κB desensitization: Implication for smokers-related osteoporosis. *Free Radic. Biol. Med.* **2018**, *117*, 6–17. [CrossRef] [PubMed]

19. Domazetovic, V.; Marcucci, G.; Iantomasi, T.; Brandi, M.L.; Vincenzini, M.T. Oxidative stress in bone remodeling: Role of antioxidants. *Clin. Cases Miner. Bone Metab.* **2017**, *14*, 209–216. [CrossRef] [PubMed]

20. Shimodaka, K.; Okura, F.; Shimizu, Y.; Saito, I.; Yanaihara, T. Osteoblast cells (MG-63 and HOS) have aromatase and 5α-reductase activities. *Biochem. Mol. Biol. Int.* **1996**, *39*, 109–116.

21. Rose, F.R.; Cyster, L.A.; Grant, D.M.; Scotchford, C.A.; Howdle, S.M.; Shakesheff, K.M. In vitro assessment of cell penetration into porous hydroxyapatite scaffolds with a central aligned channel. *Biomaterials* **2004**, *25*, 5507–5514. [CrossRef] [PubMed]

22. Kwak, J.H.; Lee, S.R.; Park, H.J.; Byun, H.E.; Sohn, E.H.; Kim, B.O.; Rhee, D.K.; Pyo, S. Kobophenol A enhances proliferation of human osteoblast-like cells with activation of the p38 pathway. *Int. Immunopharmacol.* **2013**, *17*, 704–713. [CrossRef] [PubMed]

23. Norcross, B.E.; Clement, G.; Weinstein, M. The Hantzsch pyridine synthesis: A factorial design experiment for the introductory organic laboratory. *J. Chem. Educ.* **1969**, *46*, 694–695. [CrossRef]

24. Uldrikis, J.; Preisa, I.; Duburs, G.; Zidermane, A.; Bisenieks, E.; Tirzitis, G. 2,6-Dimethyl-1,4-dihydropyridine-3,5-dicarboxylic Acid Esters and Method for Preparing Same. WO Patent 8000345 A1 19800306, 6 March 1980.

25. Pubulis, K.; Stonans, I.; Jonane-Osa, I.; Jansone, I.; Bisenieks, E.; Kalvins, I.; Vigante, B.; Bruvere, I.; Uldrikis, J.; Zuka, L.; et al. Antiviral Efficacy of Disodium 2,6-dimethyl-1,4-dihydropyridine-3,5-bis(carbonyloxyacetate) and Its Derivatives. European Patent Application EP 2 578 218 A1, 6 October 2011.

26. Bisenieks, E.; Duburs, G.; Stonans, I.; Jaschenko, E.; Domracheva, I.; Poikans, J.; Bruvere, I.; Kalvins, I.; Shestakova, I.; Uldrikis, J.; et al. Pharmaceutical Combination of 5-fluorouracil and Derivative of 1,4-dihydropyridine and Its Use in the Treatment of Cancer. EP Patent 2228365 A1, 15 September 2010.

27. Dubur, G.Y.; Uldrikis, Y.R. Preparation of 3,5-diethoxycarbonyl-2,6-dimethyl-1,4-dihydro-isonicotinic acid and 3,5-diacetyl-2,6-dimethyl-1,4-dihydroisoni-cotinic acid and their salts. *Chem. Heterocycl. Compd.* **1972**, *5*, 762–763. [CrossRef]

28. Bisenieks, E.A.; Dubur, G.Y.; Uldrikis, Y.R.; Veveris, M.M.; Kimenis, A.A.; Ivanov, E.V. 2-(2,6-Dimethyl-3,5-diethoxycarbonyl-1,4-dihydropyridine-4-carboxamide Glutaric Acid Its Disodium Salt and Method of Their Preparation. DE Patent 3337521 A1, 26 April 1984.

29. Stonans, I.; Jansone, I.; Jonane-Osa, I.; Bisenieks, E.; Duburs, G.; Kalvins, I.; Vigante, B.; Uldrikis, J.; Bruvere, I.; Zuka, L.; et al. Derivatives of 1,4-dihydropyridine Possessing Antiviral Efficacy. U.S. Patent 20130131126A1, 23 May 2013.

30. Bruvere, I.; Bisenieks, E.; Poikans, J.; Uldrikis, J.; Plotniece, A.; Pajuste, K.; Rucins, M.; Vigante, B.; Kalme, Z.; Gosteva, M.; et al. Dihydropyridine derivatives as cell growth modulators in vitro. *Oxid. Med. Cell. Longev.* **2017**, *2017*. [CrossRef] [PubMed]

31. Borovic, S.; Tirzitis, G.; Tirzite, D.; Cipak, A.; Khoschsorur, G.A.; Waeg, G.; Tatzber, F.; Scukanec-Spoljar, M.; Zarkovic, N. Bioactive 1,4-dihydroisonicotinic acid derivatives prevent oxidative damage of liver cells. *Eur. J. Pharmacol.* **2006**, *537*, 12–19. [CrossRef] [PubMed]

32. Tatzber, F.; Griebenow, S.; Wonisch, W.; Winkler, R. Dual method for the determination of peroxidase activity and total peroxides-iodide leads to a significant increase of peroxidase activity in human sera. *Anal. Biochem.* **2003**, *316*, 147–153. [CrossRef]

33. Bradford, M.M. A rapid and sensitive method for the quantitation of microgram quantities of protein utilizing the principle of protein-dye binding. *Anal. Biochem.* **1976**, *72*, 248–254. [CrossRef]

34. Rodrigues, C.; Mósca, A.F.; Martins, A.P.; Nobre, T.; Prista, C.; Antunes, F.; Gasparovic, A.C.; Soveral, G. Rat aquaporin-5 is pH-gated induced by phosphorylation and is implicated in oxidative stress. *Int. J. Mol. Sci.* **2016**, *17*, 2090. [CrossRef] [PubMed]

35. Saeed, A.; Shahzad, D.; Larik, F.A.; Channar, P.A.; Mahfooz, H.; Abbas, Q.; Hassan, M.; Raza, H.; Seo, S.-Y.; Shabir, G. Synthesis of 4-aryl-2,6-dimethyl-3,5-bis-N-(aryl)-carbamoyl-1,4-dihydropyridines as novel skin protecting and anti-aging agents. *Bangladesh J. Pharmacol.* **2017**, *12*, 25. [CrossRef]

36. Hu, R.; Li, S.-L.; Bai, H.-T.; Wang, Y.-X.; Liu, L.-B.; Lv, F.-T.; Wang, S. Regulation of oxidative stress inside living cells through polythiophene derivatives. *Chin. Chem. Lett.* **2016**, *27*, 545–549. [CrossRef]

37. Zhang, Y.; Wang, J.; Li, Y.; Wang, F.; Yang, F.; Xu, W. Synthesis and radioprotective activity of mitochondria targeted dihydropyridines in vitro. *Int. J. Mol. Sci.* **2017**, *18*, 2233. [CrossRef] [PubMed]

38. Ivanov, E.V.; Ponomarjova, T.V.; Merkusev, G.N.; Dubur, G.J.; Bisenieks, E.A.; Dauvarte, A.Z.; Pilscik, E.M. A new skin radioprotective agent Diethon (experimental study). *Radiobiol. Radiother. (Berl.)* **1990**, *31*, 69–78. [PubMed]

39. Hong, S.E.; Urahashi, S.; Kamata, R. Skin radioprotector (Diethone) modifying dermal response of radiation on rats. *J. Korean Soc. Ther. Radiol.* **1989**, *7*, 15–22.

40. Pandupuspitasari, N.S.; Khan, F.A.; Sameeullah, M.; Rehman, Z.U.; Sudjatmogo; Muktiani, A.; Yi, J. Effects of diludine on the production, oxidative status, and biochemical parameters in transition cows. *J. Environ. Agric. Sci.* **2016**, *6*, 3–9.

41. Fernandes, M.A.S.; Santos, M.S.; Vicente, J.A.F.; Moreno, A.J.M.; Velena, A.; Duburs, G.; Oliveira, C.R. Effects of 1,4-dihydropyridine derivatives (cerebrocrast, gammapyrone, glutapyrone, and diethone) on mitochondrial bioenergetics and oxidative stress: A comparative study. *Mitochondrion* **2003**, *3*, 47–59. [CrossRef]

42. Filipan-Litvić, M.; Litvić, M.; Vinković, V. An efficient, metal-free, room temperature aromatization of Hantzsch-1,4-dihydropyridines with urea-hydrogen peroxide adduct, catalyzed by molecular iodine. *Tetrahedron* **2008**, *64*, 5649–5656. [CrossRef]

43. Tirzit, G.D.; Kazush, E.Y.; Dubur, G.Y. Influence of 1,4-dihydropyridine derivatives on the generation of hydroxyl radicals. *Chem. Heterocycl. Compd.* **1992**, *28*, 435–437. [CrossRef]

44. Josephy, P.D.; Eling, T.; Mason, R.P. The horseradish peroxidase-catalyzed oxidation of 3,5,3',5'-tetramethylbenzidine. Free radical and charge-transfer complex intermediates. *J. Biol. Chem.* **1982**, *257*, 3669–3675. [PubMed]

45. Duburs, G.Y.; Kumerova, A.O.; Uldrikis, Y.R. Enzymic oxidation of hydrogenated pyridines with peroxidase-hydrogen peroxide system. *Latv. PSR ZA Vestis* **1970**, *73*, 73–77.

46. Afanasyeva, M.S.; Taraban, M.B.; Polyakov, N.E.; Purtov, P.A.; Leshina, T.V.; Grissom, C.B. Elementary steps of enzymatic oxidation of nifedipine catalyzed by horseradish peroxidase. *J. Phys. Chem. B* **2006**, *110*, 21232–21237. [CrossRef] [PubMed]

47. Huyser, E.S.; Harmony, J.A.K.; McMillian, F.L. Peroxide oxidations of dihydropyridine derivatives. *J. Am. Chem. Soc.* **1972**, *94*, 3176–3180. [CrossRef]

48. Lovaković, T.; Poljak-Blaži, M.; Duburs, G.; Cipak, A.; Cindrić, M.; Vigante, B.; Bisenieks, E.; Jaganjac, M.; Mrakovčić, L.; Dedić, A.; et al. Growth modulation of human cells in vitro by mild oxidative stress and 1,4-dihydropyridine derivative antioxidants. *Coll. Antropol.* **2011**, *35*, 137–141. [PubMed]

49. Wilson, C. Bone: Oxidative stress and osteoporosis. *Nat. Rev. Endocrinol.* **2014**, *10*, 3. [CrossRef] [PubMed]

50. Alía, M.; Ramos, S.; Mateos, R.; Bravo, L.; Goya, L. Response of the antioxidant defense system to tert-butyl hydroperoxide and hydrogen peroxide in a human hepatoma cell line (HepG2). *J. Biochem. Mol. Toxicol.* **2005**, *19*, 119–128. [CrossRef] [PubMed]

51. Zhao, W.; Feng, H.; Sun, W.; Liu, K.; Lu, J.-J.; Chen, X. Tert-butyl hydroperoxide (t-BHP) induced apoptosis and necroptosis in endothelial cells: Roles of NOX4 and mitochondrion. *Redox Biol.* **2017**, *11*, 524–534. [CrossRef] [PubMed]

52. Lee, K.C.; Teng, C.C.; Shen, C.H.; Huang, W.S.; Lu, C.C.; Kuo, H.C.; Tung, S.Y. Protective effect of black garlic extracts on tert-butyl hydroperoxide-induced injury in hepatocytes via a c-Jun N-terminal kinase-dependent mechanism. *Exp. Ther. Med.* **2018**, *15*, 2468–2474. [CrossRef] [PubMed]

53. Langston, J.W.; Li, W.; Harrison, L.; Aw, T.Y. Activation of promoter activity of the catalytic subunit of γ-glutamylcysteine ligase (GCL) in brain endothelial cells by insulin requires antioxidant response element 4 and altered glycemic status: Implication for GCL expression and GSH synthesis. *Free Radic. Biol. Med.* **2011**, *51*, 1749–1757. [CrossRef] [PubMed]

54. Aktay, G.; Emre, M.H.; Polat, A. Influence of dihydropyridine calcium antagonist nitrendipine on benzo(a)pyrene-induced oxidative stress. *Arch. Pharm. Res.* **2011**, *34*, 1171–1175. [CrossRef] [PubMed]

55. Milkovic, L.; Zarkovic, N.; Saso, L. Controversy about pharmacological modulation of Nrf2 for cancer therapy. *Redox Biol.* **2017**, *12*, 727–732. [CrossRef] [PubMed]

56. Lisk, C.; McCord, J.; Bose, S.; Sullivan, T.; Loomis, Z.; Nozik-Grayck, E.; Schroeder, T.; Hamilton, K.; Irwin, D.C. Nrf2 activation: A potential strategy for the prevention of acute mountain sickness. *Free Radic. Biol. Med.* **2013**, *63*, 264–273. [CrossRef] [PubMed]

57. Zarkovic, N. 4-hydroxynonenal as a bioactive marker of pathophysiological processes. *Mol. Aspects Med.* **2003**, *24*, 281–291. [CrossRef]

58. Zarkovic, K.; Jakovcevic, A.; Zarkovic, N. Contribution of the HNE-immunohistochemistry to modern pathological concepts of major human diseases. *Free Radic. Biol. Med.* **2017**, *111*, 110–126. [CrossRef] [PubMed]

59. Gasparovic, A.C.; Milkovic, L.; Sunjic, S.B.; Zarkovic, N. Cancer growth regulation by 4-hydroxynonenal. *Free Radic. Biol. Med.* **2017**, *111*, 226–234. [CrossRef] [PubMed]

60. Milkovic, L.; Hoppe, A.; Detsch, R.; Boccaccini, A.R.; Zarkovic, N. Effects of Cu-doped 45S5 bioactive glass on the lipid peroxidation-associated growth of human osteoblast-like cells in vitro. *J. Biomed. Mater. Res. A* **2014**, *102*, 3556–3561. [CrossRef] [PubMed]

61. Milkovic, L.; Cipak Gasparovic, A.; Zarkovic, N. Overview on major lipid peroxidation bioactive factor 4-hydroxynonenal as pluripotent growth-regulating factor. *Free Radic. Res.* **2015**, *49*, 850–860. [CrossRef] [PubMed]

62. Mouthuy, P.-A.; Snelling, S.J.B.; Dakin, S.G.; Milković, L.; Gašparović, A.Č.; Carr, A.J.; Žarković, N. Biocompatibility of implantable materials: An oxidative stress viewpoint. *Biomaterials* **2016**, *109*, 55–68. [CrossRef] [PubMed]

63. Herbert, V. Prooxidant effects of antioxidant vitamins. Introduction. *J. Nutr.* **1996**, *126*, 1197S–1200S. [CrossRef] [PubMed]

64. Tafazoli, S.; Wright, J.S.; O'Brien, P.J. Prooxidant and antioxidant activity of vitamin E analogues and troglitazone. *Chem. Res. Toxicol.* **2005**, *18*, 1567–1574. [CrossRef] [PubMed]

65. Winterbourn, C. Pro-Oxidants or Antioxidants. (Pro-oxidants and Antioxidants. What Characterizes an Antioxidant). In Proceedings of the Sunrise Free Radical School Free Radical Biology (SFRBM), San Francisco, CA, USA, 19–21 November 2009; University of Otago: Christchurch, New Zealand, 2009; pp. 1–33. Available online: https://pdfs.semanticscholar.org/presentation/83c9/974cfea4d3171558be4814673d22bbb7f857.pdf (accessed on 16 July 2018).

66. Shimizu, H.; Nakagami, H.; Yasumasa, N.; Mariana, O.K.; Kyutoku, M.; Koriyama, H.; Nakagami, F.; Shimamura, M.; Rakugi, H.; Morishita, R. Cilnidipine, but not amlodipine, ameliorates osteoporosis in ovariectomized hypertensive rats through inhibition of the N-type calcium channel. *Hypertens. Res.* **2012**, *35*, 77–81. [CrossRef] [PubMed]
67. Ushijima, K.; Liu, Y.; Maekawa, T.; Ishikawa, E.; Motosugi, Y.; Ando, H.; Tsuruoka, S.; Fujimura, A. Protective effect of amlodipine against osteoporosis in stroke-prone spontaneously hypertensive rats. *Eur. J. Pharmacol.* **2010**, *635*, 227–230. [CrossRef] [PubMed]

MDPI

St. Alban-Anlage 66

4052 Basel

Switzerland

Tel. +41 61 683 77 34

Fax +41 61 302 89 18

www.mdpi.com

Antioxidants Editorial Office

E-mail: antioxidants@mdpi.com

www.mdpi.com/journal/antioxidants

www.ingramcontent.com/pod-product-compliance
Lightning Source LLC
Chambersburg PA
CBHW051853210326
41597CB00033B/5883